ELSEVIER OCEAN ENGINEERING BOOK SERIES

VOLUME 4

LOAD AND GLOBAL RESPONSE OF SHIPS

Elsevier Science Internet Homepage

http://www.elsevier.nl (Europe)
http://www.elsevier.com (America)
http://www.elsevier.co.jp (Asia)

Consult the Elsevier homepage for full catalogue information on all books, journals and electronic products and services.

Elsevier Titles of Related Interest

BAI
Pipelines and Risers
ISBN: 008-043712-5

BJORHOVDE, COLSON & ZANDONINI
Connections in Steel Structures III.
ISBN: 008-042821-5

CHAN & TENG
ICASS '99, Advances in Steel Structures. (2 Volume Set).
ISBN: 008-043015-5

DUBINA
SDSS '99 – Stability and Ductility of Steel Structures.
ISBN: 008-043016-3

FRANGOPOL, COROTIS & RACKWITZ
Reliability and Optimization of Structural Systems.
ISBN: 008-042826-6

FUKUMOTO
Structural Stability Design.
ISBN: 008-042263-2

GODOY
Thin-Walled Structures with Structural Imperfections: Analysis and Behavior.
ISBN: 008-042266-7

GUEDES-SOARES
Advances in Safety and Reliability (3 Volume Set).
ISBN: 008-042835-5

MOAN & BERGE
13th Int Ship & Offshore Structures Congress (ISSC 1997).
ISBN: 008-042829-0

OHTSUBO & SUMI
Proceedings of the 14th International Ship and Offshore Structures Congress.
ISBN: 008-043602-1

OWENS
Steel in Construction (CD-ROM Proceedings with Printed Abstracts Volume, 268 papers).
ISBN: 008-042997-1

SRIVASTAVA
Structural Engineering World Wide 1998 (CD-ROM Proceedings with Printed Abstracts Volume 702 papers).
ISBN: 008-042845-2

USAMI & ITOH
Stability and Ductility of Steel Structures.
ISBN: 008-043320-0

VASSALOS
Contemporary Ideas on Ship Stability.
ISBN: 008-043652-8

VUGTS
BOSS '97 Behaviour of Offshore Structures (3 Volume Set).
ISBN: 008-042834-7

WATSON
Practical Ship Design.
ISBN: 008-042999-8

YOUNG
Wind Generated Ocean Waves.
ISBN: 008-043317-0

ZINGONI
Structural Engineering, Mechanics and Computation
ISBN: 008-043948-9

Related Journals

Free specimen copy gladly sent on request. Elsevier Science Ltd, The Boulevard, Langford Lane, Kidlington, Oxford, OX5 1GB, UK

Applied Ocean Research
Advances in Engineering Software
CAD
Coastal Engineering
Composite Structures
Computers and Structures
Construction and Building Materials
Engineering Failure Analysis
Engineering Fracture Mechanics
Engineering Structures
Finite Elements in Analysis and Design
International Journal of Solids and Structures
Journal of Constructional Steel Research
Marine Structures
NDT & E International
Ocean Engineering
Structural Safety
Thin-Walled Structures

To Contact the Publisher

Elsevier Science welcomes enquiries concerning publishing proposals: books, journal special issues, conference proceedings, etc. All formats and media can be considered. Should you have a publishing proposal you wish to discuss, please contact, without obligation, the publisher responsible for Elsevier's ocean engineering publishing programme:

Ian Salusbury
Senior Publishing Editor
Elsevier Science Ltd
The Boulevard, Langford Lane
Kidlington, Oxford
OX5 1GB, UK

Phone: +44 1865 843425
Fax: +44 1865 843920
E.mail: i.salusbury@elsevier.co.uk

General enquiries, including placing orders, should be directed to Elsevier's Regional Sales Offices – please access the Elsevier homepage for full contact details (homepage details at the top of this page).

ELSEVIER OCEAN ENGINEERING BOOK SERIES

VOLUME 4

LOAD AND GLOBAL RESPONSE OF SHIPS

J.J. JENSEN

Department of Naval Architecture and Offshore Engineering Technical University of Denmark

2001

ELSEVIER

Amsterdam – London – New York – Oxford – Paris – Shannon – Tokyo

ELSEVIER SCIENCE Ltd
The Boulevard, Langford Lane
Kidlington, Oxford OX5 1GB, UK

First edition 2001

Library of Congress Cataloging in Publication Data
A catalog record from the Library of Congress has been applied for.

British Library Cataloguing in Publication Data
A catalogue record from the British Library has been applied for.

ISBN: 0 08 043953 5 (hardcover)

♾ The paper used in this publication meets the requirements of ANSI/NISO Z39.48-1992 (Permanence of Paper).

Printed in The Netherlands.

SERIES PREFACE

In this day and age, humankind has come to the realization that the Earth's resources are limited. In the 19th and 20th Centuries, these resources have been exploited to such an extent that their availability to future generations is now in question. In an attempt to reverse this march towards self-destruction, we have turned out attention to the oceans, realizing that these bodies of water are both sources for potable water, food and minerals and are relied upon for World commerce. In order to help engineers more knowledgeably and constructively exploit the oceans, the **Elsevier Ocean Engineering Book Series** has been created.

The **Elsevier Ocean Engineering Book Series** gives experts in various areas of ocean technology the opportunity to relate to others their knowledge and expertise. In a continual process, we are assembling world-class technologists who have both the desire and the ability to write books. These individuals select the subjects for their books based on their educational backgrounds and professional experiences.

The series differs from other ocean engineering book series in that the books are directed more towards technology than science, with a few exceptions. Those exceptions we judge to have immediate applications to many of the ocean technology fields. Our goal is to cover the broad areas of naval architecture, coastal engineering, ocean engineering acoustics, marine systems engineering, applied oceanography, ocean energy conversion, design of offshore structures, reliability of ocean structures and systems and many others. The books are written so that readers entering the topic fields can acquire a working level of expertise from their readings.

We hope that the books in the series are well-received by the ocean engineering community.

Rameswar Bhattacharyya
Michael E. McCormick

Series Editors

Preface

The present monograph covers wave load and global structural response for ships. It is primarily written as a textbook for students with an introductory background in naval architecture and a basic knowledge of statistics and strength of materials.

The subjects are treated in detail starting from first principles. The aim has been to derive and present the necessary theoretical framework for predicting the extreme loads and the corresponding hull girder stresses a ship may be subjected to during its operational lifetime.

Although some account is given to reliability analysis, the present treatment has to be supplemented with methods for detailed stress evaluation and for structural strength assessment before a complete structural reliability analysis can be carried out.

The content of the book has developed over the years and invaluable comments and suggestions have been received from my colleagues at the department. Special thanks goes to Professor Preben Terndrup Pedersen, for his constant encouragement and friendship.

The typing of the manuscript by Vibeke Lybecker Jensen, Vivi Jessen and Linda Andersen is highly appreciated.

Jørgen Juncher Jensen

Contents

1 Introduction

A ship is a very complex structure consisting of more than 100,000 different parts. A design usually starts by a shipowner's request for a ship with the ability to carry a specific type and amount of cargo, possibly to sail on a specific route with a given speed.

Based on this small quantity of data, the naval architects make a preliminary design which satisfies these requirements. Due attention is paid to a large number of problems such as

- Restrictions on length, beam and draught of the ship imposed by relevant harbours and channels.
- Optimal hull form with respect to wave resistance.
- Economical and environmentally friendly propulsion system.
- Effective loading and unloading of cargo.
- Location of cargo, ballast and fuel oil tank to ensure proper trim in all conditions.
- Hydrostatic stability in intact and in damaged conditions.
- Electrical consumption and generator sets.
- Lifeboats with easy access and other life-saving equipment.
- Fire resistant subdivision of the ship and required fire fighting equipment.
- Heating and ventilation.
- Accommodation for crews and, perhaps, passengers.
- Navigation equipment.
- Maintenance cost.

This list contains just some of the main issues, others may be added depending on the type of ship.

The result of the preliminary design phase is a set of drawings. The main drawing is the general arrangement, showing the ship in all three projections with the location of tanks, cargo gears, life-saving equipment and machinery clearly indicated. Additional drawings show the different loading conditions, the machinery layout as well as other pertinent features of the ship. The drawings are supplemented with a specification of the various equipment on board.

When or if this design is approved by the shipowner, the next step is determination of the scantlings. This step is taken in close collaboration with a classification society. On acceptance of the structural design, the classification society issues a class notation for the ship, which makes it possible for the shipowner to insure the ship in an insurance company.

The classification societies have issued rules and regulations for a proper structural analysis of a ship and selection of the scantlings. Previously, those rules rather explicitly gave formulas for the thickness of the hull platings, the size of the stiffeners, the scantlings of internal bulkheads and so on. Such empirical rules must necessarily be rather conservative in order to apply to a large variety of ships. With the advent of powerful computers, the rules have changed. Today, the naval architect can perform the structural analysis using mainly rational methods based on first principles. The classification society then specifies proper safety factors against local and global failure modes, taking into account the consequences of failure and the analysis procedure used. A cruder method of analysis then necessitates a larger safety factor. Therefore, the effort made by the experienced naval architect to perform a detailed structural analysis will be paid back not just by a rational structural arrangement but often also in lesser lightweight of the ship and thus a higher payload throughout the operational lifetime of the ship.

The background of rational methods for ship structural design is structural and fluid mechanics with extensive use of advanced statistical methods for determining the loads due to the waves and for estimating the safety against structural failure.

Determination of the scantlings by rational methods can be divided into the following main steps:

- Determination of hydrostatic loads in a calm sea.
- Determination of wave-induced loads taking into account the statistical nature of ocean waves.
- Determination of hull girder stresses due to the above-mentioned loads.
- Assessment of the safety against hull girder failure due to excessive maximum loads. Too low safety requires a change of the scantlings of the hull plating and other longitudinal structural members (bulkheads, girders, longitudinals).
- Determination of local stresses due to pressure loads from the sea and from the cargo in association with the hull girder stresses.
- Assessment of the safety against local failure with respect to buckling, yielding and fatigue. Redesign if insufficient safety.
- Determination of the forced vibration levels due to time-varying forces generated by the machinery, the propeller and the sea. This is of less importance to normal merchant vessels, but often needed in the case of fast vessels and passenger ships in order to ensure a comfortable ship and to reduce problems with fatigue cracks.
- Optimal inspection and repair procedures to reduce docking time and to ensure that the ageing ship has sufficient strength. Especially, attention must be paid to the detection and repair of fatigue cracks and corroded areas.

- Structural assessment of the consequences of accidental loads due to collision, grounding, fire and explosion.

The present lecture notes deal with the global load and response analysis of ships, covering mainly the first four items mentioned above.

In Chapter 2, the equilibrium position of a floating body in a calm sea is determined together with the hydrostatic sectional loads in the hull girder. Moreover, a short description of hydrostatic stability is given.

Chapter 3 is devoted to a stochastic description of ocean waves. The necessary statistical methods are derived and linear as well as non-linear formulations of the wave elevation are given.

Chapter 4 describes procedures for determination of the global wave loads on ships. The linear formulation in the frequency domain is the basic method, but non-linear formulations in both frequency and time domains are also treated in order to obtain realistic extreme hull girder loads.

Chapter 5 deals with the determination of hull girder stresses due to sectional bending, shear and torsional loads. The analysis is performed within the framework of elastic thin-walled beam theory.

Hull vibrations are treated in Chapter 6 and, finally, in Chapter 7, the concept of reliability analysis is introduced.

No textbooks available today cover all parts of ship structural design. However, Hughes (1988) gives a very comprehensive discussion of many topics and can be recommended for further reading.

2 Hydrostatic Loads

When a structure is at rest in a calm sea, the submerged parts of the structure will be subjected to hydrostatic pressure forces. The *hydrostatic pressure* p acts perpendicularly to the exposed surfaces with a magnitude

$$p = -\rho g Z \tag{2.1}$$

where ρ is the mass density of sea water, g is the acceleration of gravity. Z is the vertical coordinate in a global XYZ-coordinate system, situated with the (X,Y)-coordinates in the still water surface and with Z pointing out of the water, see Figure 2.1.

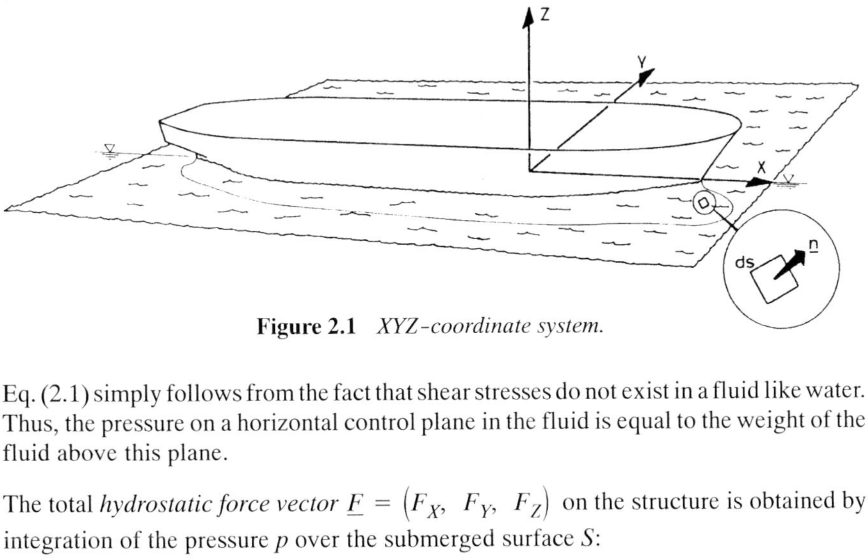

Figure 2.1 *XYZ-coordinate system.*

Eq. (2.1) simply follows from the fact that shear stresses do not exist in a fluid like water. Thus, the pressure on a horizontal control plane in the fluid is equal to the weight of the fluid above this plane.

The total *hydrostatic force vector* $\underline{F} = (F_X, F_Y, F_Z)$ on the structure is obtained by integration of the pressure p over the submerged surface S:

$$\underline{F} = -\iint_S p\,\underline{n}\,dS \tag{2.2}$$

where $\underline{n} = (n_X, n_Y, n_Z)$ denotes the normal vector to the surface, directed into the water, see Figure 2.1.

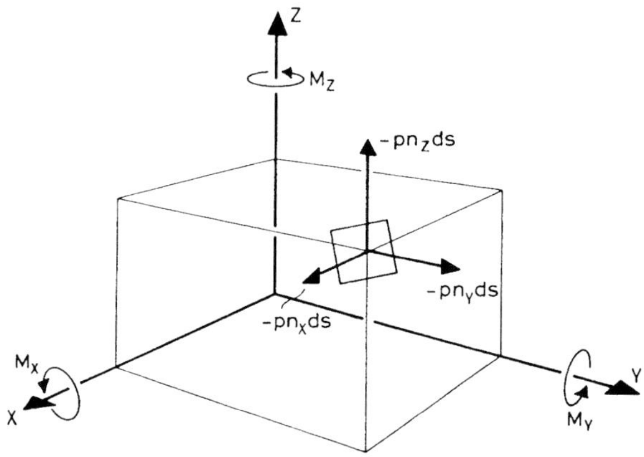

Figure 2.2 *Moments with respect to the XYZ-system.*

As the normal vector is a function of (X, Y, Z), the integration in Eq. (2.2) can be quite complicated. Later, in Section 2.2, two different procedures will be described for performing integrations over the submerged surface.

Static equilibrium requires that the net total force and moment vectors on the structure are zero. The moment vector $\underline{M} = \left(M_X,\ M_Y,\ M_Z \right)$ due to the hydrostatic pressure becomes see (Figure 2.2):

$$M_X = -\iint_S p \left(n_Z\ Y - n_Y\ Z \right)\ dS \tag{2.3}$$

$$M_Y = -\iint_S p \left(n_X\ Z - n_Z\ X \right)\ dS \tag{2.4}$$

$$M_Z = -\iint_S p \left(n_Y\ X - n_X\ Y \right)\ dS \tag{2.5}$$

Here $M_X,\ M_Y,\ M_Z$ are the moments of the hydrostatic pressure force with respect to the X-, Y- and Z-axis, respectively.

Static equilibrium requires that the force $\underline{F}$ and the moment $\underline{M}$ are counteracted by other static forces:

$$\underline{F} = -\ \underline{F}_r \tag{2.6}$$

and

$$\underline{M} = -\underline{M}_r \tag{2.7}$$

where $\underline{F}_r$ and $\underline{M}_r$ are the total force and moment vectors, respectively, of these other forces. For marine structures, these forces may be gravitational forces, reaction forces from contact with a quay or a seabed, forces due to wind and current and mooring forces.

For structures with a fixed position in the sea, as for instance bottom-supported offshore drilling rigs or grounded ships, Eqs. (2.1) - (2.5) yield $\underline{F}$, $\underline{M}$ and the reaction forces then follow from Eqs. (2.6) - (2.7).

For floating structures like ships, semi-submersibles and tension-leg platforms, the hydrostatic pressure resultants $\underline{F}$, $\underline{M}$ cannot be calculated directly from Eqs. (2.1) - (2.5) as the submerged surface S is not known in advance. The equilibrium position of the structure in the sea and thus the submerged surface S must first be determined from Eqs. (2.1) - (2.7). The solution procedure for this problem is given in Section 2.1. This procedure makes use of some hydrostatic coefficients. Two different procedures for calculation of these coefficients as well as other coefficients, necessary to obtain sectional loads and to assess the hydrostatic stability, are presented in Section 2.2. Sectional loads may then be determined as described in Section 2.3. Finally, hydrostatic stability is discussed in Section 2.4.

2.1 EQUILIBRIUM POSITION FOR A FLOATING STRUCTURE

In this section the equilibrium position of a rigid body, floating at rest in a calm sea, is determined. A rigid body is assumed which yields a position completely determined by the translation and rotation of an xyz-coordinate fixed in the body with respect to the global XYZ-system. Thus, the six equations (2.6) - (2.7) are sufficient to determine the position of the structure in a calm sea. For a very elastic body, e.g. a subsea membrane oil storage tank, the elastic deformations of the structure due to hydrostatic pressure may seriously change the hydrostatic loadings. In such cases, the equilibrium condition can only be found by solving a complete elastic structural model subjected to all external loadings (hydrostatic pressure, gravitational loads, reaction forces) and the relevant boundary conditions. The very large elastic deformations, which necessitate such an approach, also make the problem highly non-linear and therefore often difficult to solve.

However, ships and most other marine structures are so rigid that the elastic deformations due to the hydrostatic pressure do not change the hydrostatic pressure loads to any significant degree. A few per cent reduction in the sectional forces might be achieved if the hull girder flexibility is taken into account. This reduction is, however, usually neglitible considering the uncertainties in the mass distribution.

For a floating body where the submerged surface S is completely in contact with the water, the hydrostatic pressure resultants $\underline{F}$, $\underline{M}$, Eqs. (2.2) - (2.5), can be written more conveniently by applying the Gauss integral theorem, which states that in general

$$\iint_{S_0} \underline{v} \cdot \underline{n}\, dS_o = \iiint_{\nabla} div\, \underline{v}\, d\nabla \qquad (2.8)$$

where S_o is the complete surface of the *volume* ∇. The vector $\underline{v} = (v_X, v_Y, v_Z)$ is any differentiable vector and $\underline{n}$ is the normal to the surface, pointing out of the volume. Furthermore,

$$\underline{v} \cdot \underline{n} = v_X n_X + v_Y n_Y + v_Z n_Z \qquad (2.9)$$

$$div\, \underline{v} = \frac{\partial v_X}{\partial X} + \frac{\partial v_Y}{\partial Y} + \frac{\partial v_Z}{\partial Z} \qquad (2.10)$$

In the present case, S_o is chosen as the submerged surface S plus the intersection surface A_w between the structure and the still water surface, see Figure 2.3. Thus, S_o becomes a single-connected closed surface which bounds a volume equal to the volume ∇ of the submerged part of the structure.

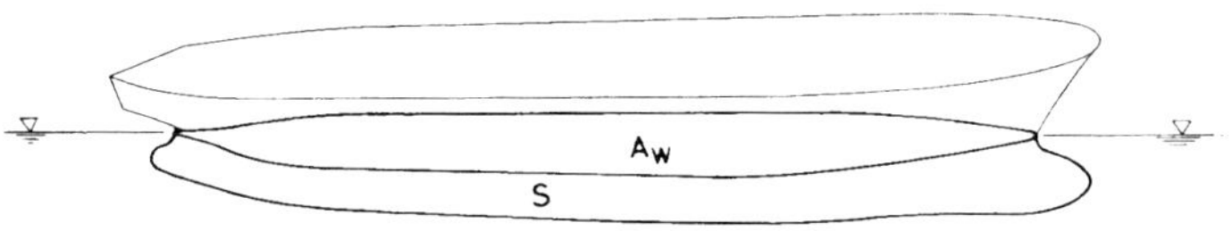

Figure 2.3 *Surfaces S_o, A_w defining the submerged volume.*

Insertion of Eq. (2.1) into Eq. (2.2) yields

$$\underline{F} = \rho g \iint_S Z\, \underline{n}\, dS = \rho g \iint_{S_o} Z\, \underline{n}\, dS_o \qquad (2.11)$$

as $Z = 0$ on A_w. For $\underline{v} = (Z, 0, 0)$, Eq. (2.8) yields

$$\iint_{S_0} Z\, n_X dS_0 = \iiint_{\nabla} \frac{\partial Z}{\partial X}\, d\nabla = 0 \qquad (2.12)$$

and similarly with $\underline{v} = (0, Z, 0)$ and $\underline{v} = (0, 0, Z)$:

$$\iint_{S_0} Z n_Y dS_0 = 0 \qquad (2.13)$$

$$\iint_{S_0} Zn_Z dS_0 = \iiint_{\nabla} \frac{\partial Z}{\partial Z} d\nabla = \iiint_{\nabla} d\nabla \equiv \nabla \tag{2.14}$$

Hence

$$F_X = \rho\, g \iint_{S_0} Zn_X dS_0 = 0 \tag{2.15}$$

$$F_Y = \rho\, g \iint_{S_0} Zn_Y dS_0 = 0 \tag{2.16}$$

$$F_Z = \rho\, g \iint_{S_0} Zn_Z dS_0 = \rho g \nabla \tag{2.17}$$

The physics behind Eqs. (2.15) - (2.17) were discovered by Archimedes, who stated that the net force on a submerged body is directed vertically upwards and equal to the weight of the displaced volume of the fluid.

Likewise, the moment vector $\underline{M}$, Eqs. (2.3) – (2.5), can be rewritten by use of Eq. (2.8)

$$M_X = \rho g \iint_{S_0} Z\left(n_Z Y - n_Y Z\right) dS_0 = \rho g \iiint_{\nabla} Y d\nabla \tag{2.18}$$

$$M_Y = \rho g \iint_{S_0} Z\left(n_X Z - n_Z X\right) dS_0 = -\rho g \iiint_{\nabla} X d\nabla \tag{2.19}$$

$$M_Z = \rho g \iint_{S_0} Z\left(n_Y X - n_X Y\right) dS_0 = 0 \tag{2.20}$$

Thus, only three out of the six force and moment resultant components of the hydrostatic pressure are different from zero. The three equilibrium equations in Eqs. (2.6) - (2.7) concerned with forces acting in the (X,Y)-directions are thus independent of the hydrostatic pressure loadings. This implies that the horizontal position in the XY-plane of the structure is independent of the hydrostatic pressure and is either indeterminate, as for a freely floating structure, or determined from an equilibrium between, for example, horizontal mooring forces and loads derived from wind and current.

$$\begin{Bmatrix} X \\ Y \\ Z \end{Bmatrix} = \begin{bmatrix} \cos\theta & 0 & -\sin\theta \\ 0 & 1 & 0 \\ \sin\theta & 0 & \cos\theta \end{bmatrix} \begin{bmatrix} 1 & 0 & 0 \\ 0 & \cos\varphi & \sin\varphi \\ 0 & -\sin\varphi & \cos\varphi \end{bmatrix} \begin{Bmatrix} x \\ y \\ z \end{Bmatrix} + \begin{Bmatrix} 0 \\ 0 \\ -T \end{Bmatrix} \tag{2.37}$$

or alternatively

$$X = x\cos\theta + y\sin\theta\sin\varphi - z\sin\theta\cos\varphi \tag{2.38}$$

$$Y = y\cos\varphi + z\sin\varphi \tag{2.39}$$

$$Z = x\sin\theta - y\cos\theta\sin\varphi + z\cos\theta\cos\varphi - T \tag{2.40}$$

The unknown quantities are: the trim angle θ, the heeling angle φ and the translation T. As the still water surface is defined by $Z = 0$, the translation T is the vertical distance between the origin of the xyz-system and the still water surface and is therefore a measure of the draught.

The three equilibrium equations (2.33) - (2.35) can now be expressed in terms of the three variables θ, φ and T. Explicit expressions are found for X_g and Y_g by substitution of $x = x_g$, $y = y_g$ and $z = z_g$ into Eqs. (2.38) and (2.39). However, the submerged volume ∇ and its volume moments L_X and L_Y also depend upon θ, φ and T through the integration domain. This makes the equilibrium equations non-linear in the variables and an iterative solution procedure must be applied in order to solve the Egs. (2.33) - (2.35).

The solution procedure starts with an initial guess (T_0, θ_0, φ_0) for the three unknowns. The corresponding values of the residuals (E_1, E_2, E_3), Eqs. (2.33) - (2.35), are probably not zero, but can be used to obtain an improved estimate T_1, θ_1, φ_1 by a Taylor series expansion about (T_0, θ_0, φ_0):

$$\begin{aligned} E_i(T_1, \theta_1, \varphi_1) \simeq E_i\big|_0 + \frac{\partial E_i}{\partial T}\bigg|_0 (T_1 - T_0) + \frac{\partial E_i}{\partial \theta}\bigg|_0 (\theta_1 - \theta_0) \\ + \frac{\partial E_i}{\partial \varphi}\bigg|_0 (\varphi_1 - \varphi_0) \quad ; \quad i = 1,2,3 \end{aligned} \tag{2.41}$$

truncated after the first order terms. The notation $|_0$ means evaluation of the expression using T_0, θ_0 and φ_0. For the sake of simplicity, this notation will be omitted in the following. On the assumption that $E_i(T_1, \theta_1, \varphi_1) = 0$ for $i = 1,2,3$, the following linear equations determine T_1, θ_1 and φ_1:

$$\begin{bmatrix} \frac{\partial E_1}{\partial T} & \frac{\partial E_1}{\partial \theta} & \frac{\partial E_1}{\partial \varphi} \\ \frac{\partial E_2}{\partial T} & \frac{\partial E_2}{\partial \theta} & \frac{\partial E_2}{\partial \varphi} \\ \frac{\partial E_3}{\partial T} & \frac{\partial E_3}{\partial \theta} & \frac{\partial E_3}{\partial \varphi} \end{bmatrix} \begin{Bmatrix} T_1 - T_0 \\ \theta_1 - \theta_0 \\ \varphi_1 - \varphi_0 \end{Bmatrix} = - \begin{Bmatrix} E_1 \\ E_2 \\ E_3 \end{Bmatrix} \tag{2.42}$$

As higher order terms in the Taylor series expansion have been omitted, $E_i\,(T_1,\ \theta_1,\ \varphi_1)$ are, however, normally not zero. Therefore, $(T_1,\ \theta_1,\ \varphi_1)$ must be used in a new iteration replacing $(T_0,\ \theta_0,\ \varphi_0)$ to yield a better estimate $(T_2,\ \theta_2,\ \varphi_2)$ and so forth. Usually, only a few iterations are needed, but for complicated structures like damaged ships several iterations may be necessary before a convergence $(E_i \simeq 0)$ is achieved.

In order to solve Eq. (2.42), the derivatives of E_i on the left hand side must be determined. This is most conveniently done by analysing the integral J of an arbitrary function $G(X,Y)$ over the submerged volume ∇:

$$J \equiv \iiint\limits_{\nabla} G(X,Y)\,d\nabla$$

The integration in the Z-direction ranges from the submerged surface S, given by $Z_S = Z_S(X,Y)$, to the still water surface $(Z = 0)$. It is easily seen that differentiation of J with respect to $\alpha = (T,\theta,\varphi)$ yields

$$\frac{\partial J}{\partial \alpha} \simeq \iint\limits_{A_W} - \frac{\partial Z_S}{\partial \alpha}\,G(X,Y)\,dXdY + \iiint\limits_{\nabla} \frac{\partial G}{\partial \alpha}\,d\nabla \tag{2.43}$$

for any single-connected submerged surface. The only assumption in Eq. (2.43) is that the *water plane area* A_W defined as the area of the plane $Z = 0$ inside the structure:

$$A_W = \iint\limits_{A_W} dX\,dY \tag{2.44}$$

does not change with α.

From Eqs. (2.38) - (2.40) it follows that

$$\frac{\partial Z}{\partial T} = -1 \quad ; \quad \frac{\partial Z}{\partial \theta} = X \quad ; \quad \frac{\partial Z}{\partial \varphi} = -Y\cos\theta \tag{2.45}$$

and, therefore, for $G(X,Y) = 1$ Eq. (2.43) yields

$$\frac{\partial \nabla}{\partial T} = A_W \quad ; \quad \frac{\partial \nabla}{\partial \theta} = -S_Y \quad ; \quad \frac{\partial \nabla}{\partial \varphi} = S_X\cos\theta \tag{2.46}$$

where the *moments of the water plane* A_W:

$$S_X = \iint_{A_W} Y \, dX \, dY \quad ; \quad S_Y = \iint_{A_W} X \, dX \, dY \tag{2.47}$$

have been introduced.

Similarly, $G(X,Y) = X$ leads to

$$\frac{\partial L_X}{\partial T} = S_Y \tag{2.48}$$

$$\frac{\partial L_X}{\partial \theta} = -\, I_{YY} - L_Z - T\nabla \tag{2.49}$$

$$\frac{\partial L_X}{\partial \varphi} = I_{XY} \cos\theta + L_Y \sin\theta \tag{2.50}$$

using

$$\frac{\partial X}{\partial \theta} = -\,(Z + T) \quad ; \quad \frac{\partial X}{\partial \varphi} = Y \sin\theta \tag{2.51}$$

and, finally, $G(X,Y) = Y$ yields

$$\frac{\partial L_Y}{\partial T} = S_X \tag{2.52}$$

$$\frac{\partial L_Y}{\partial \theta} = -\, I_{XY} \tag{2.53}$$

$$\frac{\partial L_Y}{\partial \varphi} = I_{XX} \cos\theta - L_X \sin\theta + L_Z \cos\theta + T\nabla \cos\theta \tag{2.54}$$

using

$$\frac{\partial Y}{\partial \theta} = 0 \quad ; \quad \frac{\partial Y}{\partial \varphi} = -\, X \sin\theta + (Z + T)\cos\theta \tag{2.55}$$

Here the *quadratic moments of the water plane area* are given by

$$I_{XX} = \iint_{A_W} Y^2 \, dX \, dY \tag{2.56}$$

$$I_{XY} = \iint_{A_W} XY \, dX \, dY \tag{2.57}$$

$$I_{YY} = \iint_{A_W} X^2 \, dX \, dY \tag{2.58}$$

and the *volume moment*

$$L_Z = \iiint_{\nabla} Z \, d\nabla \tag{2.59}$$

analogous to L_X and L_Y defined in Eqs. (2.24) and (2.25), respectively.

The derivatives of the residuals E_1, E_2, and E_3 can now be found from Eqs. (2.33) - (2.35), (2.46), (2.48) - (2.50) and (2.52) - (2.54) along with Eqs. (2.51), (2.55):

$$\frac{\partial E_1}{\partial T} = -A_W \;\; ; \;\; \frac{\partial E_1}{\partial \theta} = S_Y \;\; ; \;\; \frac{\partial E_1}{\partial \varphi} = -S_X \cos\theta \tag{2.60}$$

$$\frac{\partial E_2}{\partial T} = -S_Y \; ; \; \frac{\partial E_2}{\partial \theta} = I_{YY} - TE_1 - E_4 \; ; \; \frac{\partial E_2}{\partial \varphi} = -I_{XY}\cos\theta + E_3 \sin\theta \tag{2.61}$$

$$\frac{\partial E_3}{\partial T} = -S_X \; ; \; \frac{\partial E_3}{\partial \theta} = I_{XY} \; ; \; \frac{\partial E_3}{\partial \varphi} = -I_{XX}\cos\theta - E_2 \sin\theta + E_4 \cos\theta + TE_1 \cos\theta \tag{2.62}$$

where

$$E_4 \equiv Z_g V - L_Z \tag{2.63}$$

analogous to the definitions (2.34) and (2.35) of E_2 and E_3. Note, however, that usually $E_4 \neq 0$ at equilibrium.

In the next section, the calculation of the water plane coefficients A_W, S_X, S_Y I_{XX}, I_{XY} and I_{YY}, the volume ∇ and the volume moments L_X, L_Y and L_Z will be discussed.

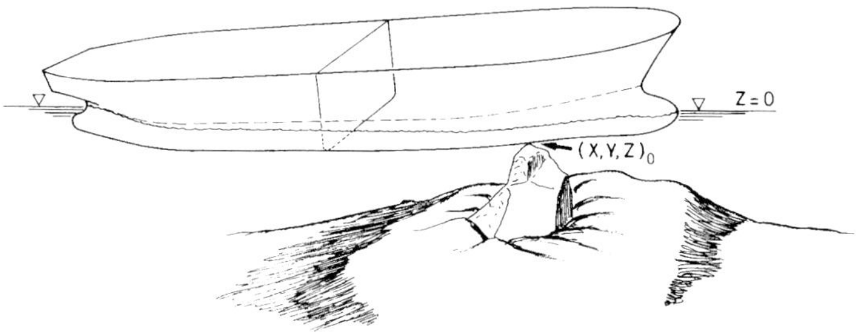

Figure 2.5 *Ship grounded on a rock.*

Example 2.1.1

If, for instance for a grounded ship (Figure 2.5), the hull is supported by the seabed at a point (x_0, y_0, z_0), then Eqs. (2.30) - (2.32) for the external forces must include the unknown reaction R at this point:

$$F_{rZ} = -gM + R \tag{2.64}$$

$$M_{rX} = -gMY_g + RY_0 \tag{2.65}$$

$$M_{rY} = gMX_g - RX_0 \tag{2.66}$$

where X_0 and Y_0 are the global coordinates for the contact point, obtained from Eqs. (2.38) - (2.40) by use of $(x, y, z) = (x_0, y_0, z_0)$. Hence, the equilibrium equations (2.21) - (2.23) can be written

$$E_1^* = E_1 - \frac{R}{\rho g} = V - \frac{R}{\rho g} - \nabla = 0 \tag{2.67}$$

$$E_2^* \equiv E_2 - X_0 E_1 = X_g V - L_X - X_0(V - \nabla) = 0 \tag{2.68}$$

$$E_3^* \equiv E_3 - Y_0 E_1 = Y_g V - L_Y - Y_0(V - \nabla) = 0 \tag{2.69}$$

where the two moment equilibrium equations are now conveniently are taken with respect to the contact point (X_0, Y_0) to eliminate the reaction force R from the equations. Thus, R only enters into the equilibrium equation (2.67). The four unknowns R, T, θ and φ are determined from Eqs. (2.67) - (2.69) and the geometrical condition for the contact point, Eq. (2.40):

$$E_0^* \equiv -T + x_0 \sin\theta - y_0 \cos\theta \sin\varphi + z_0 \cos\theta \cos\varphi - Z_0 = 0 \tag{2.70}$$

where Z_0 is the known global vertical coordinate for the contact point. The solution to Eqs. (2.67) - (2.70) can be obtained by solving the three non-linear equations (2.68) - (2.70) for T, θ and φ by using the same procedure as for the freely floating structure:

$$E_2^* = 0 \quad ; \quad E_3^* = 0 \quad ; \quad E_0^* = 0 \tag{2.71}$$

with derivatives

$$\frac{\partial E_2^*}{\partial T} = -S_Y + X_0 A_W$$

$$\frac{\partial E_2^*}{\partial \theta} = \frac{\partial E_2}{\partial \theta} - \frac{\partial X_0}{\partial \theta} E_1 - X_0 \frac{\partial E_1}{\partial \theta}$$

$$= I_{YY} - TE_1 - E_4 + (Z_0 + T)E_1 - X_0 S_Y$$

$$= I_{YY} - X_0 S_Y - E_4^*$$

$$\begin{aligned}
\frac{\partial E_2^*}{\partial \varphi} &= -(I_{XY} - X_0 S_X)\cos\theta + E_3^*\sin\theta \\
\frac{\partial E_3^*}{\partial T} &= -S_X + Y_0 A_W \\
\frac{\partial E_3^*}{\partial \theta} &= I_{XY} - Y_0 S_Y \qquad (2.72) \\
\frac{\partial E_3^*}{\partial \varphi} &= -(I_{XX} - Y_0 S_X)\cos\theta - E_2^*\sin\theta + E_4^*\cos\theta \\
\frac{\partial E_0^*}{\partial T} &= -1, \quad \frac{\partial E_0^*}{\partial \theta} = X_0, \quad \frac{\partial E_0^*}{\partial \varphi} = -Y_0\cos\theta
\end{aligned}$$

where

$$E_4^* = E_4 - Z_0 E_1 \qquad (2.73)$$

When (2.71) has been solved for the next approximation $(T, \theta, \varphi)_i$ by use of the derivatives calculated at the previous approximation $(R, T, \theta, \varphi)_{i-1}$, the new value of the reaction R_i is obtained explicitly from Eqs. (2.67).

In a grounding situation, a final value of R less than zero probably implies that the ship lifts off from the contact point and becomes freely floating again.

In the case of two contact points, only one free parameter defining the position of the ship in the water, remains, as a new geometrical restraint similar to Eq. (2.70) holds for the second contact point. With three points of contact, the equilibrium position is completely given and the three restraint quations yield the three unknown reactions.

2.2 HYDROSTATIC COEFFICIENTS

In order to calculate the equilibrium position of the structure, the following hydrostatic coefficients must be determined at each iteration step:

Volume integrals:

Volume ∇, Eq. (2.14)
Volume moments: L_X, L_Y, L_Z; Eqs. (2.24) - (2.25), (2.59)

Water plane integrals:

Water plane area: A_W; Eq. (2.44)
Moments of water plane area: S_X, S_Y; Eqs. (2.47)
I_{XX}, I_{XY}, I_{YY}; Eqs (2.56) - (2.58)

It should be noted that all moments are defined in terms of the global coordinates, but use of the transformation (2.38) - (2.40) makes it possible to express them in the local xyz-coordinate system fixed in the structure. Thus, for instance

$$L_Z = \iiint_{\nabla} Z\, d\nabla = L_x \sin\theta - L_y \cos\theta \sin\varphi + L_z \cos\theta \cos\varphi - T\nabla \tag{2.74}$$

where the volume moments are given by

$$L_x = \iiint_{\nabla} x\, d\nabla \quad ; \quad L_y = \iiint_{\nabla} y\, d\nabla \quad ; \quad L_z = \iiint_{\nabla} z\, d\nabla \tag{2.75}$$

For the integrals over the water plane area, the relation

$$dXdY = \det \begin{bmatrix} \dfrac{\partial X}{\partial x} & \dfrac{\partial X}{\partial y} \\ \dfrac{\partial Y}{\partial x} & \dfrac{\partial Y}{\partial y} \end{bmatrix} dxdy \tag{2.76}$$

is used. At the water plane $Z = 0$, Eqs. (2.38) - (2.40) imply that

$$X = \frac{x}{\cos\theta} - T\tan\theta \tag{2.77}$$

$$Y = -x\tan\theta\tan\varphi + \frac{y}{\cos\varphi} + \frac{T\tan\varphi}{\cos\theta} \tag{2.78}$$

so that

$$dXdY = \frac{dxdy}{\cos\theta\cos\varphi} \tag{2.79}$$

This yields the following expressions:

$$A_W = \iint_{A_W} dXdY = \frac{A_w}{\cos\theta\cos\varphi} \tag{2.80}$$

$$S_X = \iint_{A_w} YdXdY = -S_y \frac{\tan\theta}{\cos\theta}\frac{\tan\varphi}{\cos\varphi} + \frac{S_x}{\cos\theta\cos^2\varphi} + T\frac{\tan\varphi}{\cos\theta} A_w \tag{2.81}$$

with analogous formulas for S_Y, I_{XX}, I_{XY} and I_{YY}. Here

$$A_w = \iint_{A_w} dxdy \tag{2.82}$$

is the water plane area, projected onto the local xy-plane and

$$S_y = \iint_{A_w} x dx dy \quad ; \quad S_x = \iint_{A_w} y dx dy \tag{2.83}$$

are the corresponding area moments.

Example 2.2.1

For a structure symmetrical about the plane $y = 0$ and with the centre of gravity in this plane ($y_g = 0$), the equilibrium equation (2.35) and the relation (2.39) yield

$$z_g V \sin\varphi - L_z \sin\varphi = 0 \tag{2.84}$$

as $L_y = 0$ due to the symmetry. The vertical centre of buoyancy

$$z_b \equiv \frac{L_z}{\nabla} \tag{2.85}$$

usually differs from the vertical centre of gravity z_g and hence, the equilibrium equation (2.84) requires that $\varphi = 0$ (or $\varphi = 180°$). The two remaining equilibrium equations (2.33) - (2.34) are solved by use of the derivatives of E_1 and E_2, Eqs. (2.60) - (2.61), with respect to T and θ.

For ships with a pronounced longitudinal axis, the angle of trim θ is usually very small, so that $\cos\theta \simeq 1$ and $\sin\theta \simeq 0$ can be assumed in the derivatives of E_1 and E_2 without reducing the rate of convergence towards the equilibrium position. In addition, the term E_4, Eq. (2.63), is often also small in comparison with I_{YY}. In that case, the derivatives assume the simple form of

$$\begin{aligned} \frac{\partial E_1}{\partial T} &= -A_w \quad ; \quad \frac{\partial E_1}{\partial \theta} = S_y \\ \frac{\partial E_2}{\partial T} &= -S_y \quad ; \quad \frac{\partial E_2}{\partial \theta} = I_{yy} \end{aligned} \tag{2.86}$$

containing only water plane coefficients calculated in the local xyz-system. These formulas are often used in hand calculations for simple geometrical structures.

From Eqs. (2.42) and (2.86) it follows that the first estimate $(\Delta T, \Delta\theta)$ for the correction to the assumed equilibrium position satisfies

$$\begin{aligned} -A_w \Delta T + S_y \Delta\theta &= -E_1 \\ -S_y \Delta T + I_{yy} \Delta\theta &= -E_2 \end{aligned} \tag{2.87}$$

where E_1 and E_2 are calculated for the assumed position. The solution becomes

$$\begin{aligned} \Delta T &= \frac{E_1 I_{yy} - E_2 S_y}{I_{yy} A_w - S_y^2} \\ \Delta\theta &= \frac{E_1 S_y - E_2 A_w}{I_{yy} A_w - S_y^2} \end{aligned} \tag{2.88}$$

These formulas are also useful to determine the change in equilibrium position when a mass item m is added (or removed) from the ship. In such cases

$$E_1 = \frac{m}{\rho} \quad ; \quad E_2 = x_g^{'}\frac{m}{\rho} \tag{2.89}$$

where $x_g^{'}$ is the centre of gravity of the additional mass m.

By introduction of the longitudinal centre of flotation x_f defined by

$$S_y = \iint_{A_w} x dx dy \equiv x_f A_w \tag{2.90}$$

the quadratic moment I_{yy} can be written

$$\begin{aligned} I_{yy} &= \iint_{A_w} x^2 dx dy = \iint_{A_w} (x - x_f + x_f)^2 dx dy \\ &= \iint_{A_w} (x - x_f)^2 dx dy + x_f^2 A_w \\ &\equiv I_{fyy} + x_f^2 A_w \end{aligned} \tag{2.91}$$

Inserting Eqs. (2.89) - (2.91) into Eqs. (2.88) yields

$$\Delta\theta = \frac{m}{\rho}(x_f - x_g^{'})\frac{1}{I_{fyy}} \tag{2.92}$$

and

$$\Delta T_f \equiv \Delta T - x_f \Delta\theta = \frac{m}{\rho}\frac{1}{A_w} \tag{2.93}$$

The draught change ΔT_f is the draught change at $x = x_f$, see Figure 2.6.

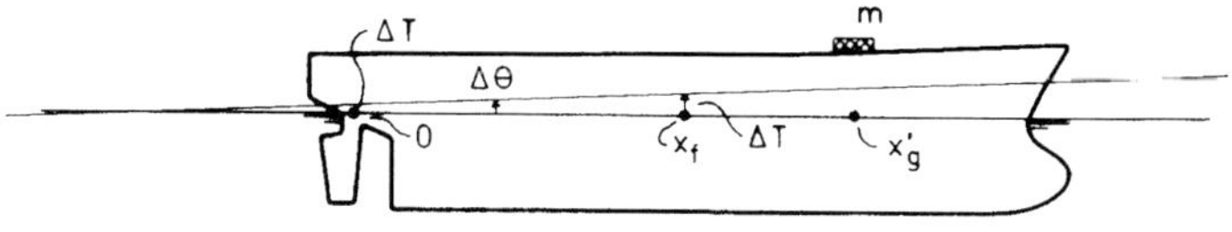

Figure 2.6 *Draught and trim changes due to an added mass item.*

The ratios

$$\frac{m}{\Delta T_f} = \rho A_w \quad ; \quad \frac{m(x_f - x_g^{'})}{\Delta\theta} = \rho I_{fyy} \tag{2.94}$$

are denoted the mass to change draught and the moment to change trim, respectively, and are very useful data for the captain during loading and unloading of the ship.

Example 2.2.2

Consider a rectangular box with the length L, the breadth B and the depth D. The total mass is M with a centre of gravity $(x_g,\ 0,\ z_g)$ in a coordinate system (x,y,z) with $y = 0$ and $x = 0$ at the intersection between the vertical planes of symmetry and with $z = 0$ in the bottom of the box. Thus, as in Example 2.2.1, the angle of heel φ is equal to zero.

On the assumption that the depth is larger than the sectional draught along the whole length of the structure, the volume coefficients become

$$\nabla = LB\tilde{T}$$

$$L_x = -\frac{1}{12}BL^3\tan\theta$$

$$L_y = 0$$

$$L_z = \frac{1}{2}BL\left(\tilde{T}^2 + \frac{1}{12}L^2\tan^2\theta\right)$$

using

$$\iiint_{\nabla} f(x,\ y,\ z)\,d\nabla = \int_{-L/2}^{L/2}\left[\int_{-B/2}^{B/2}\left\{\int_0^{\tilde{T}-x\tan\theta} f(x,\ y,\ z)\,dz\right\}dy\right]dx$$

and $\tilde{T} \equiv T/\cos\theta$. Physically, $\tilde{T}$ and T are the draught amidships measured perpendicularly to the yz-plane and YZ-plane, respectively.

The water plane coefficients are

$$A_w = BL$$

$$S_x = S_y = I_{xy} = 0$$

$$I_{xx} = \frac{1}{12}B^3L$$

$$I_{yy} = \frac{1}{12}L^3B$$

Hence, the two equilibrium equations (2.33) and (2.34) become

$$E_1 = V - BL\tilde{T} = 0$$

$$E_2 = \left(x_g\cos\theta - z_g\sin\theta\right)V + \frac{1}{12}BL^3\sin\theta + \frac{1}{2}BL\left(\tilde{T}^2 + \frac{1}{12}L^2\tan^2\theta\right)\sin\theta = 0$$

by application of Eqs. (2.38) and (2.40). The first equation yields

$$\tilde{T} = \frac{\nabla}{BL} = \frac{M}{\rho BL}$$

whereas the second equation provides an equation in $\tan\theta$:

$$\frac{1}{24}L^2\tan^3\theta + \left(- z_g\tilde{T} + \frac{1}{12}L^2 + \frac{1}{2}\tilde{T}^2 \right)\tan\theta + x_g\tilde{T} = 0$$

As a numerical example, let

$L = 100$ m, $B = 20$ m, $x_g = 10$ m, $z_g = 5$ m and $V = 14000$ m^3

Then, the equilibrium position becomes

$$\tilde{T} = \frac{\nabla}{BL} = 7m$$

and

$$\tan\theta = -0.08476 \Rightarrow \theta = -4.845^o$$

so that

$$T = \tilde{T}\cos\theta = 6.975 \text{ m}$$

With $(\tilde{T}, \theta)_0 = (V/BL, 0)_0$ as a first guess, the iteration procedure (2.42) yields

$$-BL(\tilde{T}_1 - \tilde{T}_{0)} = -E_1|_0 = 0$$

$$\left[I_{yy} - (z_gV - L_z)\right]\theta_1 = -E_2|_0 = -x_gV$$

or

$$\tilde{T}_1 = \tilde{T}_0 = \tilde{T} = 7 \text{ m}$$

and

$$\tan\theta_1 = -\frac{x_g\tilde{T}}{\frac{1}{12}L^2 - \left(z_g\tilde{T} - \frac{1}{2}\tilde{T}^2\right)} = -0.08507$$

which implies

$$\theta_1 = -4.862^o$$

In hand calculations, see Example 2.2.1, the approximate formula for the derivative may be used:

$$\frac{\partial E_2}{\partial \theta} \simeq I_{yy}$$

so that

$$\tan\theta_1 = -\frac{x_g \tilde{T}}{\frac{1}{12}L^2} = -0.08400$$

$$\theta_1 = -4.802^o$$

In both cases, the first estimate θ_1 is very close to the exact value $\theta = -4.845^o$.

The transformation of the volume and water plane area integrals from the global XYZ-system to the local xyz-system facilitates greatly the numerical integrations as the surface geometry is given in the xyz-system. However, two- or three-dimensional integrations still have to be performed. For ships, the integration are normally performed by first determination of the sectional properties, e.g. the properties for x = constant, followed by an integration along the x-axis. For more complicated structures with no specific longitudinal axis or with a very complex exterior form such as a semi-submersible or a damaged ship with flooded compartments, it may be more feasible to consider individually each panel subjected to the hydrostatic pressure. These two procedures will be discussed in the following subsection.

2.2.1 Integration of Sectional Coefficients

In this procedure, volume integrations are carried out as

$$\iiint_{\nabla} x^i y^j z^k dxdydz = \int_{x_{min}}^{x_{max}} x^i \left[\iint_{A(x)} y^j z^k dydz \right] dx \tag{2.95}$$

where $A(x)$ is the submerged sectional area at $x = x$, see Figure 2.7. By suitable choice of i, j, k the volume moments L_x, L_y and L_z as well as the volume ∇ are obtained. x_{max} and x_{min} are the maximum and minimum values of the x-coordinate of the submerged volume, respectively, see Figure 2.7. The integrand

$$S_{jk}(x) \equiv \iint_{A(x)} y^j z^k dydz \tag{2.96}$$

in Eq. (2.95) represents sectional hydrostatic coefficients often denoted *Bonjean curves*. If the heeling angle φ is equal to zero, these coefficients will be a function of only the local sectional draught and can easily be visualised as curves.

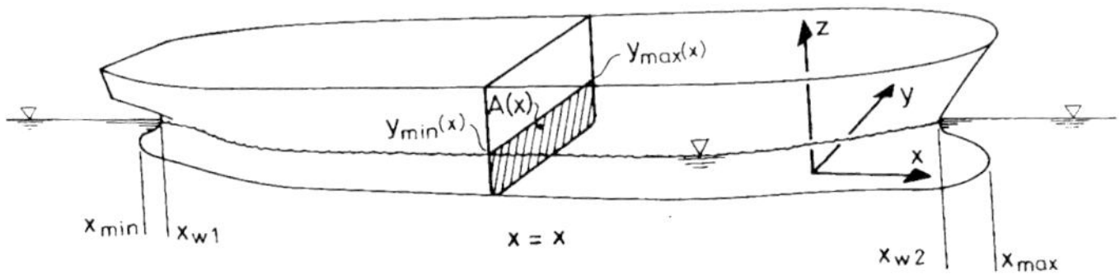

Figure 2.7 *Submerged sectional area A(x) and the integration domain.*

The sectional integration domain $A(x)$ in Eq. (2.96) is bounded by the sectional contour Ω_S:

$$z_S = z_s(y) \tag{2.97}$$

and the water line Ω_w at $x = x$:

$$z_w(y) = y \tan\varphi - \frac{x \tan\theta}{\cos\varphi} + \frac{T}{\cos\theta\cos\varphi} \tag{2.98}$$

which follows from Eq. (2.40) with $Z = 0$. Numerical integration of Eq. (2.96) is usually most easily carried out by transforming the integral over $A(x)$ to an integral along the contour Ω by use of Green's theorem:

$$\int\int_A \left[\frac{\partial P}{\partial y} - \frac{\partial Q}{\partial z}\right] dydz = \int_\Omega [Qdy + Pdz] \tag{2.99}$$

where $P = P(y, z)$ and $Q = Q(y, z)$ are any differentiable functions of y, z and where $\Omega = \Omega_s + \Omega_w$ is the closed contour of the area A. In order to avoid problems with uniqueness by use of Eqs. (2.97) and (2.98), the sectional contour is normally given as

$$y_S = y_s(t) \quad ; \quad z_S = z_s(t) \tag{2.100}$$

$$y_W = y_w(t) \quad ; \quad z_W = z_w(t) \tag{2.101}$$

where $t \in [t_1; t_2]$ is a continuous parameter. Typical piece-wise third order polynomials in t yield a sufficiently accurate description of any hull form. The coefficients in these polynomials are determined from the requirement that continuity in both the tangent and the curvature must exist between each segment, if not otherwise specified.

By use of Eqs. (2.99) - (2.101), the integral S_{jk} can be written:

$$S_{jk} = \frac{1}{j+1} \int_{t_1}^{t_2} y^{j+1}(t)\, z^k(t)\, \frac{dz}{dt}\, dt$$
$$= -\frac{1}{k+1} \int_{t_1}^{t_2} y^j(t)\, z^{k+1}(t)\, \frac{dy}{dt}\, dt \tag{2.102}$$

on the assumption that either $\frac{\partial P}{\partial y}$ or $\frac{\partial Q}{\partial z}$ is equal to $y^j z^k$. That the expressions above are identical is easily verified by partial integration using $y(t_1) = y(t_2)$; $z(t_1) = z(t_2)$ as the contour $\Omega_s + \Omega_w$ is closed. Which of the expressions is most suitable depends on the parameter descriptions (2.100) and (2.101). A polynomial form of the integrand is seen to allow for analytical integration of Eq. (2.102). Finally, the volume integrations Eq. (2.95)

$$\iiint_{\nabla} x^i y^j z^k dx\, dy\, dz = \int_{x_{min}}^{x_{max}} x^i S_{jk}(x)\, dx \tag{2.103}$$

must normally be carried out numerically, for example by means of the Simpson rule.

The water plane coefficients A_w, S_x, S_y, I_{xx}, I_{xy} and I_{yy} are obtained by integration first along the y-axis then with respect to x. In general

$$h_{ij} \equiv \iint_{A_{wp}} x^i y^j dx\, dy = \int_{x_{w2}}^{x_{w1}} x^i \left[\int_{B(x)} y^j dy \right] dx \tag{2.104}$$

with $x_{w1} \leq x \leq x_{w2}$. Here

$$\int_{B(x)} y^j\, dy = \int_{y_{min}(x)}^{y_{max}(x)} y^j\, dy = \frac{1}{j+1} \left(y_{max}^{j+1}(x) - y_{min}^{j+1}(x) \right) \tag{2.105}$$

where $y_{min}(x) \leq y\,(x|Z = 0) \leq y_{max}(x)$, see Figure 2.7.

2.2.2 Integration of Submerged Panels

For complicated submerged surfaces, definition and integration of sectional coefficients as described in the previous subsection may not be feasible. Alternatively, it may be appropriate to add directly the contributions from each submerged panel to the hydrostatic coefficients.

The water plane coefficients A_W, S_x, S_y, I_{xx}, I_{xy} and I_{yy} are obtained as before from the general expression

$$H_{ij} \equiv \iint_{A_W} X^i Y^j dX dY \tag{2.106}$$

by suitable choice of (i,j). Each submerged panel thus contributes to H_{ij} by its projection onto the still water surface. If the closed contour of this projection of a panel p is denoted Ω_p and given by $\left(X_p(t),\ Y_p(t);\ t_{1p} \le t \le t_{2p}\right)$, then Green's theorem, Eq. (2.99), yields

$$\begin{aligned} H_{ij} &= \frac{1}{i+1} \sum_{p=1}^{n} \int_{\Omega_p} X_p^{i+1}\, Y_p^j \frac{dY_p}{dt}\, dt \\ &= -\frac{1}{j+1} \sum_{p=1}^{n} \int_{\Omega_p} X_p^i\, Y_p^{j+1} \frac{dX_p}{dt}\, dt \end{aligned} \tag{2.107}$$

where n is the total number of submerged panels. Which of the two formulas to apply depends on the way $(X_p,\ Y_p)$ are given. If each panel is approximated by a polygon, the contour consists of a set of m straight lines, each given by for example

$$Y_{pq} = a_{pq} X_{pq} + b_{pq} \quad ; \quad q = 1, 2, \ldots, m \qquad X_{pq1} \le X_{pq} \le X_{pq2} \tag{2.108}$$

where the coefficients a_{pq}, b_{pq} are determined from the contour curve given in local xyz-coordinates and transformed to the XYZ-system by means of Eqs. (2.38) - (2.40). Taking $t \equiv X_{pq}$, Eq. (2.107) becomes

$$\begin{aligned} H_{ij} &= -\frac{1}{j+1} \sum_{p=1}^{n} \sum_{q=1}^{m} \int_{X_{pq1}}^{X_{pq2}} X_{pq}^i\, Y_{pq}^{j+1}\, dX_{pq} \\ &= -\frac{1}{j+1} \sum_{p=1}^{n} \sum_{q=1}^{m} \int_{X_{pq1}}^{X_{pq2}} t^i \left(a_{pq}\, t + b_{pq}\right)^{j+1} dt \end{aligned} \tag{2.109}$$

so that analytical integration can easily be performed for the water plane coefficients. Thus the numerical integration along the x-axis using sectional coefficients, as described previously, is avoided. Of course, the approximation of the submerged surface by n flat panels introduces another numerical error.

$$\overline{M}_y(x) = -g\overline{x}_g M(\overline{x}) + \rho g L_x(\overline{x}) + \overline{x}\,Q_z(\overline{x}) \tag{2.135}$$

using Eqs (2.112) and (2.118). Here

$$\overline{x}_g M(\overline{x}) = \int_{x_{\min}}^{\overline{x}} x m(x) dx \tag{2.136}$$

and

$$L_x(\overline{x}) = \int_{x_{\min}}^{\overline{x}} x A(x) dx \tag{2.137}$$

so that Eq. (2.135) takes the form

$$\begin{aligned} \overline{M}_y(\overline{x}) &= \int_{x_{\min}}^{\overline{x}} \left[(\overline{x} - x)q(x)\right] dx \\ &= \int_{x_{\min}}^{\overline{x}} \int_{x_{\min}}^{x} q(u)du\, dx = \int_{x_{\min}}^{\overline{x}} Q_z(x)dx \end{aligned} \tag{2.138}$$

by use of Eqs. (2.133) and (2.134). The formulas (2.133), (2.134) and (2.138) provide a convenient means of determining the vertical shear force and the vertical bending moment distribution in the hull by simple integrations.

2.4 HYDROSTATIC STABILITY

An important aspect related to hydrostatic loads is the concept of hydrostatic stability. The equilibrium position determined in Section 2.1 may be stable or unstable. Stable equilibrium implies that if the structure is subjected to small external forces, this will not lead to large changes in the equilibrium configuration and, also, that if these external forces are removed, then the original equilibrium position will be reestablished. If this is not the case, the equilibrium is characterised as unstable.

In addition to this initial stability analysis, information is normally also needed on how the structure will behave if subjected to large quasi-static wind and wave loads. This behaviour is determined by applying additional external forces and/or moments to Eqs. (2.30) - (2.32). The results for ships are typically presented as the so-called righting arm curves which give the relation between an external heeling moment and the resulting heeling angle.

The results of both types of hydrostatic stability analysis are the main input to the current international rules assessing the stability of floating vessels. A discussion of these rules is, however, outside the scope of this presentation, but it must be noted, that at present even dynamic load effects are modelled in these rules by the hydrostatic stability properties, due to insufficient knowledge of the parameters entering a dynamic stability analysis.

In the following, the two concepts of hydrostatic stability will be discussed.

2.4.1 Initial Hydrostatic Stability

Consider a structure at equilibrium in a calm sea. The necessary and sufficient condition for a stable equilibrium is that the potential energy Π of the structure has a minimum. If the potential energy in the equilibrium position is assumed as reference, then a stable configuration requires that

$$\Delta\Pi(\Delta T, \Delta\theta, \Delta\varphi) > 0 \tag{2.139}$$

for any small changes, given by ΔT, $\Delta\theta$ and $\Delta\varphi$, of the equilibrium position.

The potential energy is equal to the work done by the external forces and moments. In the global XYZ-system this work becomes

$$\Delta W(\Delta T, \Delta\theta, \Delta\varphi) = \Delta F_Z \Delta T + \Delta M_Y \Delta\theta + \Delta M_X \Delta\varphi \tag{2.140}$$

where the force increment, ΔF_Z, and the moment increments, ΔM_X and ΔM_Y, are found from Eqs. (2.21) - (2.23) and (2.33) - (2.35):

$$\Delta F_Z = -\rho g \left[\frac{\partial E_1}{\partial T} \Delta T + \frac{\partial E_1}{\partial \theta} \Delta\theta + \frac{\partial E_1}{\partial \varphi} \Delta\varphi \right] \tag{2.141}$$

$$\Delta M_Y = \rho g \left[\frac{\partial E_2}{\partial T} \Delta T + \frac{\partial E_2}{\partial \theta} \Delta\theta + \frac{\partial E_2}{\partial \varphi} \Delta\varphi \right] \tag{2.142}$$

$$\Delta M_X = -\rho g \left[\frac{\partial E_3}{\partial T} \Delta T + \frac{\partial E_3}{\partial \theta} \Delta\theta + \frac{\partial E_3}{\partial \varphi} \Delta\varphi \right] \tag{2.143}$$

Inserting Eqs. (2.141) - (2.143) in Eq. (2.140) and requiring

$$\Delta W = \Delta\Pi > 0 \tag{2.144}$$

yield the required condition for a stable equilibrium condition. The derivatives $\partial E_1/\partial T$ etc. are given by Eqs. (2.60) - (2.62). Evaluation of the requirement (2.144) is facilitated by a proper choice of coordinate systems. Without loss of generality, the local xyz- and global XYZ-systems are taken to coincide in the equilibrium condition (i.e. $T = \theta = \varphi = 0$) and, furthermore, to be orientated so that (see Example 2.4.1)

$$S_x = S_y = I_{xy} = 0 \tag{2.145}$$

then, as $E_1 = E_2 = E_3 = 0$ in the equilibrium position:

$$\frac{1}{\rho g}\Delta W = A_w(\Delta T)^2 + (I_{yy} - E_4)(\Delta\theta)^2 + (I_{xx} - E_4)(\Delta\varphi)^2 \quad (2.146)$$

This shows that stable equilibrium is ensured if

$$E_4 = z_g \nabla - L_z < \min(I_{xx}, I_{yy}) \quad (2.147)$$

using Eq. (2.63). By defining the *vertical centre of buoyancy* z_b as

$$z_b \equiv \frac{\iiint\limits_{\nabla} z\,d\nabla}{\iiint\limits_{\nabla} d\nabla} = \frac{L_z}{\nabla} \quad (2.148)$$

and the *metacentric height* z_m by

$$z_m \equiv z_b + \frac{\min(I_{xx}, I_{yy})}{\nabla} \quad (2.149)$$

the requirement (2.147) simply becomes

$$GM \equiv z_m - z_g > 0 \quad (2.150)$$

The metacentre $(0,0,z_m)$ can be considered as the centre of rotation of the structure for small excursions from the equilibrium position. Therefore, as for a pendulum, stable equilibrium requires that this point is situated above the centre of gravity.

Example 2.4.1

A water plane coordinate system (xy) satisfying Eq. (2.145) can always be determined. Let $x'y'$ be a given, arbitrarily orientated system in which the moments $S_{x'}$, $S_{y'}$, $I_{x'x'}$, $I_{x'y'}$ and $I_{y'y'}$ of the water plane area A_w are known. A new xy-system is defined by a translation (x_0', y_0') and a rotation (α) of the $x'y'$- system:

$$x = (x' - x_0')\cos\alpha - (y' - y_0')\sin\alpha$$

$$y = (x' - x_0')\sin\alpha + (y' - y_0')\cos\alpha$$

Substitution of this transformation into the two first requirements in (2.145) yields

$$S_x = \iint_{A_w} y dx dy = \left(S_{y'} - x_0' A_w\right) \sin\alpha + \left(S_{x'} - y_0' A_w\right) \cos\alpha = 0$$

$$S_y = \iint_{A_w} x dx dy = \left(S_{y'} - x_0' A_w\right) \cos\alpha - \left(S_{x'} - y_0' A_w\right) \sin\alpha = 0$$

implying that the centre of flotation

$$\left(x_0', y_0'\right) = \left(\frac{S_{y'}}{A_w}, \frac{S_{x'}}{A_w}\right)$$

irrespective of the rotation α. The last requirement gives:

$$I_{xy} = \iint_{A_w} xy dx dy = \left[I_{y'y'} - x_0'^2 A_w - \left(I_{x'x'} - y_0'^2 A_w\right)\right] \cos\alpha \sin\alpha$$

$$+ \left[I_{x'y'} - x_0' y_0' A_w\right]\left(\cos^2\alpha - \sin^2\alpha\right) = 0$$

or

$$\tan 2\alpha = 2 \frac{I_{x'y'} - x_0' y_0' A_w}{I_{y'y'} - x_0'^2 A_w - \left(I_{x'x'} - y_0'^2 A_w\right)}$$

After a little algebra the minimum value of I_{xx} and I_{yy} to be used in Eq. (2.149) becomes:

$$\min\left(I_{xx}, I_{yy}\right) = \frac{1}{2}\left[I_{y'y'} - x_0'^2 A_w + I_{x'x'} - y_0'^2 A_w\right]$$
$$- \frac{1}{2}\left[\left\{I_{y'y'} - x_0'^2 A_w - \left(I_{x'x'} - y_0'^2 A_w\right)\right\}^2 + 4\left(I_{x'y'} - x_0' y_0' A_w\right)^2\right]^{1/2}$$

The angle α associated with the minimum value of (I_{xx}, I_{yy}) defines the direction towards which the structure has the least metacentric height and, therefore, least stability.

2.4.2 Righting Arm

Whereas the initial stability analysis checks the hydrostatic stability for small disturbances from the equilibrium position, the determination of righting arm curves gives information on the stability when the structures are subjected to large external forces in addition to the gravitational and hydrostatic loads. These forces may represent quasi-static loads due to wind and waves.

The main problem in connection with this analysis is the definition of a proper measure of the change in equilibrium position. One possibility is always to use the angle of rotation associated with the current minimum metacentric height. However, this procedure has the drawback that this angle is measured relative to an axis (the x-axis) whose direction depends on the shape of the current water plane area, due to the requirement $I_{xy} = 0$. Thus, even for initially symmetrical structures like ships, this angle of rotation is difficult to visualise. The calculation procedure is, however, straightforward. The external moment is conservatively assumed always to act in the direction of minimum metacentric height, as the structure is free to rotate about a vertical axis. The external moment M^e is then applied in small increments ΔM_i^e ; $i = 1, 2, ...$, so that, for each increment, the corresponding change in equilibrium position can be determined from the linearised equilibrium equations (2.42). In order to apply ΔM_i^e in the current direction with the least metacentric height, it is appropriate to let the local xyz- and global XYZ-coordinates coincide at each load step and be orientated so that the requirements (2.145) are satisfied. This transformation can be done as described in Example 2.4.1. If $I_{xx} < I_{yy}$ the minimum metacentric height is associated with rotations about the current x-axis. Therefore, in the equilibrium equations (2.42), the right-hand side should be taken to be $(0,\ 0,\ -\Delta M_i^e/\rho g)$. As the current XY-system satisfies (2.145), the solution Eqs. (2.42) becomes

$$\Delta T_i = \Delta\theta_i = 0 \quad ; \quad \Delta\varphi_i = \frac{\Delta M_i^e}{\rho g \nabla (z_m - z_g)} \qquad (2.151)$$

where z_m $(= I_{xx}/\nabla)$ and z_g are the vertical coordinates for the metacentre and the centre of gravity, respectively, in the current xyz-system. For the next load increment ΔM_{i+1}^e, the equilibrium position $\left(X_{i+1},\ Y_{i+1},\ Z_{i+1}\right)$ is given by the transformation (2.38) - (2.40) with T, θ and φ given by Eqs. (2.151):

$$X_{i+1} = x_i \quad ; \quad Y_{i+1} = y_i + z_i \Delta\varphi_i \quad ; \quad Z_{i+1} = z_i - y_i \Delta\varphi_i \qquad (2.152)$$

The $(XYZ)_{i+1}$-system must then be transformed so that it satisfies Eq. (2.145) before solving for the change in equilibrium position due to the load increment ΔM_{i+1}^e.

Although this procedure yields the equilibrium position as a function of the applied moment, the result is difficult to interpret if the direction corresponding to the minimum metacentric height changes with the magnitude of applied moment. In such cases, no simple measure of the change in equilibrium position with applied moment normally exists. However, successive use of Eq. (2.152) together with the transformation required to satisfy Eq. (2.145) will yield the current equilibrium position relative to an $x'y'z'$-system fixed in the structure.

To avoid the difficulties described above, another procedure is normally used, especially for ships. The local xyz-system is kept fixed with respect to the structure and, rather than specifying an external moment M^e, the angle of heel φ is prescribed in increments $\Delta\varphi_i$. As no change in the xyz-system is made, the accumulated angle φ is simply the algebraic sum of $\Delta\varphi_i$:

$$\varphi_n = \sum_{i=1}^{n} \Delta\varphi_i \tag{2.153}$$

In the procedure, the external moment M^e necessary to generate an angle of heel φ is assumed to act about the global X-axis, i.e. $M^e_X = M^e$, $M^e_Y = 0$ and $F^e_Z = 0$. The equilibrium equations (2.21) - (2.23) can then be written

$$E_1 = 0 \quad ; \quad E_2 = 0 \quad ; \quad E_3 = -\frac{M^e}{\rho g} \tag{2.154}$$

with the definitions (2.33) - (2.35) of E_1, E_2 and E_3. For a given increment $\Delta\varphi_i$, the equations (2.154) are solved by use of the iterative scheme, Eqs. (2.42), as follows. First, the associated changes ΔT_i and $\Delta\theta_i$ are determined iteratively by

$$\begin{bmatrix} \frac{\partial E_1}{\partial T} & \frac{\partial E_1}{\partial \theta} \\ \frac{\partial E_2}{\partial T} & \frac{\partial E_2}{\partial \theta} \end{bmatrix} \begin{Bmatrix} \Delta T_i^{(j)} \\ \Delta\theta_i^{(j)} \end{Bmatrix} = -\begin{Bmatrix} E_1 + \frac{\partial E_1}{\partial \varphi}\Delta\varphi \\ E_2 + \frac{\partial E_2}{\partial \varphi}\Delta\varphi \end{Bmatrix} \tag{2.155}$$

The coefficients in (2.155) are evaluated in the current equilibrium position:

$$T_{i-1} + \sum_{k=1}^{j-1} \Delta T_i^{(k)} \quad ; \quad \theta_{i-1} + \sum_{k=1}^{j-1} \Delta\theta_i^{(k)} \quad ; \quad \varphi_i \tag{2.156}$$

where T_{i-1} and θ_{i-1}, are the equilibrium values obtained with $\varphi = \varphi_{i-1}$. The iteration scheme usually converges very fast, so only a few iterations (m) are needed to get ΔT_i and $\Delta\theta_i$:

$$\Delta T_i = \sum_{j=1}^{m} \Delta T_i^{(j)} \quad ; \quad \Delta\theta_i = \sum_{j=1}^{m} \Delta\theta_i^{(j)} \tag{2.157}$$

and, hence, the new equilibrium is found:

$$T_i = T_{i-1} + \Delta T_i \quad ; \quad \theta_i = \theta_{i-1} + \Delta\theta_i \quad ; \quad \varphi_i = \varphi_{i-1} + \Delta\varphi_i \tag{2.158}$$

After determination of the equilibrium positions $(T, \theta, \varphi)_i$; $i = 1, 2, ..., n$ the external moments M^e_i needed to create these equilibrium positions are found from Eq. (2.154):

$$M^e_i = -\rho g E_3(T_i, \theta_i, \varphi_i) \tag{2.159}$$

As $T_i = T_i(\varphi_i)$ and $\theta_i = \theta_i(\varphi_i)$ the result can also be written

$$GZ(\varphi) \equiv \frac{M^e(\varphi)}{\rho g \nabla} \tag{2.160}$$

by introduction of the righting arm GZ. If $GZ(\varphi)$ and φ have the same sign, the structure is hydrostatically stable at this angle of heel φ. $GZ(\varphi) = 0$ corresponds to an equilibrium configuration without any additional external moment, whereas opposite signs on $GZ(\varphi)$ and φ signify an unstable state.

The result (2.160) gives a simple relation between an external moment and the corresponding change of equilibrium position, which is a clear advantage as compared to the result from the first procedure. If the direction associated with the lowest metacentric height does not change, the two procedures will yield identical results. Otherwise, the first procedure yields larger changes in the equilibrium position for a given external moment than the second procedure. Therefore, as regards the stability of the structure, the second method must be considered as non-conservative. This must implicitly be taken into account in the formulation of stability requirements based on the GZ-curve, Eq. (2.160).

The procedure leading to Eq. (2.160) is equally valid for intact and damage stability analysis. In intact stability analysis, the submerged volume ∇ is constant whereas, for damage stability cases, it may change with the angle φ due to successive flooding of compartments. The calculation of $M^e = M^e(\varphi)$ properly takes flooding into account, but the definition of GZ, Eq. (2.160), can be made in different ways. The definition applied in the current stability rules uses Eq. (2.160) with ∇ equal to the intact submerged volume. This definition, in which ∇ is independent of φ, is called the "added weight" method as the water in the flooded compartments can be considered as additional weights. A more rational definition is the "lost buoyancy" method, where the lost buoyancy in the flooded compartments is subtracted from the intact submerged volume to get the real submerged volume $\nabla = \nabla(\varphi)$ to be used in Eq. (2.160). In the present context, the submerged volume to be used in Eq. (2.160) is the only difference between the "added weight" and "lost buoyancy" methods. However, for hand calculations, the two methods look quite different.

Finally, a relation between the initial stability analysis and the GZ-curve should be mentioned.

For small angles of heel φ ($\varphi \ll 1$), both procedures for the righting arm, Eq. (2.151) and Eq. (2.159), lead to the same result:

$$\varphi = \frac{M^e}{\rho g \nabla (z_m - z_g)} \tag{2.161}$$

as

$$E_3(\varphi) \simeq E_3\Big|_0 + \frac{\partial E_3}{\partial \varphi}\Big|_0 \varphi = (-I_{xx} + E_4)\varphi = \nabla(z_g - z_m)\varphi \tag{2.162}$$

By introduction of GM, Eq. (2.150), from the initial stability analysis and GZ, defined by Eq. (2.160), Eq. (2.161) can be written

$$GZ(\varphi) = GM \cdot \varphi \quad (\varphi \ll 1) \tag{2.163}$$

GM is seen to be the slope of the GZ-curve at the equilibrium configuration ($\varphi = 0$).

Example 2.4.2

The calculation of the initial stability, Eq. (2.150), and the GZ-curve, Eq. (2.160), is well suited for numerical calculation. The results obtained by use of the I-ship code (see e.g. Jensen, Baatrup and Andersen, 1995) for a Ro-Ro ship are given in Figure 2.12 as an illustrative example.

```
Intact equilibrium position

Total mass of this loading condition .............. :  16900.000 [t]
X-coordinate for center-of-gravity ................ :     84.853 [m]
Y-coordinate for center-of-gravity ................ :       .000 [m]
Z-coordinate for center-of-gravity ................ :     10.000 [m]


Draught at Lpp/2, measured perp. to water plane ... :      6.501 [m]
Draught at AP, measured perp. to water plane ...... :      6.544 [m]
Draught at FP, measured perp. to water plane ...... :      6.458 [m]
Angle of heel ..................................... :       .000 [deg]
Calculated displacement ........................... :  16899.990 [t]
LCF measured positive forward Lpp/2 ............... :     -1.720 [m]
BMT measured upwards from center of bouyancy ...... :      7.628 [m]
GM (upright) without free surface corrections ..... :      1.121 [m]
GM (upright) with free surface corrections ........ :      1.121 [m]
Corresponding roll period (IMO A.685(17)) ......... :     18.984 [s]
Moment to change trim (MCT) ....................... :  24529.332 [t*m/m]
Mass to change immersion .......................... :   3040.281 [t/m]


Moulded volumen ................................... :  16422.107 [m^3]
Block coefficient ................................. :       .597
LCB measured positive forward Lpp/2 ............... :      2.229 [m]
KB measured in CL from the base line .............. :      3.493 [m]
Wetted surface (moulded) .......................... :   4324.060 [m^2]
Water plane area .................................. :   2954.311 [m^2]
Water plane area coefficient ...................... :       .698
BML measured upwards from center of bouyancy ...... :    239.836 [m]
```

Statical stability results

Angle of heel [deg]	Draught AP perp. to WP [m]	Draught FP perp. to WP [m]	MS [m]	KN [m]	GZ (corrected) [m]
.000	6.544	6.458	.000	.000	.000
10.000	6.296	6.430	.022	1.953	.216
20.000	5.570	6.332	.156	3.960	.539
30.000	4.438	6.036	.394	5.955	.955
40.000	2.922	5.322	.563	7.711	1.283

Figure 2.12 *Hydrostatic stability information sheet.*

3 Ocean Waves

In the previous chapter, the hydrostatic and the gravitational loads acting on a structure at rest in a calm sea were derived. However, the real ocean environment always differs from this ideal static equilibrium state due to continuous changes in the meteorological conditions. As it is well known from weather forecasts, we can, with our present knowledge, only estimate average values of atmospheric pressure and wind speed. Local variations in space or time cannot be predicted. The reason is not just lack of a complete physical model of the earth climate, but also that local variations depend on the past detailed history of the global climate. Weather conditions with the same macroscopic (average) parameters therefore show different and unpredictable local variations in the key variables of the system. All these variables may thus be said to be random or stochastic processes, for which only average (expected) values can be predicted, leaving a detailed description out of reach. The average values can, however, be used to estimate the probability that a variable is within certain given bounds. This is very important as we can thus determine the probability that a load derived from the climatic model exceeds a given design value. These loads may be directly related to the wind speed at the location of the structure, but may also be derived loads such as those due to waves, current and ice. For fixed offshore structures, all four types are important but, for floating structures, which can move freely in the horizontal plane, the forces due to wind, current and ice are usually negligible compared to those generated by the ocean waves.

As ocean waves are generated by local variation in wind speed and atmospheric pressure, they are to be considered a stochastic process. The magnitude of the waves is the result of an energy transfer between the wind and the ocean, which takes place in a narrow boundary layer on the surface of the sea. A complete description of this process is not available. Instead, the average parameters describing the state of the sea may be related to the average wind speed, direction, duration and fetch over the considered area of the sea.

In this chapter, the statistical properties of wind-driven ocean waves will be described, including the necessary fundamentals of stochastic processes. This will facilitate the load and response analysis described in the two following chapters, as the statistical analysis of these processes is basically the same as for the ocean waves.

The literature on stochastic processes is extensive and only a small fraction of the procedures available is discussed here. For further reference, the following textbooks can be recommended: Price and Bishop (1974), Madsen, Krenk and Lind (1986) and, especially, Ochi (1990).

3.1 RANDOM VARIABLES

The probability $P(X \leq x)$ that a random variable X is less than or equal to a particular value x is denoted $F(x)$:

$$F(x) = P(X \leq x) \tag{3.1}$$

By definition, $F(x) = 0$ if the event $X \leq x$ can never take place whereas $F(x) = 1$ if X always is less than or equal to x. Hence, if the random variable X is bounded, $a \leq X \leq b$, then

$$F(a) = 0 \quad ; \quad F(b) = 1 \tag{3.2}$$

Furthermore, for $a \leq x_1 < x_2 \leq b$:

$$P(X \leq x_2) = P(X \leq x_1) + P(x_1 < X \leq x_2)$$

$$\geq P(X \leq x_1)$$

as the event $P(x_1 \leq X \leq x_2)$ is non-negative. Thus

$$F(x_2) \geq F(x_1) \quad \text{for } x_2 > x_1 \tag{3.3}$$

so that $F(x)$ is a never decreasing function of x. The definition given above hold for all types of random variables representing discrete as well as continuous events. In the present treatment, only continuous random variables are considered, and for such variables a *probability density function* $p(x)$ can be defined by the integral

$$F(x) = \int_a^x p(u)\, du \tag{3.4}$$

Provided that $F(x)$ is differentiable, Eq. (3.4) yields

$$p(x) = \frac{dF}{dx} \tag{3.5}$$

From definitions (3.1) and (3.4) it follows that the probability that X falls within a small range dx about x is

$$P(x < X \leq x + dx) = p(x)dx \tag{3.6}$$

which shows that $p(x)$ is a non-negative function of x.

Due to the relation (3.4) between $F(x)$ and $p(x)$, $F(x)$ is denoted the *cumulative probability density function* or the *probability distribution function*.

As

$$P(x_1 < X \leq x_2) = \int_{x_1}^{x_2} p(u)du$$

it is seen that

$$P(X = x) = \int_{x_0}^{x_0} p(u)du = 0 \tag{3.7}$$

Thus, the probability that a continuous random variable assumes a specific value is zero.

The stochastic nature of the variable X is completely described by either $F(x)$ or $p(x)$. However, often neither $F(x)$ nor $p(x)$ can be derived from the physical process represented by X. Instead, average values of certain functions $G(X)$ may be available from previous outcomes of X. These average or *expected values* $E[G(X)]$ are defined as

$$E[G(X)] \equiv \int_a^b G(x)p(x)dx \tag{3.8}$$

The most useful average values are the *moments*:

$$\mu_n \equiv E[X^n] \quad ; \quad n = 1, 2, 3, ..., \tag{3.9}$$

or the *central moments*:

$$\zeta_n \equiv E[(X - \mu)^n] \quad ; \quad n = 2, 3, ..., \tag{3.10}$$

where the *mean value*

$$\mu \equiv \mu_1 = E[X] \tag{3.11}$$

The second central moment ζ_2 is termed the *variance*

$$\zeta_2 = E[(X - \mu)^2] \tag{3.12}$$

The *standard deviation* s defined as

$$s = \sqrt{\zeta_2} \tag{3.13}$$

has, like the mean value μ, the same physical dimension as the variable X.

The moments μ_n and ζ_n are clearly related. From Eq. (3.10) it follows that

$$\zeta_2 = E[X^2] - 2\mu E[X] + \mu^2 = \mu_2 - \mu_1^2$$

$$\zeta_3 = \mu_3 - 3\mu_2\mu_1 + 2\mu_1^3$$

$$\zeta_4 = \mu_4 - 4\mu_3\mu_1 + 6\mu_2\mu_1^2 - 3\mu_1^4$$

and so on. Obviously, $\zeta_n = \mu_n$ if $\mu = 0$.

The following non-dimensional values of the central moments are often used:

$$\textit{Coefficient of variation:} \quad v = \frac{\sqrt{\zeta_2}}{\mu} = \frac{s}{\mu}$$

$$\textit{Skewness:} \quad \gamma_1 = \frac{\zeta_3}{\zeta_2^{3/2}} \tag{3.14}$$

$$\textit{Kurtosis:} \quad \gamma_2 = \frac{\zeta_4}{\zeta_2^2}$$

From the knowledge of the moments, approximations to the probability density function can be constructed, either as series expansions or by tranformation, see Sections 3.1.2 and 3.1.3. First, however, it is appropriate to introduce the *normal distribution*.

The normal distribution plays a significant role in the theory of stochastic processes as the statistical properties of the majority of physical phenomena can be described by this distribution. Non-linear effects can, to a certain extent, also be included by modifications to the normal distribution, as shown in Sections 3.1.2 and 3.1.3.

However, if we want to analyse the statistical behaviour of discrete events, like the occurrence of peaks or zero upcrossings, other probability distributions are more useful. For that purpose the Weibull distribution and the Gumbel distribution are described in Sections 3.1.4 and 3.1.5 respectively. Finally, probability distributions for several random variables are introduced in Section 3.1.6 and applied in Section 3.1.7 in order to derive the very important *central limit theorem*.

3.1.1 The Normal Distribution

The normal distribution, also called the *Gaussian distribution*, has the probability density function

$$p(x) = \frac{1}{\sqrt{2\pi}\, s} \exp\left[-\frac{1}{2}\left(\frac{x-\mu}{s}\right)^2\right] \quad ; \quad -\infty < x < \infty \tag{3.15}$$

where the two parameters μ, s are the mean value and the standard deviation, respectively:

$$\int_{-\infty}^{\infty} xp(x)dx = \mu$$

$$\int_{-\infty}^{\infty} (x - \mu)^2 p(x)dx = s^2$$

An analytical expression for the probability distribution function F(x) is not possible. Rather than using a numerical table for each combination of (μ, s), it is appropriate to introduce the standard normal distribution with a mean value zero and a unit standard deviation. Its probability density function is

$$\varphi(u) = \frac{1}{\sqrt{2\pi}} e^{-\frac{1}{2}u^2} \qquad -\infty < u < \infty \tag{3.16}$$

and the associated probability distribution function:

$$\Phi(u) = \frac{1}{\sqrt{2\pi}} \int_{-\infty}^{u} e^{-\frac{1}{2}v^2 dv} \tag{3.17}$$

Hence, by the transformation $u = (x - \mu)/s$

$$F(x) = \Phi\left(\frac{x - \mu}{s}\right) \tag{3.18}$$

The *standard normal distribution* $\Phi(u)$ is tabulated in Table 3.1 and shown in Figure 3.1 together with the probability density function $\varphi(u)$. It is seen from Eq. (3.17) that

$$\Phi(-u) = 1 - \Phi(u) \tag{3.19}$$

so that only tabular values for $u > 0$ are necessary.

Asymtotically, Abramowitz and Stegun (1966):

$$\Phi(-u) \simeq \frac{1}{\sqrt{2\pi}\,u} e^{-\frac{1}{2}u^2} \quad ; u \gg 1 \tag{3.20}$$

The moments μ_n of the standard normal distribution become

$$\mu_n = \zeta_n = \frac{1}{\sqrt{2\pi}} \int_{-\infty}^{\infty} u^n e^{-\frac{1}{2}u^2} du = \begin{cases} 0 & ; \quad n \text{ odd} \\ 1 \cdot 3 \cdot 5 \cdots (n-1) & ; \quad n \text{ even} \end{cases} \tag{3.21}$$

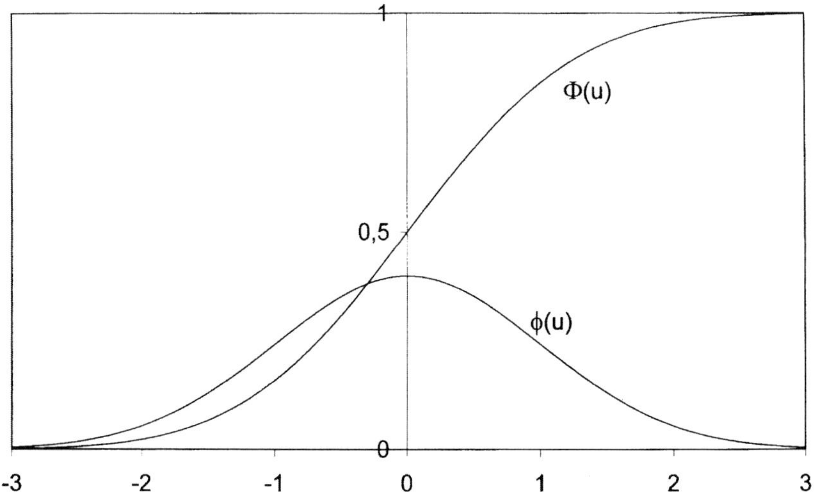

Figure 3.1 *Probability density function φ(u) and probability distribution function Φ(x) of the standard normal distribution.*

Table 3.1 *Standard normal distribution Φ(z).*

z	.00	.01	.02	.03	.04	.05	.06	.07	.08	.09
.0	.5000	.5040	.5080	.5120	.5160	.5199	.5239	.5279	.5319	.5359
.1	.5398	.5438	.5478	.5517	.5557	.5596	.5636	.5675	.5714	.5753
.2	.5793	.5832	.5871	.5910	.5948	.5987	.6026	.6064	.6103	.6141
.3	.6179	.6217	.6255	.6293	.6331	.6368	.6406	.6443	.6480	.6517
.4	.6554	.6591	.6628	.6664	.6700	.6736	.6772	.6808	.6844	.6879
.5	.6915	.6950	.6985	.7019	.7054	.7088	.7123	.7157	.7190	.7224
.6	.7257	.7291	.7324	.7357	.7389	.7422	.7454	.7486	.7517	.7549
.7	.7580	.7611	.7642	.7673	.7704	.7734	.7764	.7794	.7823	.7852
.8	.7881	.7910	.7939	.7967	.7995	.8023	.8051	.8078	.8106	.8133
.9	.8159	.8186	.8212	.8238	.8264	.8289	.8315	.8340	.8365	.8389
1.0	.8413	.8438	.8461	.8485	.8508	.8531	.8554	.8577	.8599	.8621
1.1	.8643	.8665	.8686	.8708	.8729	.8749	.8770	.8790	.8810	.8830
1.2	.8849	.8869	.8888	.8907	.8925	.8944	.8962	.8980	.8997	.9015
1.3	.9032	.9049	.9066	.9082	.9099	.9115	.9131	.9147	.9162	.9177
1.4	.9192	.9207	.9222	.9236	.9251	.9265	.9279	.9292	.9306	.9319
1.5	.9332	.9345	.9357	.9370	.9382	.9394	.9406	.9418	.9429	.9441
1.6	.9452	.9463	.9474	.9484	.9495	.9505	.9515	.9525	.9535	.9545
1.7	.9554	.9564	.9573	.9582	.9591	.9599	.9608	.9616	.9625	.9633
1.8	.9641	.9649	.9656	.9664	.9671	.9678	.9686	.9693	.9699	.9706
1.9	.9713	.9719	.9726	.9732	.9738	.9744	.9750	.9756	.9761	.9767
2.0	.9772	.9778	.9783	.9788	.9793	.9798	.9803	.9808	.9812	.9817
2.1	.9821	.9826	.9830	.9834	.9838	.9842	.9846	.9850	.9854	.9857
2.2	.9861	.9864	.9868	.9871	.9875	.9878	.9881	.9884	.9887	.9890
2.3	.9893	.9896	.9898	.9901	.9904	.9906	.9909	.9911	.9913	.9916
2.4	.9918	.9920	.9922	.9925	.9927	.9929	.9931	.9932	.9934	.9936
2.5	.9938	.9940	.9941	.9943	.9945	.9946	.9948	.9949	.9951	.9952
2.6	.9953	.9955	.9956	.9957	.9959	.9960	.9961	.9962	.9963	.9964
2.7	.9965	.9966	.9967	.9968	.9969	.9970	.9971	.9972	.9973	.9974
2.8	.9974	.9975	.9976	.9977	.9977	.9978	.9979	.9979	.9980	.9981
2.9	.9981	.9982	.9982	.9983	.9984	.9984	.9985	.9985	.9986	.9986
3.0	.9987	.9987	.9987	.9988	.9988	.9989	.9989	.9989	.9990	.9990
3.1	.9990	.9991	.9991	.9991	.9992	.9992	.9992	.9992	.9993	.9993
3.2	.9993	.9993	.9994	.9994	.9994	.9994	.9994	.9995	.9995	.9995
3.3	.9995	.9995	.9995	.9996	.9996	.9996	.9996	.9996	.9996	.9997
3.4	.9997	.9997	.9997	.9997	.9997	.9997	.9997	.9997	.9997	.9998

3.1.2 Series Expansion of the Probability Density Function

Consider an unbounded random variable $-\infty < X < \infty$ and define the so-called characteristic function

$$\phi(it) = \int_{-\infty}^{\infty} p(x)e^{itx}dx = E\left[e^{itX}\right] \tag{3.22}$$

where i is the imaginary unit (with the property $i^2 = -1$). Substitution of the expansion

$$e^{itx} = 1 + \sum_{n=1}^{\infty} \frac{(itx)^n}{n!}$$

into Eq. (3.22) yields

$$\phi(it) = 1 + \sum_{n=1}^{\infty} \frac{\mu_n}{n!}(it)^n \tag{3.23}$$

From Eq. (3.22) or Eq. (3.23) it is seen that $\phi(it)$ has the property

$$\left.\frac{d^n\phi}{dt^n}\right|_{t=0} = i^n\mu_n \tag{3.24}$$

When the moments μ_n are known, a formal solution for $p(x)$ is given by the Fourier transformation of Eq. (3.22):

$$p(x) = \frac{1}{2\pi}\int_{-\infty}^{\infty} \phi(it)e^{-itx}dt \tag{3.25}$$

However, substitution of Eq. (3.23) into Eq. (3.25) yields a meaningless result as each of the terms

$$\frac{1}{2\pi}\int_{-\infty}^{\infty} (it)^n e^{-itx}dt$$

is real but infinite. To avoid this problem, other expansions can be constructed by use of different moments. An expansion in the *cumulants* κ_n defined by

$$\ln\phi(it) \equiv \sum_{n=1}^{\infty} \frac{\kappa_n}{n!}(it)^n \tag{3.26}$$

will be shown to be very useful for generating probability density functions deviating only slightly from a normal distribution. By use of the expansion

$$\ln(1 + u) = u - \frac{1}{2}u^2 + \frac{1}{3}u^3 - \frac{1}{4}u^4 + \dots$$

it follows from Eqs. (3.23) and (3.26) that

$$\begin{aligned} \kappa_1 &= \mu_1 \\ \kappa_2 &= \mu_2 - \mu_1^2 = \zeta_2 \\ \kappa_3 &= \mu_3 - 3\mu_2\mu_1 + 2\mu_1^3 = \zeta_3 \\ \kappa_4 &= \mu_4 - 4\mu_3\mu_1 - 3\mu_2^2 + 12\mu_2\mu_1^2 - 6\mu_1^4 = \zeta_4 - 3\zeta_2^2 \end{aligned} \tag{3.27}$$

Moments with n greater than 4 are seldom used as their estimates from e.g. measurements are connected with great uncertainty.

The series expansion of *p(x)* in terms of the cumulants follows from Eqs. (3.25) and (3.26):

$$p(x) = \frac{1}{2\pi} \int_{-\infty}^{\infty} e^{\ln\phi(it) - itx} dx$$

$$= \frac{1}{2\pi} \int_{-\infty}^{\infty} \exp\left[\sum_{n=1}^{\infty} \frac{\kappa_n}{n!}(it)^n - itx \right] dx$$

Now, consider the case where the cumulants κ_n are small for n > 2 . Then *p(x)* can be written

$$p(x) = \frac{1}{2\pi} \int_{-\infty}^{\infty} \exp\left[(\kappa_1 - x)it - \frac{1}{2}\kappa_2 t^2 \right]$$

$$\cdot \left\{ 1 + \frac{\kappa_3}{3!}(it)^3 + \frac{\kappa_4}{4!}(it)^4 + \dots + \frac{1}{2!}\left(\frac{\kappa_3}{3!}(it)^3 + \frac{\kappa_4}{4!}(it)^4 + \dots \right)^2 + \dots \right\} dt$$

Introducing

$$t = \frac{u}{\sqrt{\kappa_2}} \quad , \quad (x - \kappa_1) = f\sqrt{\kappa_2} \quad , \quad \lambda_n = \frac{\kappa_n}{\kappa_2^{n/2}} \tag{3.28}$$

we obtain

$$p(x) = \frac{1}{2\pi\sqrt{\kappa_2}} \int_{-\infty}^{\infty} e^{-\frac{1}{2}(u^2+2ifu)} \left\{ 1 + \frac{\lambda_3}{6}(iu)^3 + \frac{\lambda_4}{24}(iu)^4 + \ldots \right.$$

$$\left. + \frac{1}{2!}\left(\frac{\lambda_3}{6}(iu)^3 + \frac{\lambda_4}{24}(iu)^4 + \ldots\right)^2 + \ldots \right\} du$$

The integration is carried out by means of the identity:

$$\frac{1}{\sqrt{2\pi}} \int_{-\infty}^{\infty} e^{-\frac{1}{2}(u^2+2ifu)}(iu)^n \, du = \frac{(-1)^n}{\sqrt{2\pi}} \frac{d^n}{df^n} \int_{-\infty}^{\infty} e^{-\frac{1}{2}(u^2+2ifu)} \, du$$
$$= (-1)^n \frac{d^n}{df^n} e^{-\frac{1}{2}f^2} \equiv He_n(f)\, e^{-\frac{1}{2}f^2} \tag{3.29}$$

where $He_n(f)$ are denoted the *Hermite polynomials* of order n. The values of $He_n(f)$ follow from (3.29):

$$He_1(f) = f \; ; \; He_2(f) = f^2 - 1$$
$$He_3(f) = f^3 - 3f \; ; \; He_4(f) = f^4 - 6f^2 + 3 \tag{3.30}$$

and so forth.

Use of the identity (3.29) leads to the final result for $p(x)$:

$$p(x) = \frac{1}{\sqrt{2\pi\kappa_2}} e^{-\frac{1}{2}f^2} \left[1 + \frac{1}{6}\lambda_3 He_3(f) + \frac{1}{24}\lambda_4 He_4(f) + \ldots \right.$$
$$\left. + \frac{1}{2!}\left(\frac{1}{36}\lambda_3^2 He_6(f) + \frac{1}{576}\lambda_4^2 He_8(f) + \frac{1}{72}\lambda_3\lambda_4 He_7(f) + \ldots\right) + \ldots \right] \tag{3.31}$$

which is the so-called *Gram-Charlier series expansion.* It is seen that if $\kappa_n = 0$ for $n \geq 3$, then Eq. (3.31) reduces to the normal distribution, which is thus characterised by having all cumulants $\kappa_n = 0, \; n > 2$. This makes the use of cumulants to describe slightly non-Gaussian processes very convenient. Of course, even for small cumulants $\kappa_n, n > 2$, the expansion breaks down in the tail ($f \to \pm \infty$) as the Hermite polynomials, Eq. (3.30), go to infinity. The expansion (3.31) is only useful if the value inside the square brackets does not deviate to much from unity. A negative value signifies a meaningless result as *p(x)* must by definition be non-negative.

Note that, according to Eqs. (3.14), (3.27) and (3.28), λ_3 and λ_4 are closely related to the skewness γ_1 and the kurtosis γ_2:

$$\lambda_3 = \gamma_1$$
$$\lambda_4 = \gamma_2 - 3 \tag{3.32}$$

For a normal distribution $\gamma_2 = 3$, and λ_4 is therefore often denoted the *coefficient of excess*.

We will return to Eq. (3.31) later in this and in the next chapter. The derivation of Eq. (3.31) is given in the classical paper by Longuet-Higgins (1963) on the statistical properties of sea waves.

3.1.3 Transformation of a Random Variable

In the previous section, a rigorous series expansion was derived for the probability density function of a slightly non-Gaussian variable. The drawback of this expansion is its limited applicability, especially in the tails of the distribution.

A much simpler result can be obtained, if a monotonic, deterministic relationship $x = g(u)$ between the random variable X with an unknown probability density function $p_x(x)$ and another random variable U with a known distribution $p_u(u)$ can be assumed. Hence

$$P(X \leq x) = P(X \leq g(u)) = P(U \leq g^{-1}(x))$$

and, from Eq. (3.6):

$$p_x(x)\, dx = p_u(u)\, du$$

or

$$p_x(x) = \left[p_u(u)\, \frac{du}{dx}\right]_{u=g^{-1}(x)} \tag{3.33}$$

The validity of the assumption $x = g(u)$ must of course be demonstrated, preferably by a physical model as for instance shown in Example 3.1.1.

<u>Example 3.1.1</u>

Consider the drag term in Morison's equation for wave loads on cylinders. This term is proportional to $u|u|$ where u is the wave particle velocity, which is usually taken to be normally distributed with the zero mean and standard deviation s_u. In this case Eq. (3.33) yields the probability density function $p_x(x)$ for the drag term $x \equiv u\,|u|$:

$$p_x(x) = \frac{1}{\sqrt{2\pi}\, s_u} \exp\left(-\frac{1}{2}\frac{|x|}{s_u^2}\right) \frac{1}{2\sqrt{|x|}} \quad ; \quad -\infty < x < \infty$$

as

$$u = g^{-1}(x) = \sqrt{|x|} \cdot sign(x)$$

However, in many cases an analytical relation between x and u is not available, but instead the lowest order central moments ζ_n are known for the random variable X. A power series expansion of X in terms of U may then be appropriate:

$$X = g(U) = \sum_{j=0}^{m} c_j U^j \tag{3.34}$$

where the deterministic coefficient c_i is determined so that X has the prescribed central moments:

$$\zeta_n = E\left[(X - \mu)^n \right] = E\left[\left\{ \sum_{j=0}^{m} c_j \left(U^j - E\left[U^j \right] \right) \right\}^n \right] \tag{3.35}$$

This is a non-linear algebraic system of equations in the unknown coefficients c_j. For the number of coefficients to be equal to the number N of known moments, m must be taken to be N - 1. Furthermore, the moments $E[U^j]$, $j = 1,2,\ldots,N(N-1)$ should be available for the random variable U. If U is standard and normally distributed these moments are given by Eq. (3.21).

Example 3.1.2

As mentioned previously, often only the four lowest moments μ, ζ_2, ζ_3 and ζ_4 are known with sufficient accuracy. In such cases Eq. (3.34) yields a cubic power series. If U is taken to be standard normally distributed, the coefficients c_i in Eq. (3.34) are determined from

$$\begin{aligned}
\mu &= c_0 + c_2 \\
\zeta_2 &= c_1^2 + 6\, c_1\, c_3 + 2\, c_2^2 + 15\, c_3^2 \\
\zeta_3 &= c_2 \left(6\, c_1^2 + 8c_2^2 + 72\, c_1\, c_3 + 270\, c_3^2\right) \\
\zeta_4 &= 60\, c_2^4 + 3\, c_1^4 + 10395\, c_3^4 + 60\, c_2^2\, c_1^2 + 4500\, c_2^2\, c_3^2 + 630 c_1^2\, c_3^2 \\
&\quad + 936 c_1 c_2^2 c_3 + 3780 c_1 c_3^3 + 60 c_1^3 c_3
\end{aligned} \tag{3.36}$$

The coefficients c_1, c_2 and c_3 must be determined numerically, for instance by application of the Newton-Raphson method. It should be mentioned that a good initial guess for c_1, c_2 and c_3 is required. Here, the approximate values given by Winterstein (1985) are useful.

The result is only valid if the transformation (3.34) is monotonic, which requires that dx/du does not change sign for any value of u. Differentiation of Eq. (3.34), m = 3, yields

$$c_1 + 2\, c_2\, u + 3\, c_3\, u^2 > 0 \quad \text{for all } u$$

or

$$c_2^2 < 3\, c_1\, c_3 \tag{3.37}$$

The accuracy of the *transformation method* given by Eqs. (3.34) and (3.35) depends completely on the assumption that the higher order moments of X can be neglected in the transformation.

This may seem a dubious assumption but several examples, e.g. Winterstein (1985), show that very accurate results are obtained even for the tail of the distribution. As long as Eq. (3.37) is satisfied no unphysical (i.e. negative) values of the probability density function are obtained.

3.1.4 The Weibull Distribution

The probability distribution $F(x)$ for the *Weibull distribution* is given by

$$F(x) = 1 - e^{-(x/\alpha)^{\beta}} \quad ; \quad x > 0, \ \alpha > 0, \ \beta > 0 \tag{3.38}$$

This distribution is purely empirical but has proved to be a very versatile distribution for descriptions of the statistical properties of many practical problems. It is also one of the three true asymptotic extreme value distributions, see Section 3.2.5.

The two coefficients (α, β) can be related to the mean value μ and the variance ζ_2 through

$$\mu = \int_0^\infty x\, p(x)\, dx = \alpha\, \Gamma\left(1 + \frac{1}{\beta}\right)$$

$$\zeta_2 = \int_0^\infty (x - \mu)^2\, p(x)\, dx = \alpha^2 \left\{ \Gamma\left(1 + \frac{2}{\beta}\right) - \left[\Gamma\left(1 + \frac{1}{\beta}\right)\right]^2 \right\} \tag{3.39}$$

where $\Gamma(\ )$ is the *Gamma function*:

$$\Gamma(x) = \int_0^\infty t^{x-1} e^{-t}\, dt \tag{3.40}$$

The moments μ_n, Eq. (3.9), become

$$\mu_n = \int_0^\infty x^n\, p(x)\, dx = \alpha^n\, \Gamma\left(1 + \frac{n}{\beta}\right)$$

If the argument of the Gamma function is an integer n, then

$$\Gamma(n + 1) = n! \tag{3.41}$$

and in general

$$\Gamma(x + 1) = x\,\Gamma(x) \tag{3.42}$$

Thus, only tabular values of $\Gamma(x)$ are needed for, say, $1 \leq x < 2$, see Table 3.2.

3.1.4.1 The Rayleigh Distribution ($\beta = 2$)

A special case of the Weibull distribution is the Rayleigh distribution where the coefficient $\beta = 2$ yields the following distribution:

$$F(x) = 1 - e^{-(x/\alpha)^2}, \quad x \geq 0 \tag{3.43}$$

with the mean value

$$\mu = \alpha\,\Gamma\left(\frac{3}{2}\right) = \frac{\sqrt{\pi}}{2}\,\alpha \tag{3.44}$$

the variance

$$\zeta_2 = \left(1 - \frac{\pi}{4}\right)\alpha^2 \tag{3.45}$$

and the skewness

$$\gamma_1 = \frac{\zeta_3}{\zeta_2^{3/2}} = \frac{\sqrt{\pi}\,(\pi - 3)}{4\left(\sqrt{1 - \frac{\pi}{4}}\right)^3} \simeq 0.6311$$

The probability density function *p(x)* becomes

$$p(x) = \frac{dF}{dx} = \frac{2x}{\alpha^2}\,e^{-(x/\alpha)^2}\ ; \quad x \geq 0 \tag{3.46}$$

and is shown in non-dimensional form in Figure 3.2.

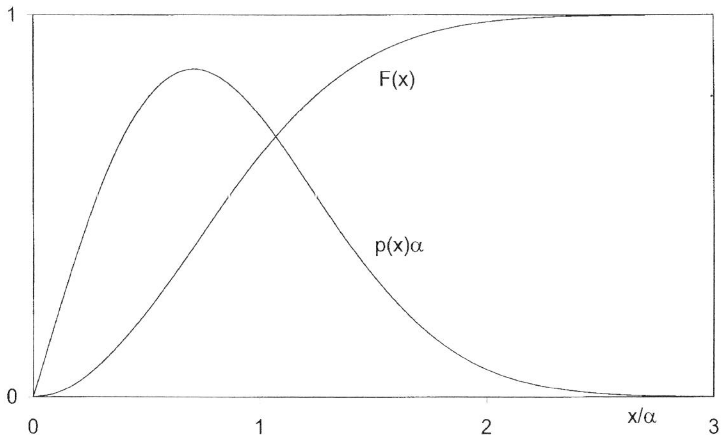

Figure 3.2 *The non-dimensional Rayleigh distribution.*

At this point, it may be convenient to illustrate different characteristic values which can be derived from a given probability distribution. The mean value, Eq. (3.44), is

$$\mu = \frac{\sqrt{\pi}}{2}\,\alpha \simeq 0.8862\,\alpha \tag{3.47}$$

Alternatively, the *most probable value* $\tilde{\mu}$, i.e. the value with the largest value of the probability density function, could be considered:

$$\left.\frac{d\,p(x)}{dx}\right|_{x=\tilde{\mu}} = 0 \Rightarrow \tilde{\mu} = \frac{\alpha}{\sqrt{2}} \simeq 0.7071\,\alpha \tag{3.48}$$

Finally, the *50 per cent fractile* μ_{50} defined by

$$P(X > \mu_{50}) = \frac{1}{2}$$

can also be regarded as a typical value. From Eq. (3.43)

$$\mu_{50} = \alpha\,\sqrt{\ln 2} \simeq 0.8326\,\alpha \tag{3.49}$$

Due to the difference between μ, $\tilde{\mu}$ and μ_{50}, care must be taken in interpretations of a cited characteristic value of a random variable.

3.1.4.2 *The Exponential Distribution* ($\beta = 1$)

Another special case of the Weibull distribution which deserves to be mentioned is the *exponential distribution* where $\beta = 1$ giving

$$F(x) = 1 - e^{-x/\alpha} \tag{3.50}$$

This results in a mean value of

$$\mu = \alpha$$

a variance

$$\zeta_2 = \alpha^2$$

and a skewness

$$\gamma_1 = \frac{\zeta_3}{\zeta_2^{3/2}} = 2$$

Table 3.2 *Gamma function $\Gamma(x)$.*

x	Γ(x)	x	Γ(x)	x	Γ(x)
1	1	1.5	0.8862269	2	1
1.01	0.9943259	1.51	0.8865917	2.1	1.0464858
1.02	0.9888442	1.52	0.8870388	2.2	1.1018025
1.03	0.9835500	1.53	0.8875676	2.3	1.1667119
1.04	0.9784382	1.54	0.8881777	2.4	1.2421693
1.05	0.9735043	1.55	0.8888683	2.5	1.3293404
1.06	0.9687436	1.56	0.8896392	2.6	1.4296246
1.07	0.9641520	1.57	0.8904897	2.7	1.5446858
1.08	0.9597253	1.58	0.8914196	2.8	1.6764908
1.09	0.9554595	1.59	0.8924282	2.9	1.8273551
1.1	0.9513508	1.6	0.8935153	3	2
1.11	0.9473955	1.61	0.8946806	3.1	2.1976203
1.12	0.9435902	1.62	0.8959237	3.2	2.4239655
1.13	0.9399314	1.63	0.8972442	3.3	2.6834374
1.14	0.9364161	1.64	0.8986420	3.4	2.9812064
1.15	0.9330409	1.65	0.9001168	3.5	3.3233510
1.16	0.9298031	1.66	0.9016684	3.6	3.7170239
1.17	0.9266996	1.67	0.9032965	3.7	4.1706518
1.18	0.9237278	1.68	0.9050010	3.8	4.6941742
1.19	0.9208850	1.69	0.9067818	3.9	5.2993297
1.2	0.9181687	1.7	0.9086387	4	6
1.21	0.9155765	1.71	0.9105717	4.1	6.8126229
1.22	0.9131059	1.72	0.9125806	4.2	7.7566895
1.23	0.9107549	1.73	0.9146654	4.3	8.8553434
1.24	0.9085211	1.74	0.9168260	4.4	10.136102
1.25	0.9064025	1.75	0.9190625	4.5	11.631728
1.26	0.9043971	1.76	0.9213749	4.6	13.381286
1.27	0.9025031	1.77	0.9237631	4.7	15.431412
1.28	0.9007185	1.78	0.9262273	4.8	17.837862
1.29	0.8990416	1.79	0.9287675	4.9	20.667386
1.3	0.8974707	1.8	0.9313838	5	24
1.31	0.8960042	1.81	0.9340763	5.1	27.931754
1.32	0.8946405	1.82	0.9368451	5.2	32.578096
1.33	0.8933781	1.83	0.9396904	5.3	38.077976
1.34	0.8922155	1.84	0.9426124	5.4	44.598848
1.35	0.8911514	1.85	0.9456112	5.5	52.342778
1.36	0.8901845	1.86	0.9486870	5.6	61.553915
1.37	0.8893135	1.87	0.9518402	5.7	72.527635
1.38	0.8885371	1.88	0.9550709	5.8	85.621737
1.39	0.8878543	1.89	0.9583793	5.9	101.27019
1.4	0.8872638	1.9	0.9617658	6	120
1.41	0.8867647	1.91	0.9652307	6.1	142.45194
1.42	0.8863558	1.92	0.9687743	6.2	169.40610
1.43	0.8860362	1.93	0.9723969	6.3	201.81328
1.44	0.8858051	1.94	0.9760989	6.4	240.83378
1.45	0.8856614	1.95	0.9798807	6.5	287.88528
1.46	0.8856043	1.96	0.9837425	6.6	344.70192
1.47	0.8856331	1.97	0.9876850	6.7	413.40752
1.48	0.8857470	1.98	0.9917084	6.8	496.60608
1.49	0.8859451	1.99	0.9958133	6.9	597.49413
1.5	0.8862269	2	1	7	720

The probability density function and the probability distribution function are shown in non-dimensional form in Figure 3.3. For this distribution the most probable value is found at the lower limit $x = 0$ for the distribution:

$$\tilde{\mu} = 0$$

and finally

$$\mu_{50} = \alpha \ln 2 \simeq 0.6931\, \alpha$$

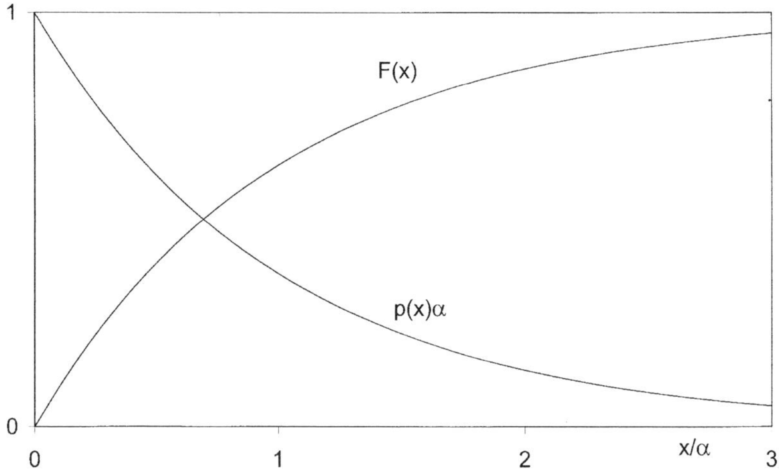

Figure 3.3 *The non-dimensional exponential distribution.*

3.1.5 The Gumbel Distribution

The *Gumbel distribution* is characterised by the probability distribution function

$$F(x) = \exp\left(-e^{-(x-\alpha)/\beta}\right) \quad ; \quad -\infty < x < \infty,\ \beta > 0 \tag{3.51}$$

and, by differentiation:

$$p(x) = \frac{1}{\beta}\, e^{-(x-\alpha)/\beta}\, F(x) \tag{3.52}$$

The mean value is found to be

$$\mu = \alpha + C\beta \tag{3.53}$$

and the variance

$$\zeta_2 = \frac{\pi^2}{6}\, \beta^2 \tag{3.54}$$

where C is Euler's constant, $C \simeq 0.577215...$*. The most probable value $\tilde{\mu}$ is obtained from

$$\left.\frac{d\,p(x)}{dx}\right|_{x=\tilde{\mu}} = 0 \Rightarrow \tilde{\mu} = \alpha$$

The 50 per cent fractile is given as

$$\tilde{\mu}_{50} = \alpha - \beta \ln(\ln 2) \simeq \alpha + 0.3665\,\beta$$

Application of the Gumbel distribution to extreme value prediction is discussed in Section 3.2.5.

3.1.6 Probability Distributions of Several Variables

The probability distribution $F(x_1, x_2,..., x_n)$ of n random variables X_i, $i = 1, 2,..., n$, is defined by

$$F(x_1, x_2, x_3,..., x_n) = P(X_1 \leq x_1, X_2 \leq x_2,..., X_n \leq x_n) \tag{3.55}$$

as a generalisation of Eq. (3.1). For continuous variables, the *joint probability density function* $p(x_1, x_2,..., x_n)$ is given by

$$F(x_1, x_2,..., x_n) = \int_{a_1}^{x_1}\int_{a_2}^{x_2} \cdots \int_{a_n}^{x_n} p(u_1, u_2,..., u_n)\, du_1\, du_2 ... du_n \tag{3.56}$$

where a_i is the lower boundary on X_i. Provided that $F(x_1, x_2,..., x_n)$ is differentiable for all values of x_i, then

$$p(x_1, x_2,..., x_n) = \frac{\partial^n F(x_1, x_2,..., x_n)}{\partial x_1\, \partial x_2 \ldots \partial x_n} \tag{3.57}$$

If the individual random variables X_i are *statistically independent*, then Eqs. (3.55) - (3.57) yield

$$F(x_1, x_2,..., x_n) = F_{X_1}(x_1)\, F_{X_2}(x_2) ... F_{X_n}(x_n) \tag{3.58}$$

and

* These results follow from

$$\int_0^\infty \ln v\, e^{-v} dv = -C \quad \text{and} \quad \int_0^\infty (\ln v)^2 e^{-v} dv = \frac{\pi^2}{6} + C^2$$

$$p(x_1, x_2, \ldots, x_n) = p_{X_1}(x_1)\, p_{X_2}(x_2) \ldots p_{X_n}(x_n) \tag{3.59}$$

where $F_{X_i}(x_i)$ and $p_{X_i}(x_i)$ are denoted the *marginal distribution* and *marginal density functions*, respectively. For statistically dependent variables, the marginal density function for X_i becomes

$$p_{X_i}(x_i) \equiv \int_{a_1}^{b_1} \ldots \int_{a_{i-1}}^{b_{i-1}} \int_{a_{i+1}}^{b_{i+1}} \int_{a_n}^{b_n} p(x_1, x_2, \ldots, x_n)\, dx_1 \ldots dx_{i-1}\, dx_{i+1} \ldots dx_n \tag{3.60}$$

where b_j is the upper boundary on X_j. The marginal distribution is obtained by integration of Eq. (3.60):

$$F_{X_i}(x_i) = \int_{a_i}^{x_i} p_{X_i}(u_i)\, du_i \tag{3.61}$$

The marginal distribution function $F_{X_i}(x_i)$ expresses the probability that a variable X_i is less than or equal to x_i, irrespective of the values of all the other variables X_j, $j = 1, .., n\,;\ j \neq i$. If, on the contrary, the values of X_j are known, i.e. $X_j = x_j$, $j = 1, 2, .., n\,;\ j \neq i$, then the *conditional probability distribution* for X_i is defined as

$$P(X_i | X_j = x_j\,; j = 1, 2, .., n\,;\ j \neq i) =$$

$$F(x_i | x_1, x_2, .., x_{i-1}, x_{i+1}, .., x_n) = \frac{\int_{a_i}^{x_i} p(x_1, x_2, .., x_{i-1}, u_i, x_{i+1}, .., x_n)\, du_i}{\int_{a_i}^{b_i} p(x_1, x_2, .., x_{i-1}, u_i, x_{i+1}, .., x_n)\, du_i} \tag{3.62}$$

where the denominator is the marginal distribution of $X_1, X_2, .., X_{i-1}, X_{i+1}, .., X_n$. The conditional probability density function is obtained by replacing the numerator on the right-hand side with $p(x_1, x_2, .., x_n)$, i.e. by differentiation of Eq. (3.62) with respect to x_i.

For two random variables *(X, Y)* Eqs. (3.60) - (3.62) yield the following relation between joint, marginal and conditional probability densities:

$$p(x, y) = p(x|y)\, p_Y(y) \tag{3.63}$$

The moments $E[G(X_1, X_2, .., X_n)]$ of any combination $G(X_1, X_2, .., X_n)$ of the random variables are defined by

$$E[G(X_1, X_2, .., X_n)] = \int_{a_1}^{b_1}\int_{a_2}^{b_2} .. \int_{a_n}^{b_n} G(x_1, x_2, .., x_n)\, p(x_1, x_2, .., x_n)\, dx_1 dx_2 .. dx_n \tag{3.64}$$

The most useful moments are the central moments:

$$\zeta_{m_1, m_2, ..., m_n} = E[(X_1 - \mu_1)^{m_1} (X_2 - \mu_2)^{m_2} ... (X_n - \mu_n)^{m_n}] \tag{3.65}$$

where the mean values are given as

$$\begin{aligned} \mu_i = E[X_i] &= \int_{a_1}^{b_1}\int_{a_2}^{b_2} .. \int_{a_n}^{b_n} x_i\, p(x_1, x_2, .., x_n)\, dx_1 dx_2 .. dx_n \\ &= \int_{a_i}^{b_i} x_i\, p_{xi}(x_i)\, dx_i \end{aligned} \tag{3.66}$$

Of special importance is the *covariance matrix* $\underline{\underline{\Sigma}}$:

$$\underline{\underline{\Sigma}} = \begin{bmatrix} \sigma_{11} & \sigma_{12} & \cdots & \sigma_{1n} \\ \cdot & & & \cdot \\ \cdot & & & \cdot \\ \sigma_{n1} & & & \sigma_{nn} \end{bmatrix} \tag{3.67}$$

with the components

$$\sigma_{ij} \equiv cov(x_i, x_j) = E[(X_i - \mu_i)(X_j - \mu_j)] \tag{3.68}$$

The diagonal term σ_{ii} is seen to be the variance of the variable X_i.

A non-dimensional measure of covariance is the *correlation matrix* $\underline{\underline{\rho}}$:

$$\underline{\underline{\rho}} = \begin{bmatrix} \rho_{11} & \rho_{12} & \cdots & \rho_{1n} \\ \cdot & & & \cdot \\ \cdot & & & \cdot \\ \rho_{n1} & & & \rho_{nn} \end{bmatrix} \tag{3.69}$$

where each of the correlation coefficients is defined as

$$\rho_{ij} = \frac{\sigma_{ij}}{s_i\, s_j} \tag{3.70}$$

by use of the standard deviations

$$s_i = \sqrt{E[(X_i - \mu_i)^2]} = \sqrt{\sigma_{ii}} \tag{3.71}$$

as normalisation factors. If the variables X_i and X_j are independent, then

$$\sigma_{ij} = E[(X_i - \mu_i)]\, E[(X_j - \mu_j)] = 0 \tag{3.72}$$

due to Eq. (3.66). Thus, statistically independent pairs of random variables have zero off-diagonal covariance and correlation coefficients. The reverse is not always true but holds for instance for the *multivariate normal distribution* defined by the joint probability density function:

$$p(x_1, x_2, \ldots, x_n) = \frac{1}{\left(\sqrt{2\pi}\right)^n \sqrt{|\underline{\underline{\Sigma}}|}} \exp\left\{ -\frac{1}{2} \left(\underline{x} - \underline{\mu}\right)^T \underline{\underline{\Sigma}}^{-1} \left(\underline{x} - \underline{\mu}\right) \right\} \tag{3.73}$$

where $\underline{\underline{\Sigma}}$ is given by Eq. (3.67). $|\ |$ denotes the matrix determinant and the subscript bar denotes a vector.

This joint distribution clearly becomes the product of the density functions of each of the variables x_i if $\sigma_{ij} = 0$ for $i \neq j$. Thus, in this case, zero correlation also implies statistical independence.

Example 3.1.3

Consider two variables (X_1, X_2) with the joint probability density function

$$p(x_1, x_2) = \frac{6}{7}\left(x_1 + x_2\right) + \frac{1}{7} \quad ; \quad \left(x_1, x_2\right) \in [0, 1]$$

It is seen that $p(x_1, x_2)$ satisfies

$$\int_0^1 \int_0^1 p(x_1, x_2)\, dx_1\, dx_2 = 1$$

as well as

$$p(x_1, x_2) \geq 0 \quad \text{for} \quad \left(x_1, x_2\right) \in [0, 1]$$

as required for a probability density function. Furthermore, the mean values are found to be

$$\mu_i = \int_0^1 \int_0^1 x_i\, p(x_1, x_2)\, dx_1\, dx_2 = \frac{9}{14} \quad ; \quad i = 1, 2$$

and the covariance to be

$$cov(x_1, x_2) = E\left[(X_1 - \mu_1)(X_2 - \mu_2)\right] =$$

$$\int_0^1\int_0^1 (x_1 - \mu_1)(x_2 - \mu_2)\, p(x_1, x_2)\, dx_1\, dx_2 = 0$$

The variables X_1 and X_2 are thus uncorrelated. By application of Eq. (3.60) the marginal probability densities become

$$p_{x_1}(x_1) = \int_0^1 p(x_1, x_2)\, dx_2 = \frac{6}{7}x_1 + \frac{4}{7}$$

$$p_{x_2}(x_2) = \int_0^1 p(x_1, x_2)\, dx_1 = \frac{6}{7}x_2 + \frac{4}{7}$$

As

$$p(x_1, x_2) \neq p_{x_1}(x_1)p_{x_2}(x_2)$$

the two variables are, however, not statistically independent.

The correlation coefficients ρ_{ij}, Eq. (3.70), are bounded:

$$-1 \leq \rho_{ij} \leq 1 \tag{3.74}$$

as

$$E\left[(X_i - \mu_i)(X_j - \mu_j)\right]^2 \leq E\left[(X_i - \mu_i)^2\right]E\left[(X_j - \mu_j)^2\right]$$

due to Schwarz' inequality*.

A correlation coefficient equal to zero signifies uncorrelated variables. For $\rho_{ij} = \pm 1$, it follows from Eq. (3.70) that

$$E\left[(X_i - \mu_i)(X_j - \mu_j)\right] = \pm\, s_i s_j$$

This can only be satisfied for

$$\frac{X_i - \mu_i}{s_i} = \pm\frac{X_j - \mu_j}{s_j}$$

---

* $E[(aX - Y)^2] = a^2E[X^2] - 2aE[XY] + E[Y^2] \geq 0$ for all a $\Rightarrow E[XY]^2 \leq E[X^2]\,E[Y^2]$

which is a complete correlation between X_i and X_j. The value of ρ_{ij} therefore serves as a convenient measure of the correlation between two random variables.

3.1.7 Central Limit Theorem

In the analysis of stochastic processes, the *central limit theorem* plays an important role. The theorem states that the sum X of n independent random variables X_i, $i = 1, 2, \ldots, n$ tends toward a normal distribution for large values of n, irrespectively of the distributions of X_i. The mean value μ_X and the standard deviation s_X are for any n given by

$$\mu_X = E[X] = \sum_{i=1}^{n} E[X_i] = \sum_{i=1}^{n} \mu_{X_i} \tag{3.75}$$

$$s_X^2 = E\left[(X - \mu_X)^2\right] = E\left[\sum_{i=1}^{n}(X_i - \mu_i)^2\right] = \sum_{i=1}^{n} E\left[(X_i - \mu_i)^2\right] = \sum_{i=1}^{n} s_{X_i}^2 \tag{3.76}$$

using Eqs. (3.65) - (3.66). Thus, according to the central limit theorem, for large n, the distribution of X is simply given by the normal distribution, Eq. (3.15), with mean and standard deviation as calculated from Eqs. (3.75) - (3.76).

The proof of this theorem is given below for the special case where all X_i are identically distributed with mean μ and standard deviation s. Then

$$\mu_X = n\mu \tag{3.77}$$

and

$$s_X = \sqrt{n}\, s \tag{3.78}$$

It is convenient to introduce the normalised variables Y and U_j by

$$Y = \frac{X - \mu_X}{s_X} = \frac{X - n\mu}{\sqrt{n}\, s} = \frac{1}{\sqrt{n}} \sum_{j=1}^{n} \frac{X_j - \mu}{s} \equiv \frac{1}{\sqrt{n}} \sum_{j=1}^{n} U_j \tag{3.79}$$

Now, consider the characteristic function $\phi(it)$, Eq. (3.22), introduced in Section 3.1.2. With $i^2 = -1$ this function becomes

$$\begin{aligned}
\phi(it) &\equiv E\left[e^{itY}\right] = E\left[\exp\left\{\frac{it}{\sqrt{n}} \sum_{j=1}^{n} U_j\right\}\right] = \prod_{j=1}^{n} E\left[\exp\left(\frac{it}{\sqrt{n}} U_j\right)\right] \\
&= \prod_{j=1}^{n} E\left[1 + \frac{it}{\sqrt{n}} U_j + \frac{1}{2!}\left(\frac{it}{\sqrt{n}}\right)^2 U_j^2 + \frac{1}{3!}\left(\frac{it}{\sqrt{n}}\right)^3 U_j^3 + \ldots\right] \\
&= \left\{1 - \frac{1}{2}\frac{t^2}{n} - \frac{it^3}{6n\sqrt{n}} \mu_{3U} + \ldots\right\}^n
\end{aligned} \tag{3.80}$$

as

$$E\left[U_j\right] = 0 \quad ; \quad E\left[U_j^2\right] = 1 \quad ; \quad E\left[U_j^k\right] \equiv \mu_{kU} \tag{3.81}$$

for $j = 1, 2, ..., n$. The moments μ_{kU}, $k = 3, 4, 5, ...$ are independent of n and bounded. Therefore, for large n*

$$\phi(it) \rightarrow \left(1 - \frac{1}{2}\frac{t^2}{n}\right)^n \rightarrow e^{-t^2/2} \tag{3.82}$$

The probability density function for Y can then be found from Eq. (3.25):

$$p(y) = \frac{1}{2\pi}\int_{-\infty}^{\infty}\phi(it)e^{-ity}dt = \frac{1}{2\pi}\int_{-\infty}^{\infty}e^{-t^2/2-ity}dt = \frac{1}{\sqrt{2\pi}}e^{-y^2/2} \tag{3.83}$$

so that finally

$$p(x) = p(y)\frac{dy}{dx} = \frac{1}{\sqrt{2\pi}\,s_x}e^{-\frac{1}{2}\left(\frac{x-\mu_x}{s_x}\right)^2} \tag{3.84}$$

which is the normal distribution.

Example 3.1.4

In order to illustrate how many terms n are necessary in order to obtain a reasonable approximation by a normal distribution, consider the case where the individual random variables X_j are uniformly distributed over the range [0,1]:

$$p(x_j) = \begin{cases} 1 & 0 \leq x_j \leq 1 \\ 0 & \text{otherwise} \end{cases} \qquad j = 1, 2, ..., n$$

implying that

$$\mu = \frac{1}{2} \quad ; \quad s = \frac{1}{\sqrt{12}}$$

and, from Eqs. (3.77) - (3.78):

$$\mu_x = \frac{n}{2} \quad ; \quad s_x = \sqrt{\frac{n}{12}}$$

* In general

$$\lim_{n\rightarrow\infty}\left(1 - \frac{a}{n}\right)^n \rightarrow e^{-a}$$

The probability distribution function $F^{(n)}(x)$ and the probability density function $p^{(n)}(x)$ for $X = X_1 + X_2 + ... + X_n$ can be determined from the recurrence relations:

$$F^{(n)}(x) = P(X \leq x) = P(X_1 + X_2 + ... + X_n \leq x)$$

$$= \sum_u \lim_{du \to 0} P(X_1 + X_2 + ... + X_{n-1} \leq x - u) \cdot P(u < X_n \leq u + du)$$

$$= \int F^{(n-1)}(x - u) f(u)\, du = \int_0^1 F^{(n-1)}(x - u)\, du = \int_{x-1}^{x} F^{(n-1)}(v)\, dv$$

and

$$p^{(n)}(x) = \frac{dF^{(n)}(x)}{dx} = F^{(n-1)}(x) - F^{(n-1)}(x - 1)$$

The recurrence starts with

$$F^{(1)}(x) = \begin{cases} 0 & x \leq 0 \\ x & 0 \leq x \leq 1 \\ 1 & x \geq 1 \end{cases}$$

and, after some algebra:

$$p^{(2)}(x) = \begin{cases} x & 0 \leq x \leq 1 \\ 2 - x & 1 \leq x \leq 2 \\ 0 & \text{otherwise} \end{cases}$$

$$p^{(3)}(x) = \begin{cases} \frac{1}{2}x^2 & 0 \leq x \leq 1 \\ -\frac{3}{2} + 3x - x^2 & 1 \leq x \leq 2 \\ \frac{9}{2} - 3x + \frac{1}{2}x^2 & 2 \leq x \leq 3 \\ 0 & \text{otherwise} \end{cases}$$

In Figure 3.4, the dashed curves show the exact probability density functions $p^{(n)}(x)$, $n = 1, 2, 3$ and 4, whereas the solid curves show the corresponding normal distributions with the same mean values and standard deviations. Although the initial uniform distribution deviates significantly from the normal distribution, it is clearly seen that already from $n \geq 3$, the normal distribution yields a good approximation to the exact distributions. Of course, the exact distributions are bounded by $x \in [0, n]$ whereas for the normal distribution $x \in]-\infty, \infty[$ for all n.

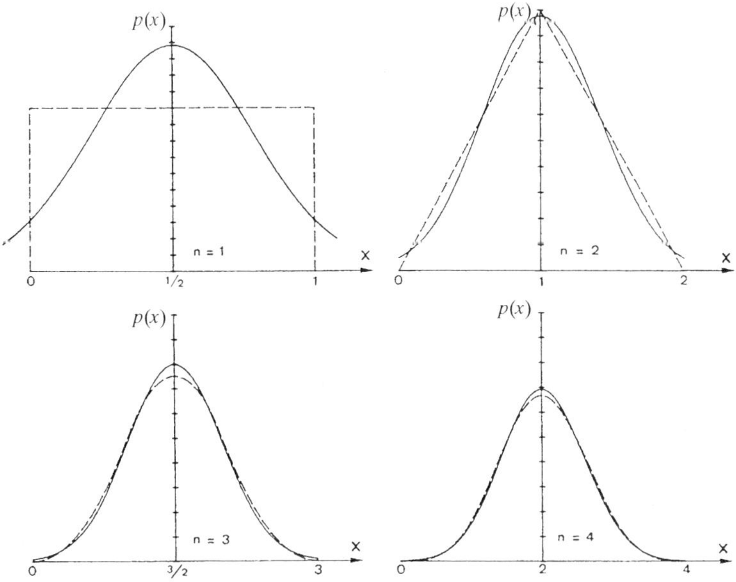

Figure 3.4 *Comparison between exact probability density distributions and normal distributions based on the central limit theorem.*

3.2 STOCHASTIC PROCESSES

The random variables used in the description of the ocean waves and the derived loads are often continuous functions of time t. If, for instance, the wave elevation is measured at a given location, wave elevation records as shown in Figure 3.5 may be obtained. However, as discussed in the introduction to this chapter, it is not possible to make a precise a priori description of this time history. Hence, if the measurements are made again under apparently identical conditions, other quite different time histories may be found. Each record $X_i(t)$ will constitute a sample in the ensemble of possible time histories which might be the outcome of the measurements.

For design purposes, from an infinite number of samples $X_i(t)$, $i = 1, 2, 3, ..$, those samples must be selected which represent the severest loading on the structure, and the probability of occurrence of these samples must be taken into acount. This analysis is performed by considering the complete ensemble as a *stochastic* (or *random) process* $X(t)$, so that each sample $X_i(t)$ is just an outcome of $X(t)$.

For any value of time t, $X(t)$ is a random variable and the results given in Section 3.1 can be used to define the probabilistic behaviour of $X(t)$. For example, the probability that $X(t)$ is less than or equal to x is

$$P(X(t) < x) = F(x; t) \tag{3.85}$$

where $F(x; t)$ is the probability distribution of X at time t. The relation between values of $X(t)$ at two different instants of time $X(t_1)$ and $X(t_2)$ can be characterised by the *covariance* $cov(t_1, t_2)$, Eq. (3.68):

$$cov(t_1, t_2) = E\left[\left(X(t_1) - \mu_1\right)\left(X(t_2) - \mu_2\right)\right] \tag{3.86}$$

where

$$\mu_i = E\left[X(t_i)\right] \tag{3.87}$$

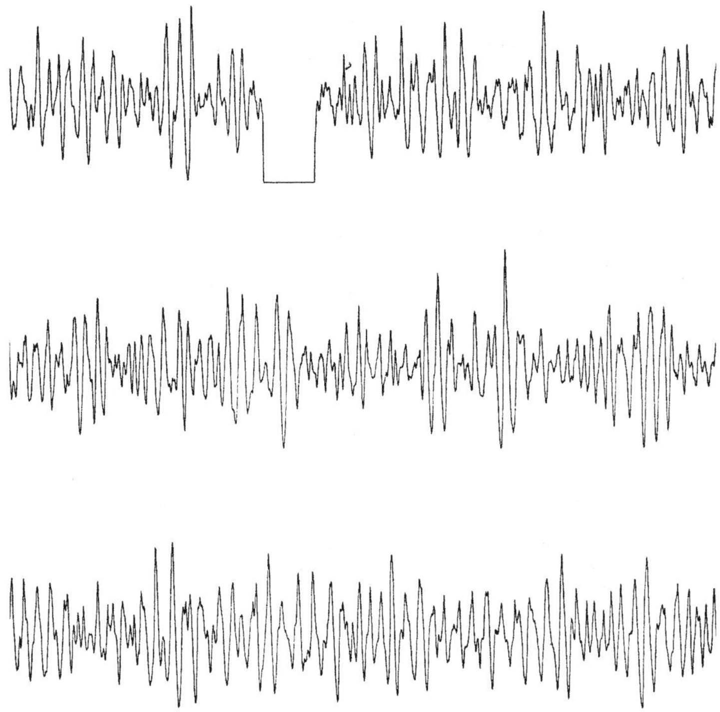

Figure 3.5 *Three 15 minutes wave elevation records, taken at Ekofisk Field, 1500 hrs 12/12/90. Courtesy Maersk Oil and Gas.*

Alternatively, the *autocorrelation* $R(t_1, t_2)$ defined as

$$R(t_1, t_2) \equiv E\left[X(t_1)X(t_2)\right] \tag{3.88}$$

may be used. It is seen that

$$R(t_1, t_2) = \text{cov}(t_1, t_2) + \mu_1\mu_2 \tag{3.89}$$

The closer the two instants of time, t_1 and t_2, are, the closer the correlation must be between $X(t_1)$ and $X(t_2)$ as $X(t)$ is a continuous process.

For $t_1 = t_2$ the autocorrelation becomes

$$R(t_1, t_1) = s_1^2 + \mu_1^2 \tag{3.90}$$

where s_1 is the standard deviation of $X(t_1)$.

A significant reduction in the complexity of the statistical predictions for $X(t)$ occurs, if the absolute value of time t does not enter the probability distribution, i.e. $F(x; t) = F(x)$. On the basis of this assumption, the stochastic process is said to be *stationary*. Only *stationary processes* will be considered here, as non-stationary processes can often be treated as a weighted sum of stationary processes. This is discussed later in Section 3.3.3

3.2.1 Stationary Stochastic Processes

On the assumption that $F(x; t) = F(x)$, the probability density function is written

$$p(x; t) = \frac{dF}{dx} = p(x)$$

Hence, all statistical moments $E[G(X)]$, for instance $\mu = E[X(t)]$ and $s^2 = E\left[(X(t) - \mu)^2\right]$, also become independent of t. The autocorrelation and the covariance become functions only of the time difference $\tau = t_2 - t_1$ and the autocorrelation is an even function of τ as

$$\begin{aligned} R(\tau) &= E[X(t)X(t + \tau)] \\ &= E[X(t - \tau)X(t)] = R(-\tau) \end{aligned} \tag{3.91}$$

Stationarity implies that the statistical moments of the process do not change with time. This criterion may be used to divide a stochastic process into a series of stationary processes. For ocean waves and derived wave load processes each stationary process may have a duration of between 30 minutes and three hours.

Based on a large set of samples $X_i(t)$, $i = 1, 2, \ldots, n$, approximate values of the statistical moments can be obtained by averaging over the whole ensemble:

$$\begin{aligned} \mu &\simeq \frac{1}{n}\sum_{i=1}^{n} X_i(t) \\ s^2 &\simeq \frac{1}{n}\sum_{i=1}^{n} (X_i(t) - \mu)^2 \\ R(\tau) &\simeq \frac{1}{n}\sum_{i=1}^{n} X_i(t)X_i(t + \tau) \end{aligned} \tag{3.92}$$

for a fixed value of t.

However, often only a single sample $X_0(t)$ is known. If this sample can be considered typical in an average sense of all possible samples, then the statistical moments of $X_0(t)$ obtained by averaging over time t:

$$\mu_0 = \frac{1}{T} \int_{-T/2}^{T/2} X_o(t)\, dt$$

$$s_0^2 = \frac{1}{T} \int_{-T/2}^{T/2} \left(X_o(t) - \mu_0\right)^2 dt \tag{3.93}$$

$$R_0(\tau) = \frac{1}{T} \int_{-T/2}^{T/2} X_o(t)\, X_o(t + \tau)\, dt$$

will often provide good estimates for μ, s, $R(\tau)$ etc. If this is the case the stationary process is also an *ergodic process*. Obviously, from Eq. (3.93), stationarity is a necessary condition for an ergodic process but the reverse does not have to be true.

3.2.2 Spectral Density

In the analysis of many physical problems, it is convenient to apply a Fourier transformation to a function defined in the time domain, in order to obtain a corresponding function defined in the frequency domain. Here the Fourier transformation is given in the following form:

$$A(t) = \int_{-\infty}^{\infty} B(\omega)\, e^{i\omega t} d\omega \tag{3.94}$$

and

$$B(\omega) = \frac{1}{2\pi} \int_{-\infty}^{\infty} A(t)\, e^{-i\omega t} dt \tag{3.95}$$

with $i^2 = -1$.

Eqs. (3.94) and (3.95) are called the Fourier transform pair. It should be noted that this is not a unique definition of the Fourier transform pair. Other, albeit very similar, forms do exist (see for example Ochi (1990) for details). Here $B(\omega)$ is called the Fourier transform of $A(t)$ while $A(t)$ is the Fourier inverse transform of $B(\omega)$.

For stationary processes, a Fourier transformation of the autocorrelation function $R(\tau)$ is particularly useful:

$$R(\tau) = \int_{-\infty}^{\infty} S_d(\omega)\, e^{i\omega\tau} d\omega \tag{3.96}$$

$$S_d(\omega) = \frac{1}{2\pi} \int_{-\infty}^{\infty} R(\tau)\, e^{-i\omega\tau} d\tau \tag{3.97}$$

The function $S_d(\omega)$ is denoted the *double sided spectral density*. However, $S_d(\omega)$ is an even function of ω as

$$S_d(-\omega) = \frac{1}{2\pi} \int_{-\infty}^{\infty} R(\tau)\, e^{i\omega\tau} d\tau$$

$$= \frac{1}{2\pi} \int_{-\infty}^{\infty} R(-u)\, e^{-i\omega u} du = S_d(\omega)$$

by use of $R(\tau) = R(-\tau)$, Eq. (3.91). The integrals (3.96) - (3.97) can then be reduced to

$$R(\tau) = \int_{0}^{\infty} S(\omega) \cos \omega\tau d\omega \tag{3.98}$$

$$S(\omega) = \frac{2}{\pi} \int_{0}^{\infty} R(\tau) \cos \omega\tau d\tau \tag{3.99}$$

by application of $e^{ix} \equiv \cos x + i \sin x$. Here

$$S(\omega) = 2\, S_d(\omega) \qquad \omega \geq 0 \tag{3.100}$$

is the *one-sided spectral density*. Whereas $R(\tau)$ varies between positive and negative values, it can be shown that $S(\omega)$ is always a non-negative function. This is illustrated later for ocean waves.

From Eqs. (3.91) and (3.98) it follows that

$$\int_{0}^{\infty} S(\omega) d\omega = R(0) = E[X(t)^2] = s^2 + \mu^2 \tag{3.101}$$

Thus, for $\mu = 0$, the area below the spectral density curve is equal to the variance of the process $E[X(t)^2]$ and therefore a measure of the average magnitude of X. In Section 3.3.2 spectral densities representing ocean waves will be discussed. Here, only a general discussion is given.

From the spectral density $S(\omega)$, the *spectral moments* can be derived:

$$m_n = \int_0^\infty \omega^n S(\omega) d\omega \tag{3.102}$$

It is seen that

$$m_0 = s^2 + \mu^2 \tag{3.103}$$

and that average frequencies $\overline{\omega}_n$ can be defined as

$$\overline{\omega}_n = \left(\frac{m_n}{m_0}\right)^{1/n} \tag{3.104}$$

If the spectral density is *narrow-banded*, that is if $S(\omega)$ has only non-zero values in a small frequency band $[\omega_0 - \Delta\omega, \omega_0 + \Delta\omega]$, then

$$\begin{aligned} m_n &\simeq \omega_0^n m_0 \\ \overline{\omega}_n &\simeq \omega_0 \\ R(\tau) &\simeq m_0 \cos\omega_0\tau \frac{\sin\Delta\omega\tau}{\Delta\omega\tau} \underset{\Delta\omega\to 0}{\to} m_0 \cos\omega_0\tau \end{aligned} \tag{3.105}$$

Thus, it can be seen that a strong correlation exists in a narrow-banded process between values of $X(t)$ separated by a large time difference $\tau \lesssim (\Delta\omega)^{-1}$. In the limit $\Delta\omega \to 0$, the process $X(t)$ becomes deterministic with a sinusoidal shape, frequency ω_0 and amplitude $\sqrt{m_0}$.

Generally, the spectral densities derived for physical processes related to ocean waves are rather narrow-banded. The *bandwidth* ε is usually characterised in terms of the spectral moments. One measure is

$$\varepsilon = \sqrt{1 - \frac{m_2^2}{m_0 m_4}} \tag{3.106}$$

It is seen from Eq. (3.105) that $\varepsilon \to 0$ for a narrow-banded spectrum. A *broad-banded spectrum* with $m_4 \to \infty$ yields the upper limit $\varepsilon = 1$.

Example 3.2.1

Consider the spectral density

$$S(\omega) = \begin{cases} 0 & \omega < \omega_1 \\ S_0 & \omega_1 \le \omega \le \omega_2 \\ 0 & \omega > \omega_2 \end{cases}$$

From Eq. (3.102) the spectral moments become

$$m_n = S_0 \frac{\omega_2^{n+1} - \omega_1^{n+1}}{n+1}$$

$$\overline{\omega}_1 = \frac{\omega_2 + \omega_1}{2} \quad ; \quad \overline{\omega}_2 = \sqrt{\frac{\omega_2^2 + \omega_1\omega_2 + \omega_1^2}{3}}$$

$$\varepsilon = \sqrt{1 - \frac{5}{9}\frac{1 + 2\alpha + 3\alpha^2 + 2\alpha^3 + \alpha^4}{1 + \alpha + \alpha^2 + \alpha^3 + \alpha^4}}$$

where $\alpha \equiv \frac{\omega_1}{\omega_2}$. Finally

$$R(\tau) = \frac{2\,m_0}{(\omega_2 - \omega_1)\tau}\cos\left(\frac{\omega_2 + \omega_1}{2}\tau\right)\sin\left(\frac{\omega_2 - \omega_1}{2}\tau\right)$$

For $\alpha \to 1$, the results tend to the narrow-banded solution (3.105). The other limit is $\alpha = 0$, i.e. $\omega_1 = 0$ or $\omega_2 \to \infty$. Here the bandwidth $\varepsilon = 2/3$.

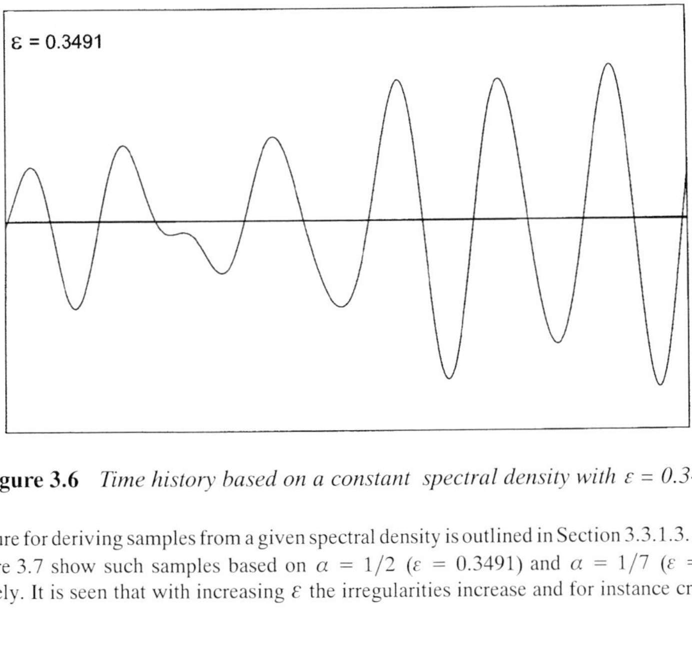

Figure 3.6 *Time history based on a constant spectral density with* $\varepsilon = 0.3491$.

A procedure for deriving samples from a given spectral density is outlined in Section 3.3.1.3. Figure 3.6 and Figure 3.7 show such samples based on $\alpha = 1/2$ ($\varepsilon = 0.3491$) and $\alpha = 1/7$ ($\varepsilon = 0.5963$), respectively. It is seen that with increasing ε the irregularities increase and for instance crests below the mean level appear more freqently. Note the similarity between the measured wave elevation records in Figure 3.5 and the simulation in Figure 3.7

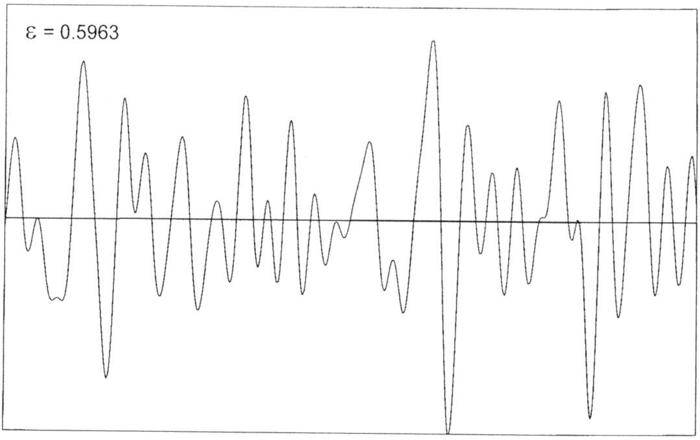

Figure 3.7 *Time history based on a constant spectral density with ε = 0.5963.*

3.2.3 Upcrossing and Peak Rates

When the stochastic process $X(t)$ represents a load or response process, it is obvious that the largest values of $X(t)$ are especially important in the design phase. Generally, the distribution of peaks (and troughs) is needed, but also the analysis of upcrossing rates gives useful information for design purposes, especially for narrow-banded processes. The results for upcrossing rates and peak distribution were first derived by Rice (1945).

For a stochastic process $X(t)$, the probability that the process during the time increment dt crosses the level x from below can be written

$$\int_{\dot{x}=0}^{\infty} p(x, \dot{x})\, dx\, d\dot{x}$$

as an upcrossing requires $\dot{x} > 0$. Here, $p(x, \dot{x})$ is the joint probability density of $(X, \dot{X})$. The *upcrossing rate* $\nu(x)$ is obtained by dividing this expression by dt. As $\frac{dx}{dt} = \dot{x}$:

$$\nu(x) = \int_{0}^{\infty} p(x, \dot{x})\, \dot{x}\, d\dot{x} \tag{3.107}$$

There is no correlation between a stationary process $X(t)$ and its time derivative $\dot{X}(t)$ since

including positive as well as negative local maxima $x \in]-\infty, \infty[$. Eq. (3.131) is the general formula derived by Rice (1945).

For a Gaussian process, the joint probability density $p(x, \dot{x}, \ddot{x})$ is a multivariate normal distribution given by Eq. (3.73). Its covariance matrix $\underline{\underline{\Sigma}}$ is of the order 3 x 3 with the coefficients:

$$\sigma_{11} = s_x^2 = m_0$$
$$\sigma_{22} = s_{\dot{x}}^2 = m_2$$
$$\sigma_{33} = s_{\ddot{x}}^2 = m_4$$
$$\sigma_{12} = \sigma_{21} = E[X(t)\,\dot{X}(t)] = 0$$
$$\sigma_{23} = \sigma_{32} = E[\dot{X}(t)\,\ddot{X}(t)] = 0$$
$$\sigma_{13} = \sigma_{31} = E[X(t)\,\ddot{X}(t)] = -m_2$$

assuming that $\mu_x = 0$*. Thus

$$\underline{\underline{\Sigma}} = \begin{bmatrix} m_0 & 0 & -m_2 \\ 0 & m_2 & 0 \\ -m_2 & 0 & m_4 \end{bmatrix} \tag{3.132}$$

The determinant becomes

$$|\underline{\underline{\Sigma}}| = m_2\left(m_0\, m_4 - m_2^2\right) \tag{3.133}$$

and the inverse matrix takes the value

$$\underline{\underline{\Sigma}}^{-1} = \frac{1}{m_2\left(m_0\, m_4 - m_2^2\right)} \begin{bmatrix} m_2 m_4 & 0 & m_2^2 \\ 0 & m_0 m_4 - m_2^2 & 0 \\ m_2^2 & 0 & m_0 m_2 \end{bmatrix} \tag{3.134}$$

Insertion of Eqs. (3.133) - (3.134) into Eq. (3.73) yields

$$p(x, \dot{x}, \ddot{x}) = p_{\dot{x}}(\dot{x})\, p_{x\ddot{x}}(x, \ddot{x}) \tag{3.135}$$

where $p_{\dot{x}}(\dot{x})$ is the marginal normal distribution of $\dot{x}$ given by Eq. (3.113). $p_{x\ddot{x}}(x, \ddot{x})$ is the joint distribution of $(X, \ddot{X})$:

$$p_{x\ddot{x}}(x, \ddot{x}) = \frac{1}{2\pi\sqrt{m_0 m_4 - m_2^2}} \exp\left\{-\frac{1}{2}\,\frac{m_4 x^2 + 2 m_2 x\ddot{x} + m_0 \ddot{x}^2}{m_0 m_4 - m_2^2}\right\} \tag{3.136}$$

* This can be done without loss of generality by a change of variable from $X(t)$ to $X(t) - \mu_x$.

Due to the non-zero covariance $\sigma_{13} = -m_2$, the stochastic process $X(t)$ and its second derivative $\ddot{X}(t)$ are correlated and hence also statistically dependent.

After some lengthy algebra substitution of Eqs. (3.135) - (3.136) into Eq. (3.131) followed by integration with respect to $\ddot{x}$ yields

$$p_p(x) = \frac{\int_{-\infty}^{0} p_{x\ddot{x}}(x,\ddot{x})\,|\ddot{x}|\,d\ddot{x}}{\int_{-\infty}^{\infty}\int_{-\infty}^{0} p_{x\ddot{x}}(x,\ddot{x})\,|\ddot{x}|\,d\ddot{x}\,dx} \tag{3.137}$$

$$= \left[\frac{\varepsilon}{\sqrt{2\pi}}\, e^{-\frac{1}{2}(u/\varepsilon)^2} + \sqrt{1-\varepsilon^2}\; u\, e^{-\frac{1}{2}u^2}\, \Phi\left(\frac{\sqrt{1-\varepsilon^2}}{\varepsilon}\, u\right)\right]\frac{1}{\sqrt{m_0}}$$

where

$$u \equiv \frac{x}{\sqrt{m_0}} \tag{3.138}$$

and ε is the bandwidth, Eq. (3.106). The above derivation was first carried out by Cartwright and Longuet-Higgins (1956).

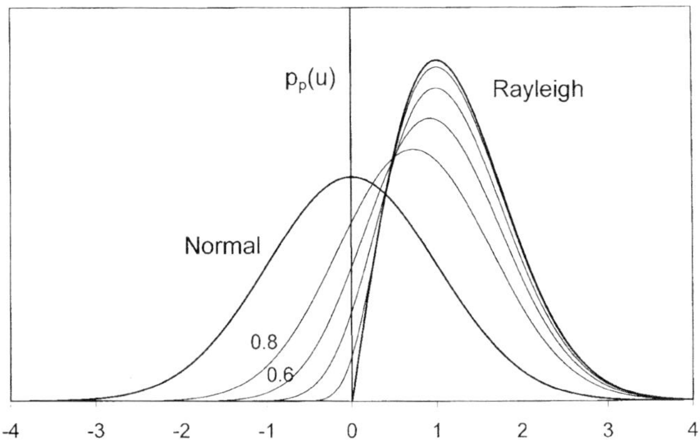

Figure 3.8 *Probability density functions of individual maxima for different bandwidths* $\varepsilon = 0$ *(Rayleigh),* 0.2, 0.4, 0.6, 0.8 *and* 1.0 *(normal).*

Figure 3.8 shows the probability density function $p_p(u) = p_p(x)\sqrt{m_0}$ for various values of the only parameter ε which enters into the expression for the probability distribution. In the narrow banded case ($\varepsilon \to 0$):

$$p_p(u) \rightarrow \begin{cases} u\, e^{-\frac{1}{2}u^2} & ; \quad u \geq 0 \\ 0 & ; \quad u < 0 \end{cases}$$

which is the Rayleigh distribution, Eq. (3.46), with $\alpha = \sqrt{2}$. This is in accordance with the result, Eq. (3.120), derived in the previous section, relating the peak distribution to the upcrossing rate for narrow-banded processes. It is seen that only positive peaks exist for $\varepsilon = 0$.

For extreme broad-banded cases ($\varepsilon = 1$), the probability density function for the maxima becomes

$$p_p(u) = \frac{1}{\sqrt{2\pi}}\, e^{-\frac{1}{2}u^2}$$

which is the normal distribution. In this case, the maxima are distributed just like the process itself and hence there is no reason to consider the distribution of maxima separately.

The probability density function, Eq. (3.137), can be integrated to yield the *probability distribution for the local maxima*:

$$\begin{aligned} F_p(u) &= \int_{-\infty}^{u} p_p(u)\, du \\ &= \Phi\left(\frac{u}{\varepsilon}\right) - \sqrt{1-\varepsilon^2}\, \Phi\left(\frac{u}{\varepsilon}\sqrt{1-\varepsilon^2}\right) e^{-\frac{1}{2}u^2} \\ &\rightarrow 1 - e^{-\frac{1}{2}u^2} \qquad \text{for } \varepsilon = 0 \end{aligned} \tag{3.139}$$

From the probability density function $p_p(x)$, Eq. (3.137), the lowest statistical moments may be derived:

$$\begin{aligned} &\textit{Mean} && \mu_x = \sqrt{m_0}\,\sqrt{\frac{\pi}{2}\left(1-\varepsilon^2\right)} \\ &\textit{Standard deviation} && s_x = \sqrt{m_0}\,\sqrt{1 - \left(\frac{\pi}{2} - 1\right)\left(1-\varepsilon^2\right)} \\ &\textit{Skewness} && \gamma_1 = \sqrt{\frac{\pi}{2}}\,(\pi - 3)\left[\frac{1-\varepsilon^2}{1 - \left(\frac{\pi}{2} - 1\right)\left(1-\varepsilon^2\right)}\right]^{3/2} \end{aligned} \tag{3.140}$$

The limiting cases of $\varepsilon = 0$ and $\varepsilon = 1$ are again seen to correspond to the results for the Rayleigh and the normal distribution, respectively. Extension to slightly non-Gaussian processes given by the Gram-Charlier series expansion (3.31) is presented by Longuet-Higgins (1964).

3.2.5 Extreme Value Predictions

Consider a stationary stochastic process $X(t)$. One of the most important questions is often how to determine the expected maximum (or minimum) value of this process during a given period of time T.

As the process is assumed to be continuous in time t, the maximum value is the largest peak. Therefore, the probability distribution $F_T(x)$ for the maximum value can be expressed as

$$F_T(x) = P(X_{p1} < x, X_{p2} < x, ..., X_{pN} < x) \tag{3.141}$$

where X_{pi}, $i = 1, 2, ..., N$ are the peaks in $t \in [0, T]$ and N is the expected number of peaks:

$$N = \nu_p T \tag{3.142}$$

The peak rate ν_p is given by Eq. (3.121) or, for a Gaussian process, by Eq. (3.129).

If the peaks X_{pi}, $i = 1, 2, ..., N$ can be assumed to be statistically independent, then

$$F_T(x) \equiv F_N(x) = [F_p(x)]^N \tag{3.143}$$

as all peaks in a stationary process have the same probability distribution. The corresponding probability density function $p_N(x)$ becomes

$$p_N(x) = \frac{dF_N}{dx} = N\left[F_p(x)\right]^{N-1} p_p(x) \tag{3.144}$$

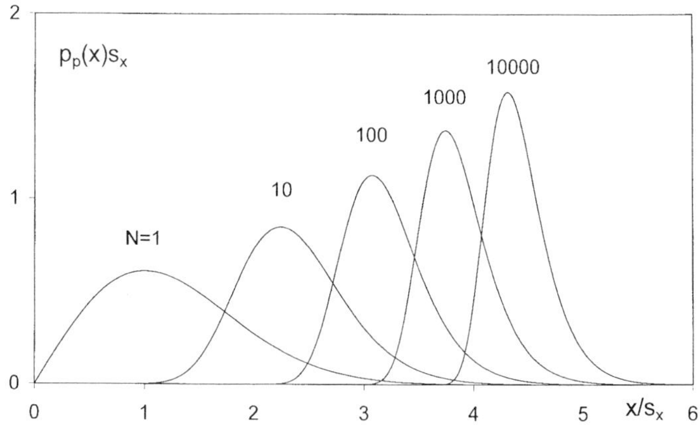

Figure 3.9 *Probability density functions $p_N(x)$ for the largest peaks among N peaks assuming a narrow-banded Gaussian process, $\varepsilon = 0$.*

where $p_p(x)$ is the probability density function of the individual maxima, Eq. (3.131). Knowing the probability distribution $F_p(x)$ of the individual maxima makes it thus possible to calculate the probability distribution of the largest peak among N peaks. For a narrow-banded Gaussian process such results are shown in Figure 3.9.

Any statistical moment or characteristic value of the largest peak may be calculated from Eq. (3.144). For example, the mean value μ_N:

$$\mu_N = \int_{-\infty}^{\infty} x\, p_N(x)\, dx = N \int_{-\infty}^{\infty} x \left[F_p(x)\right]^{N-1} p_p(x)\, dx \tag{3.145}$$

the most probable value $\tilde{\mu}_N$:

$$\left.\frac{dp_N(x)}{dx}\right|_{x=\tilde{\mu}_N} = 0 \Rightarrow (N-1)\, p_p^2(\tilde{\mu}_N) + F_p(\tilde{\mu}_N) \left.\frac{dp_p}{dx}\right|_{x=\tilde{\mu}_N} = 0 \tag{3.146}$$

and the 50 per cent fractile $\tilde{\mu}_{N,\,50}$:

$$F_N\left(\tilde{\mu}_{N,\,50}\right) = \frac{1}{2} \Rightarrow F_p\left(\tilde{\mu}_{N,\,50}\right) = \left(\frac{1}{2}\right)^{1/N} \tag{3.147}$$

Closed-form expressions for μ_N and $\tilde{\mu}_N$ are normally difficult to obtain. If, however, N is sufficiently large, approximate values for μ_N and $\tilde{\mu}_N$ can be determined. First $\tilde{\mu}_N$ is considered. Later, in Example 3.2.3, μ_N is derived for a Gaussian process.

For large values of N, $\tilde{\mu}_N$ will correspond to a value in the tail of the distribution function $F_p(x)$. This implies that $F_p(\tilde{\mu}_N) \simeq 1$ and $p_p(\tilde{\mu}_N) \simeq 0$. From l'Hôpital's rule, it follows that

$$\frac{p_p\left(\tilde{\mu}_N\right)}{1 - F_p\left(\tilde{\mu}_N\right)} \simeq \frac{\left(dp_p/dx\right)_{x=\tilde{\mu}_N}}{-\, p_p\left(\tilde{\mu}_N\right)}$$

Insertion of this result into Eq. (3.146) yields

$$\frac{F_p\left(\tilde{\mu}_N\right)}{1 - F_p\left(\tilde{\mu}_N\right)} \simeq N - 1$$

or

$$F_p\left(\tilde{\mu}_N\right) \simeq 1 - \frac{1}{N} \tag{3.148}$$

As

$$P\left(X_p > \tilde{\mu}_N\right) = 1 - F_p\left(\tilde{\mu}_N\right) \simeq \frac{1}{N}$$

the probability that each individual peak exceeds $\tilde{\mu}_N$ is approximately equal to $1/N$. With an average total of N peaks, only 1 in N peaks during the time T will exceed $\tilde{\mu}_N$ in average. Hence, T may be considered as the *return period* for the extreme peak $\tilde{\mu}_N$.

The probability that the largest peak among N peaks exceeds $\tilde{\mu}_N$ follows from Eqs. (3.141), (3.143) and (3.148):

$$
\begin{aligned}
P\left(\max_{i=1,2,\ldots N} X_{pi} > \tilde{\mu}_N\right) &= 1 - F_N\left(\tilde{\mu}_N\right) \\
&= 1 - \left[F_p\left(\tilde{\mu}_N\right)\right]^N \\
&\simeq 1 - \left[1 - \frac{1}{N}\right]^N \simeq 1 - \frac{1}{e} \simeq 0.632
\end{aligned}
\tag{3.149}
$$

as $\lim_{N\to\infty}\left(1 - \frac{a}{N}\right)^N = e^{-a}$.

For design purposes, $\tilde{\mu}_N$ may therefore not be a very appropriate value because the high probability of exceeding this value during the period T (e.g. the lifetime of the structure).

The formula for the 50 per cent fractile $\tilde{\mu}_{N,\,50}$ may also be simplified if N is large. In that case

$$
\left(\frac{1}{2}\right)^{1/N} = e^{-\frac{1}{N}\ln 2} \simeq 1 - \frac{1}{N}\ln 2
$$

so that Eq. (3.147) becomes

$$
F_p\left(\tilde{\mu}_{N,\,50}\right) \simeq 1 - \frac{1}{N}\ln 2 \tag{3.150}
$$

The probability that the maximum peak among the N peaks exceeds $\tilde{\mu}_{N,\,50}$ is per definition equal to 50 per cent so that $\tilde{\mu}_N < \tilde{\mu}_{N,\,50}$. For design purposes, it may be preferable to use the *q-fractile* $\tilde{\mu}_{Nq}$ defined as the value which the largest peak exceeds with a given small probability q:

$$
F_N\left(\tilde{\mu}_{Nq}\right) = 1 - q
$$

or

$$
F_p\left(\tilde{\mu}_{Nq}\right) = (1 - q)^{1/N} \simeq 1 - \frac{q}{N} \tag{3.151}
$$

For a Gaussian process, the probability distribution $F_p(x)$ and the associated probability density function $p_p(x)$ are given by Eqs. (3.139) and (3.137), respectively. For large values of N, $\tilde{\mu}_N$ will be much greater than $s = \sqrt{m_0}$, see Figure 3.9, and if, furthermore, the bandwidth measure ε is less than about 0.9, the standard normal distribution functions in these equations are very close to unity so that, omitting for simplicity the subscript x on s_x,

$$
F_p(x) \simeq 1 - \sqrt{1 - \varepsilon^2}\, e^{-\frac{1}{2}(x/s)^2} \tag{3.152}
$$

and

$$p_p(x) \simeq \frac{x}{s^2}\sqrt{1-\varepsilon^2}\; e^{-\frac{1}{2}(x/s)^2} \tag{3.153}$$

The most probable value $\tilde{\mu}_N$ of the largest peak is obtained from Eq. (3.148):

$$\tilde{\mu}_N \simeq s\;\sqrt{2\ln\left(\sqrt{1-\varepsilon^2}\,N\right)} \tag{3.154}$$

From Eqs. (3.130) and (3.142):

$$\sqrt{1-\varepsilon^2}\;N = \sqrt{1-\varepsilon^2}\;\nu_p\,T = \nu(0)\,T \equiv N_Z \tag{3.155}$$

so that $\tilde{\mu}_N$ can be expressed more simply in terms of the number N_Z of zero-upcrossings:

$$\tilde{\mu}_N \simeq s\;\sqrt{2\ln N_Z} \tag{3.156}$$

The 50 per cent fractile $\tilde{\mu}_{N,\,50}$ follows from Eq. (3.150):

$$\tilde{\mu}_{N,\,50} \simeq s\;\sqrt{2\ln\left(N_Z\,/\ln 2\right)} \tag{3.157}$$

using Eq. (3.155) and, finally, the q-fractile $\tilde{\mu}_{N,\,q}$, Eq. (3.151), becomes

$$\tilde{\mu}_{N,\,q} \simeq s\;\sqrt{2\ln\left(N_Z\,/q\right)} \tag{3.158}$$

for $N_Z >> 1, \;\; q << 1$.

Example 3.2.2

For stochastic processes where no analytical probability distribution $F_p(x)$ for the individual peak values is available, an empirical fit of the individual maxima to a Weibull distribution, Eq. (3.38):

$$F_p(x) = 1 - e^{-(x/\alpha)^\beta}$$

often proves useful. The most probable largest value $\tilde{\mu}_N$ is then found from Eq. (3.148):

$$\tilde{\mu}_N = \alpha\,(\ln N)^{1/\beta} \tag{3.159}$$

The fundamental assumption in the previous derivations is that the peaks are statistically independent and therefore can be rearranged in ascending order. This analysis is called *order statistics*. Another procedure based on so-called *Poisson processes* assumes statistically independent upcrossings of any given level x. A discussion of Poisson processes is beyond the scope of this treatment of stochastic processes, but it can be shown, see e.g. Ochi (1990), that the probability distribution for the upcrossing of the level x during a period of time T becomes

$$F_T(x) = e^{-\nu(x)\,T} \tag{3.160}$$

where $\nu(x)$ is the upcrossing rate, Eq. (3.107). For Gaussian processes, substitution of Eq. (3.115) by $\mu_x = 0$ into Eq. (3.160) yields

$$F_T(x) \equiv F_N(x) = \exp\left(-N_Z\, e^{-\frac{1}{2}(x/s)^2}\right) \tag{3.161}$$

with $N_Z = \nu(0)T$ and $s \equiv s_x$. For upcrossings of extreme levels, Eq. (3.161) can be approximated by

$$F_N(x) \simeq 1 - N_Z\, e^{-\frac{1}{2}(x/s)^2} \tag{3.162}$$

which is the same result as obtained previously by use of order statistics, Eqs. (3.143), (3.152):

$$\begin{aligned} F_N(x) &\simeq \left(1 - \sqrt{1-\varepsilon^2}\, e^{-\frac{1}{2}(x/s)^2}\right)^N \\ &\simeq 1 - N\sqrt{1-\varepsilon^2}\, e^{-\frac{1}{2}(x/s)^2} \\ &= 1 - N_Z\, e^{-\frac{1}{2}(x/s)^2} \end{aligned}$$

Therefore, whether to use order statistics or Poisson upcrossing is not so important in the present context as they asymptotically lead to the same results. It should, however, be mentioned that both procedures rely on some assumptions about statistically independent peaks which are only satisfied approximately in real physical processes. For estimations of wave loads on ships, the results presented here are generally sufficiently accurate.

Example 3.2.3

From the probability distribution function $F_N(x)$ given by Eq. (3.161), the mean value μ_N can be derived.

The monotonic transformation

$$y \equiv -\ln N_Z + \frac{1}{2}\left(\frac{x}{s}\right)^2 \tag{3.163}$$

implies that y is Gumbel distributed, Eq. (3.51):

$$F_N(y) = \exp\left(-e^{-y}\right)$$

Hence, from Eq. (3.53), the mean value of y becomes

$$E[y] = C\ (= 0.577215....)$$

In order to determine the mean value $\mu_N = E[x]$ of x, the transformation (3.163) is linearised around the most probable largest value $\tilde{\mu}_N$ given by (3.156), assuming μ_N to be close to $\tilde{\mu}_N$.

If $x = \tilde{\mu}_N$ the corresponding value of y is equal to zero by use of (3.163) such that

$$y \simeq \frac{dy}{dx}\Big|_{x=\tilde{\mu}_N} (x - \tilde{\mu}_N) = \frac{\tilde{\mu}_N}{s^2} (x - \tilde{\mu}_N)$$

Hence

$$E[y] \simeq \frac{\tilde{\mu}_N}{s^2} (\mu_N - \tilde{\mu}_N) = C$$

yielding

$$\mu_N \simeq \tilde{\mu}_N + C \frac{s^2}{\tilde{\mu}_N}$$

or

$$\mu_N \simeq s\sqrt{2 \ln N_Z} \left(1 + \frac{C}{\sqrt{2 \ln N_Z}} \right] \tag{3.164}$$

It is seen that $\mu_N > \tilde{\mu}_N$ and, furthermore, from Eq. (3.157) that

$$\tilde{\mu}_N < \tilde{\mu}_{N,50} < \mu_N$$

To apply order statistics or Poisson upcrossing a knowledge of the individual peak distribution or upcrossing rates is needed. For linear and slightly non-linear problems they can usually be determined theoretically but for highly non-linear problems or when actual measurements are used, information in the form of samples $X_i(t)$ of the process may only be available. In such cases the theory of *asymptotic distributions of extreme values* may be very useful. The theory does not make use of the individual peak distribution or upcrossing rate but relies on a proof, see e.g. Ochi (1990), that the extreme peaks follow a specific distribution, depending mainly on whether the process is bounded or not. For unbounded processes, this distribution is the Gumbel distribution, Eq. (3.51), provided all statistical moments of the initial distribution exist*. Its two parameters, (α, β), must be determined numerically. This can be done in a least square sense by relating (α, β) to the mean value and the standard deviation of the set of known extreme peaks, by application of Eqs. (3.53) - (3.54).

Another method which usually provides better results is the *maximum likelihood procedure*, in which it is assumed that the set of known extreme peaks $\bar{x}_1, \bar{x}_2, \ldots, \bar{x}_n$ is the most probable set of peaks to be found. This implies that the joint probability density $p(x_1, x_2, \ldots, x_n)$ should have its maximum value for $x_i = \bar{x}_i$, 1, 2,..., n. On the assumption that the peaks are statistically independent and that each follows the same Gumbel distribution

---

* The Weibull distribution, Eq. (3.38), is another extreme value distribution applicable to processes without an upper limit.

$$p(\bar{x}_1, \bar{x}_2, ..., \bar{x}_n) = \prod_{i=0}^{n} p(\bar{x}_i; \alpha, \beta) \tag{3.165}$$

where $p(\bar{x}_i; \alpha, \beta)$ is the Gumbel probability density function, Eq. (3.52), with $x = \bar{x}_i$. The parameters (α, β) are then determined so that $p(\bar{x}_1, \bar{x}_2, ..., \bar{x}_n)$ is maximised:

$$\frac{\partial p}{\partial \alpha} = 0 \quad ; \quad \frac{\partial p}{\partial \beta} = 0 \tag{3.166}$$

As $p > 0$ Eq. (3.166) can be replaced by

$$\frac{\partial \ln p}{\partial \alpha} = 0 \quad ; \quad \frac{\partial \ln p}{\partial \beta} = 0 \tag{3.167}$$

The use of $\ln p$ rather than p makes the solution for (α, β) easier, because the products in Eq. (3.165) are replaced by a summation:

$$\frac{\partial \ln p}{\partial \alpha} = \sum_{i=1}^{n} \frac{\partial \ln p(\bar{x}_i; \alpha, \beta)}{\partial \alpha} = 0 \tag{3.168}$$

Differentiation with respect to β is done analogously. This results in two non-linear equations in (α, β) which must be solved numerically.

The special property inherent in the Gumbel distribution is that its type is invariant with respect to the number of peaks. Consider a Gumbel distribution

$$F_n(x) = \exp\left(-e^{-(x-\alpha_n)/\beta_n}\right) \tag{3.169}$$

derived for the largest peak among n peaks. The distribution of the largest peak among $N = m \cdot n$ peaks becomes

$$F_N(x) = F_n(x)^{N/n} = F_n(x)^m = \exp\left(-e^{-(x-\alpha_N)/\beta_N}\right) \tag{3.170}$$

where $\alpha_N = \alpha_n + \beta \ln m$ and $\beta_N = \beta_n$. $F_N(x)$ is therefore also a Gumbel distribution, only the shift parameter α has been modified. The most probable largest peak value $\tilde{\mu}_N = \alpha_N$ as shown in Section 3.1.5.

Comparing Eq. (3.161) with Eq. (3.170) seems to indicate a difference in exponent on x, making the two distributions incompatible. However, this disagreement disappears when $n \to \infty$ as shown below.

Eq.(3.161) can be written

$$F_N(x) = \exp\left[-\exp\left[-\frac{1}{2}\left(\frac{x}{s}\right)^2 + \frac{1}{2}\left(\frac{\tilde{\mu}_N}{s}\right)^2\right]\right]$$

using Eq. (3.156). If N is large then $\tilde{\mu}_N = s\sqrt{2\ln N_Z}$ is also large and, hence, it can be expected that $|x - \tilde{\mu}_N| \ll \tilde{\mu}_N$. Thereby,

$$\left(\frac{x}{s}\right)^2 = \left(\frac{\mu_N}{s}\right)^2\left(1 + \frac{x - \tilde{\mu}_N}{\tilde{\mu}_N}\right)^2 \simeq \left(\frac{\tilde{\mu}_N}{s}\right)^2\left(1 + 2\frac{x - \tilde{\mu}_N}{\tilde{\mu}_N}\right)$$

such that

$$F_N(x) \simeq \exp\left[-\exp\left(-\frac{x - \tilde{\mu}_N}{\tilde{\mu}_N}\right)\right]$$

which is the Gumbel distribution with

$$\alpha_N = \beta_N = \tilde{\mu}_N$$

The same asymptotic behaviour can easily be proven for any individual peak distribution satisfying

$$F_p(x) = 1 - e^{-q(x)} \quad ; \quad -\infty < x < \infty \tag{3.171}$$

provided Eq. (3.148) also holds. Again the Gumbel distribution is strictly only valid for $N \rightarrow \infty$, because of the approximation

$$q(x) \simeq q(\tilde{\mu}_N) + \frac{dq}{dx}\Big|_{\tilde{\mu}_N}(x - \tilde{\mu}_N) \tag{3.172}$$

The Gumbel distribution can be applied to a realization (sample) $X(t)$, $0 \leq t \leq T$ of a stochastic process by dividing the sample into M equidistant time epochs

$$X_i(t)\,;\ \frac{i-1}{M}T < T \leq \frac{i}{M}T \quad ; \quad i = 1, 2, ...M$$

and then extract the maximum value $\bar{x}_i$ in each epoch

$$\bar{x}_i = \max\left(X_i(t)\right) \quad ; \quad i = 1, 2, ...M$$

These M maxima then determine α_M, β_M in the Gumbel distribution by either Eqs. (3.53)-(3.54) or the maximum likelihood method, Eq. (3.167). For a given sample length T it might be difficult to choose a proper value of M. If M is too high then $\bar{x}_i \ll \tilde{\mu}_M$ in several of the then small equidistant time epochs, violating the assumption $|x_i - \tilde{\mu}_M| \ll \tilde{\mu}_M$. On the other hand, if M is too small, $\tilde{\mu}_M \ll \max(\bar{x}_i)$, also contradicting this assumption. A remedy can be to transform the variable x. As the approximation (3.172) holds exactly if $q(x)$ is a linear function of x, the optimal transformation is

$$x = q^{-1}(y) \tag{3.173}$$

Thus, for a Gaussian process, the transformation $x = \sqrt{2y}/s$ will transform Eq. (3.161) to a Gumbel distribution with $\alpha_M = 0.5\left(\tilde{\mu}_M/s\right)^2$ and $\beta_M = 1$ in the variable y, a property also used in Example 3.2.3. In real cases $q(x)$ is of course not known (because then there is no need for the Gumbel distribution as $F_N(x)$ is known!). However, some knowledge of the underlying physical processes can be apply to define a transformation which extend the validity of the Gumbel distribution significantly, Naess, Storli and Storm (1996).

The main disadvantage with the application of the Gumbel distribution is that only the largest maxima in each time epoch is utilized. Thus many large maxima may be omitted from the analysis, resulting in a large loss of information. An alternative is to assume that all data above a certain threshold value u are important for the extreme value distribution. An appropriate model for the probability distribution function $F_u(v)$ for the excess $V = X - u$ is the *Generalized Pareto (GP) distribution*. Each of the three asymptotic extreme value distributions has a corresponding GP distribution, Pickands (1975). For the Gumbel distribution this GP distribution is

$$F_u(v) = 1 - e^{v/\gamma} \quad ; \quad v > 0 \tag{3.174}$$

The relation between the parameter in the two distributions, Eq. (3.51) and (3.174) can be determined from the assumption that the number of peaks exceeding the threshold is Poisson distributed, Pickands (1975), and becomes

$$\alpha = u + \gamma \ln n_u \quad ; \quad \beta = \gamma \tag{3.175}$$

where n_u is the number of peaks above the threshold level u.

In order to derive extreme value predictions from a sample $X(t)$; $0 \leq t \leq T$ of a stochastic process, the following procedure based on the Peak-Over Threshold method can be used:

- Select a moderately high threshold u and collect all peaks $\bar{x}_i$; $i = 1, 2, ..., n_u$ above u. The number n_u depends clearly on both u, T and the actual sample. Thus n_u is a random number.
- Fit the excesses $\bar{v}_i = \bar{x}_i - u$; $i = 1, 2, ..., n_u$ to the GP distribution, Eq. (3.174). For instance the maximum likelyhood procedure, Eq. (3.168), with $p(v\,;\,\gamma)$ derived from Eq. (3.174) yields simply

$$\gamma = \frac{1}{n_u}\sum_{i=1}^{n_u} \bar{v}_i \tag{3.176}$$

- Calculate α, β in the Gumbel distribution from Eq. (3.175).
- Determine the cumulative probability distributions for the largest peak over a period $T_e > T$ from Eq. (3.170) with

$$\alpha_N = \alpha + \beta \ln (T_e/T) = u + \gamma \ln (n_u N)$$
$$\beta_N = \beta = \gamma$$

where $N = T_e/T$.

- The mean value μ_N, the most probable value $\tilde{\mu}_N$ and the 50 per cent fractile of the largest peak becomes

$$\begin{aligned} \mu_N &= \alpha_N + C\beta_N \simeq \alpha_N + 0.5772\,\beta_N \\ \tilde{\mu}_N &= \alpha_N \\ \tilde{\mu}_{50,N} &= \alpha_N - \beta_N \ln(\ln 2) \simeq \alpha_N + 0.3665\,\beta_N \end{aligned} \tag{3.177}$$

according to the results derived in Section 3.1.5. The value with a return period T_e follows from $F_N(x_N) = 1 - \frac{1}{N}$:

$$\begin{aligned} x_N &= \alpha - \beta \ln\left[-\ln\left(1 - \frac{1}{N}\right)\right] \\ &\simeq \alpha + \beta\,\ln N = \alpha_N = \tilde{\mu}_N \end{aligned} \tag{3.178}$$

- The result should be rather insensitive to the choice of threshold u. Therefore the analysis should be done with different values of u to check the validity of the GP distribution for the excess.

The procedure outlined above assumes that the extreme value distribution is of the Gumbel type. This is so for many of the stochastic processes related to wave loads. However, in some cases the two other asymptotic extreme value distributions (Fréchet and Weibull) and their associated GP distributions might be useful. A detailed description of these distributions and their applications is given in Castillo (1988).

3.2.6 Conditional Mean Processes

The expected (mean) variation $< X(t) >$ of a stochastic process $X(t)$ in the vicinity of a large peak x_0 at $t=0$ is also of interest. It is given by

$$\begin{aligned} < X(t) > &\equiv E\left[X(t)\,|\,X(0) = x_0, \dot{X}(0) = \dot{x}_0 = 0\right] \\ &= \int_{-\infty}^{\infty} u p\,(u\,|\,x_0,\,\dot{x}_0 = 0)\,dx \end{aligned} \tag{3.179}$$

where $p\,(x\,|\,x_0, 0)$ is the conditional probability density function of $X(t)$, given $X(0) = x_0$ and $\dot{X}(0)=0$. From Eq. (3.62) it follows that

$$p\,(x\,|\,x_0, \dot{x}_0) = \frac{p\,(x_0, \dot{x}_0, x)}{p\,(x_0, \dot{x}_0)} \tag{3.180}$$

where $p\,(x_0, \dot{x}, x)$ and $p\,(x_0, \dot{x}_0)$ are the joint probability density functions of $X(0)$, $\dot{X}(0)$, $X(t)$ and $X(0)$, $\dot{X}(0)$, respectively. For Gaussian processes these functions become multivariate normal distributed, Eq. (3.73). The result for $p\,(x_0, \dot{x}_0)$ is given by Eqs. (3.109), (3.112) and (3.113)

$$p\left(x_0, \dot{x}_0 = 0\right) = \frac{1}{2\pi\, s_x\, s_{\dot{x}}} \exp\left[-\frac{1}{2}\left(\frac{x_0}{s_x}\right)^2\right] \tag{3.181}$$

assuming $E\,[X(t)] = 0$. The elements σ_i in the covariance matrix $\underline{\underline{\Sigma}}$ of $p\,(x_0, \dot{x}_0, x)$ are

$$\begin{aligned}
\sigma_{11} &= E\left[X(0)^2\right] = s_x^2 = m_0 \\
\sigma_{22} &= E\left[\dot{X}(0)^2\right] = s_{\dot{x}}^2 = m_2 \\
\sigma_{12} &= \sigma_{21} = E\left[X(0)\,\dot{X}(0)\right] = 0 \\
\sigma_{33} &= E\left[X(t)^2\right] = s_x^2 = m_0 \\
\sigma_{13} &= \sigma_{31} = E\left[X(0)\,X(t)\right] = R(t) \\
\sigma_{23} &= \sigma_{32} = E\left[\dot{X}(0)\,X(t)\right] = S(t)
\end{aligned} \tag{3.182}$$

using the properties of stationary processes. The autocorrelations $R(t)$ and $S(t)$ are the only functions of t in Eq. (3.182). The determinant $|\underline{\underline{\Sigma}}|$ becomes

$$|\underline{\underline{\Sigma}}| = m_0^2\, m_2 - R(t)^2 m_2 - S(t)^2\, m_0 \tag{3.183}$$

and the inverse matrix takes the form

$$\underline{\underline{\Sigma}}^{-1} = \frac{1}{|\underline{\underline{\Sigma}}|}\begin{bmatrix} m_0\, m_2 - S(t)^2 & R(t)\, S(t) & -R(t)\, m_2 \\ R(t)\, S(t) & m_0^2 - R(t)^2 & -S(t)\, m_0 \\ -R(t) m_2 & -S(t)\, m_0 & m_0 m_2 \end{bmatrix} \tag{3.184}$$

Carrying out the matrix multiplications in Eq. (3.73) using $\underline{x} = \left\{x_0, \dot{x}_0 = 0, x\right\}^T$ yield

$$p\left(x_0, 0, x(t)\right) = \frac{1}{\left(\sqrt{2\pi}\right)^3 \sqrt{|\underline{\underline{\Sigma}}|}} \times \tag{3.185}$$
$$\exp\left[-\frac{1}{2}\frac{\left(m_0 m_2 - S(t)^2\right) x_0^2 + m_0 m_2\, x(t)^2 - 2\,R(t)\, m_2 x_0\, x(t)}{|\underline{\underline{\Sigma}}|}\right]$$

and hence Eq. (3.180) becomes

$$p\,(x(t)|x_0, \dot{x}_0 = 0) = \frac{1}{\sqrt{2\pi}\; u(t)} \exp\left[-\frac{1}{2}\left(\frac{x(t) - r(t) x_0}{u(t)}\right)^2\right] \tag{3.186}$$

after a little algebra. Here

$$u(t) = \sqrt{m_0\left(1 - r^2(t) - s^2(t)\right)} \tag{3.187}$$

with

$$r(t) = \frac{R(t)}{m_0}$$
$$s(t) \equiv \frac{S(t)}{\sqrt{m_0 m_2}}$$

From Eq. (3.186) it is seen that $X(t)$, given $X(0) = x_0$ and $\dot{X}(0) = 0$, is a normal distributed, non-stationary process with mean value

$$< X(t) > = r(t)\, x_0 \tag{3.188}$$

and standard deviation $u(t)$. As $u(t)$ is independent of x_0, the conditional process $<X(t)>$tends toward the deterministic mean value, Eq. (3.188), for large peak values x_0. Thus the very simple result

$$< X(t) > = \frac{E\,[X(t)\,X(0)]}{E[X^2(0)]} x_0 \tag{3.189}$$

provides a good estimate of the variation of a stochastic process around a large peak x_0 at $t=0$. This result can for instance be used to define critical wave episodes, as shown later in Example 3.3.2, for application in time-domain simulations of ship responses in waves.

The result, Eq. (3.189), can be generalized to processes conditional on several parameters, Friis Hansen and Nielsen (1995) and slightly non-Gaussian processes, Jensen (1996).

3.3 RANDOM SEA WAVES

Measurements of the surface elevation in the open sea have shown that over a relative short period, in the order of hours, the elevation can be considered as a stationary stochastic process with a probability distribution close to a normal distribution and a relatively narrow-banded spectral density.

For most practical purposes, the wave elevation can therefore be taken to be normally distributed but in large storm-generated waves, the non-linearities in the free surface conditions result in some skewness in the distribution. This phenomenon is equivalent to the behaviour seen in regular waves where small amplitude waves have a nearly sinusoidal profile. If the wave amplitude increases with a constant wave length the profile changes form so that it exhibits vertical asymmetry, as described by for example a Stokes fifth order wave.

The measured observations can be explained mathematically by potential flow modelling of surface waves. This is done in Section 3.3.1.

As shown in Section 3.2, a Gaussian process is completely described by the spectral density $S(\omega)$. Unfortunately, it is not possible to determine the spectral density for ocean waves analytically due to its basis on a complex energy transfer from the atmospheric condition to the waves. Section 3.3.2 deals with the spectral model, normally applied for ocean-going ships.

In the prediction of design wave loads, the limitation to stationary processes of a duration measured in hours has to be relaxed. In Section 3.3.3, the standard procedure where the long-term extreme values are obtained by proper weighting of all stationary (short-term) processes is described.

This ends the description of ocean waves as the necessary results are available for the prediction of hull girder loads and, to some extent, local sea pressure loads required for fatigue analysis. However, for the analysis of bottom-supported offshore structures, extensions regarding the wave kinematics (velocities, accelerations) are needed together with methods of accounting for strong non-linearities in the load description for drag-dominated slender structural members. Several textbooks are available on this subject, see for instance Chakrabarti (1987) and Clauss, Lehmann and Östergaard (1992).

3.3.1 Surface Waves

The theory of surface waves is based on the assumption that the fluid, i.e. the sea water, can be considered as incompressible and irrotational and that any viscous effects can be neglected. Thus, the kinematics in the complete domain of water can be derived from a velocity potential $\phi(X, Y, Z, t)$ satisfying the *Laplace equation*:

$$\nabla^2\phi = \frac{\partial^2\phi}{\partial X^2} + \frac{\partial^2\phi}{\partial Y^2} + \frac{\partial^2\phi}{\partial Z^2} = 0 \tag{3.190}$$

where the global XYZ-coordinate system can conveniently be positioned as in Chapter 2 with the Z-axis pointing upwards and with the XY-plane coinciding with the position of the still water surface.* From the velocity potential ϕ, wave particle velocities v_X, v_Y, v_Z become

$$v_X = \frac{\partial\phi}{\partial X} \quad ; \quad v_Y = \frac{\partial\phi}{\partial Y} \quad ; \quad v_Z = \frac{\partial\phi}{\partial Z} \tag{3.191}$$

and the pressure p follows from *Bernouilli's equation*:

$$p + \rho\,\frac{\partial\phi}{\partial t} + \rho g Z + \frac{1}{2}\,\rho v^2 = p_{atm} \tag{3.192}$$

where ρ is the mass density of sea water, g the acceleration of gravity, v the scalar velocity:

* From here on X, Y, Z do not any longer represent statistical processes but global coordinates as in Chapter 2!

$$v^2 = v_X^2 + v_Y^2 + v_Z^2 \tag{3.193}$$

and p_{atm} the atmospheric pressure just above the water surface. The *velocity potential* $\phi(X, Y, Z, t)$ and the wave elevation $h(X, Y, t)$ follow from the linear field equation, (3.190) with appropriate boundary conditions. Assuming periodicity in the horizontal X-, Y-directions and zero vertical velocity at the seabed leaves only the boundary conditions at the free, but unknown, water surface h. Here two conditions apply. The first is Eq. (3.192) evaluated at the free surface $Z = h(X, Y, t)$:

$$\left[\frac{\partial \phi}{\partial t} + gZ + \frac{1}{2}v^2\right]_{Z=h} = 0 \tag{3.194}$$

and the second requirement is that the vertical velocity at the free surface should satisfy

$$\left.\frac{\partial \phi}{\partial Z}\right|_{Z=h} = \frac{Dh}{Dt} \tag{3.195}$$

because water particles on the free surface stay there. The total derivative D/Dt is needed as the waves propagate with respect to the fixed XYZ-system. Hence Eq. (3.195) can be written

$$\left.\frac{\partial \phi}{\partial Z}\right|_{Z=h} = \frac{\partial h}{\partial t} + \frac{\partial h}{\partial X}\frac{\partial \phi}{\partial X} + \left.\frac{\partial h}{\partial Y}\frac{\partial \phi}{\partial Y}\right|_{Z=h} \tag{3.196}$$

by application of Eq. (3.191). It is seen that both free-surface boundary conditions are non-linear in the unknown functions ϕ, h. This makes a general analytical solution impossible and various approximate procedures have been developed. The standard method is a perturbational procedure in which ϕ, h are expressed as power series in a small parameter ε:

$$\begin{aligned} \phi(X, Y, Z, t) &= \varepsilon\phi^{(1)}(X, Y, Z, t) + \varepsilon^2\phi^{(2)}(X, Y, Z, t) + \ldots \\ h(X, Y, t) &= \varepsilon h^{(1)}(X, Y, t) + \varepsilon^2 h^{(2)}(X, Y, t) + \ldots \end{aligned} \tag{3.197}$$

Substitution of these series expansions into the governing equations makes it, at least in principle, possible to set up equations which can be solved analytically. For deterministic waves Stokes' fifth order wave theory is such a solution including terms up to $\phi^{(5)}$, $h^{(5)}$ using the wave slope as the small parameter. Solutions which can describe the wave elevation h as a stochastic process have been derived to second order.

Higher than second order solutions are very difficult to obtain, not only because of the complexities of solving for higher order terms but also, due to lack of physical knowledge of the interaction between different wave components.

3.3.1.1 Linear (Airy) Waves

Substitution of the expansion Eq. (3.197) into the free-surface conditions Eqs. (3.194) and (3.196) and retainment only of terms of the order $0(\varepsilon)$ yield

$$\left.\frac{\partial \phi}{\partial t}\right|_{Z=0} + gh = 0 \tag{3.198}$$

and

$$\left.\frac{\partial \phi}{\partial Z}\right|_{Z=0} = \frac{\partial h}{\partial t} \tag{3.199}$$

These two equations can be reduced to one equation in ϕ alone:

$$\left[\frac{\partial^2 \phi}{\partial t^2} + g\,\frac{\partial \phi}{\partial Z}\right]_{Z=0} = 0 \tag{3.200}$$

The field equation (3.190), horizontal periodicity conditions, the bottom condition and the free-surface condition (3.200) constitute a homogeneous, linear system. Depending on the periodicity conditions, various but unique solutions are easily found. In the present case two-dimensional (*long-crested*) waves travelling in the X-direction are considered. Furthermore, infinite water depth is assumed to be most relevant for sea loads on ships. Thus, the velocity potential ϕ becomes

$$\phi^{(1)}(X,Z,t) = a\,\frac{\omega}{k}\,e^{kZ}\sin(kX - \omega t + \theta) \tag{3.201}$$

where the frequency ω and the wave number k are related by the so-called *dispersion relation*:

$$\omega^2 = kg \tag{3.202}$$

derrived from Eq. (3.200). The solution contains three undetermined parameters a, ω (or k) and θ. The wave elevation h follows from Eq. (3.198) and Eqs. (3.201) - (3.202):

$$h^{(1)}(X,t) = a\cos(kX - \omega t + \theta) \tag{3.203}$$

from which it is seen that a is the amplitude of the sinusoidal wave elevation and θ a phase lag, relative to the choice of origo of time and space.

Finally, the pressure $p = p(X,Z,t)$ in the water becomes

$$p = p_s + p^{(1)}$$

where p_s is the *hydrostatic pressure* in a calm sea:

$$p_s(Z) = p_{atm} - \rho g Z$$

and

$$p^{(1)}(X,Z,t) = -\rho\frac{\partial\phi^{(1)}}{\partial t} = \rho g a e^{kZ}\cos(kX - \omega t + \theta) \tag{3.204}$$

is the *first order pressure* due to the water particle motions described by $\phi^{(1)}$.

For application to a stochastic wave description, it is important to realise that the first order solution is obtained from a homogeneous linear differential equation with homogeneous boundary conditions and, therefore, any values of a, ω and θ can be chosen and, furthermore, any sum of n such solutions will also be a solution. A general solution can thus be written

$$h^{(1)}(X,t) = \sum_{i=1}^{n} a_i\cos(k_i X - \omega_i t + \theta_i) \tag{3.205}$$

$$\phi^{(1)}(X,Z,t) = \sum_{i=1}^{n} a_i\frac{\omega_i}{k_i}e^{k_i Z}\sin(k_i X - \omega_i t + \theta_i) \tag{3.206}$$

$$p^{(1)}(X,Z,t) = \rho g\sum_{i=1}^{n} a_i e^{k_i Z}\cos(k_i X - \omega_i t + \theta_i) \tag{3.207}$$

where each pair of ω_i, k_i must satisfy Eq. (3.202).

3.3.1.2 Second Order Waves

The second order solutions $\phi^{(2)}$, $h^{(2)}$ follow from solution of the same linear differential equation (3.190), but with non-homogeneous boundary conditions at the free surface, see e.g. Longuet-Higgins (1963). The non-homogeneous terms consist of quadratic terms in the first order solution, as such terms will be of the order ε^2 like the second order solutions. So, $\phi^{(2)}$ and $h^{(2)}$ depend on the first order solution $\phi^{(1)}$, $h^{(1)}$, and thus on the undetermined coefficients a_i, ω_i and θ_i. The second order solutions become (see e.g. Jensen and Pedersen (1979)):

$$h^{(2)}(X,t) = \frac{1}{4}\sum_{i=1}^{n}\sum_{j=1}^{n} a_i a_j \left[\left(k_i + k_j\right) \cos\left(\psi_i + \psi_j\right) - \left|k_i - k_j\right| \cos\left(\psi_i - \psi_j\right) \right] \tag{3.208}$$

$$\phi^{(2)}(X,Z,t) = \frac{1}{2}\sum_{i=1}^{n}\sum_{j=1}^{n} a_i a_j \max\left(-\omega_i, \omega_j\right) e^{|k_i - k_j|Z} \sin\left(\psi_i - \psi_j\right) \tag{3.209}$$

$$p^{(2)}(X,Z,t) = -\frac{1}{2}\rho \sum_{i=1}^{n}\sum_{j=1}^{n} a_i a_j \left[\max\left(\omega_i, \omega_j\right) \left|\omega_i - \omega_j\right| e^{|k_i - k_j|Z} + \omega_i \omega_j e^{(k_i + k_j)Z}\right] \cos\left(\psi_i - \psi_j\right) \tag{3.210}$$

where

$$\max\left(-\omega_i, \omega_j\right) \equiv \begin{cases} -\omega_i & \text{if } \omega_i > \omega_j \\ \omega_j & \text{if } \omega_i < \omega_j \end{cases} \tag{3.211}$$

and

$$\psi_i \equiv k_i X - \omega_i t + \theta_i \tag{3.212}$$

It should be noted that, due to the non-homogeneous equations governing the second order solution, no new constants appear. The second order solution is uniquely given by the first order solution. It should also be noted that, in the present case of unidirectional deep water waves, the second order velocity potential and pressure only contain difference frequencies

$$\psi_i - \psi_j = \left(k_i - k_j\right) X - \left(\omega_i - \omega_j\right) t + \theta_i - \theta_j$$

in the trigonometric functions.

3.3.1.3 Random Linear Surface Waves

Consider the linear solution, Eq. (3.205), for the wave elevation. Both the amplitudes a_i and the phase lags θ_i can be chosen freely. A stochastic wave elevation $H(X,t)$ satisfying the linearised wave equations can therefore be obtained by taking one or both of these parameters to be random variables.

Here the phase lags θ_i, $i = 1, 2, ..., h$ are taken to be statistically independent random variables, each uniformly distributed between 0 and 2π*:

$$p(\theta_i) = \begin{cases} \dfrac{1}{2\pi} & 0 \le \theta_i \le 2\pi \\ 0 & \text{otherwise} \end{cases} \tag{3.213}$$

From the wave elevation process

$$H(X, t) = \sum_{i=1}^{n} a_i \cos\left(k_i X - \omega_i t + \theta_i\right) \tag{3.214}$$

and Eq. (3.213), statistical moments $E[G(H)]$ of $H(X, t)$ can be derived:

$$\begin{aligned} E[G(H)] &= \int_0^{2\pi} \cdots \int_0^{2\pi} G\left(h\left(\theta_1, \theta_2, ..., \theta_n\right)\right) p\left(\theta_1\right) p\left(\theta_2\right) ... p(\theta_n)\, d\theta_1 d\theta_2 ... d\theta_n \\ &= \frac{1}{(2\pi)^n} \int_0^{2\pi} \cdots \int_0^{2\pi} G\left(h\left(\theta_1, \theta_2, ..., \theta_n\right)\right) d\theta_1 d\theta_2 ... d\theta_n \end{aligned} \tag{3.215}$$

It is easily seen that, due to the integration over a full period of the trigonometric functions in Eq. (3.214), $E[G(H)]$ will be independent of X, t. Thus, the wave elevation $H(X, t)$ given by Eq. (3.214) is a stationary stochastic process and $X = 0$, $t = 0$ can be used in Eq. (3.214) without loss of generality in calculations of the statistical moments.

The mean value μ_H becomes

$$\mu_H = E\left[\sum_{i=1}^{n} a_i \cos\theta_i\right] = \sum_{i=1}^{n} a_i E\left[\cos\theta_i\right] = 0 \tag{3.216}$$

as

$$E\left[\cos\theta_i\right] = \frac{1}{2\pi} \int_0^{2\pi} \cos\theta_i \, d\theta_i = 0$$

* This is a fundamental assumption, based partly on measurements. However, the possibility of wave groupings (coupling between phases) are thereby ignored for the linear terms.

With zero mean value the variance is found to be

$$\zeta_2 = s_H^2 = E\left[\left(\sum_{i=1}^{n} a_i \cos\theta_i\right)^2\right]$$
$$= \sum_{i=1}^{n} a_i^2 E\left[\cos^2\theta_i\right] = \frac{1}{2}\sum_{i=1}^{n} a_i^2 \tag{3.217}$$

as

$$\frac{1}{2\pi}\int_0^{2\pi} \cos^2\theta_i \, d\theta_i = \frac{1}{2}$$

If the number of terms n is large, further moments are not needed since the central limit theorem, Section 3.1.7, implies that the wave elevation H is normally distributed with the mean value $\mu_H = 0$, the standard deviation

$$s_H = \sqrt{\frac{1}{2}\sum_{i=1}^{n} a_i^2} \tag{3.218}$$

and the probability density function $p(h)$

$$p(h) = \frac{1}{\sqrt{2\pi}\, s_H} e^{-\frac{1}{2}(h/s_H)^2} \tag{3.219}$$

Thus, the wave elevation process, Eq. (3.214), satisfies the measured observations that over a short period of time the wave elevation can be considered approximately as a stationary Gaussian process.

The individual amplitudes a_i needed to compute s_H by means of Eq. (3.218) can be related to the spectral density $S(\omega) \equiv S_H(\omega)$ of the wave elevation through the autocorrelation Eq. (3.91):

$$R(\tau) = E\left[\left(\sum_{i=1}^{n} a_i \cos\theta_i\right)\left(\sum_{j=1}^{n} a_j \cos\left(\theta_j - \omega_j\tau\right)\right)\right]$$
$$= \sum_{i=1}^{n} a_i^2 E\left[\cos\theta_i \cos\left(\theta_i - \omega_i\tau\right)\right] \tag{3.220}$$
$$= \frac{1}{2}\sum_{i=1}^{n} a_i^2 \cos\omega_i\tau$$

as

$$\frac{1}{2\pi}\int_0^{2\pi} \cos\theta_i\left(\cos\theta_i\cos\omega_i\tau + \sin\theta_i\sin\omega_i\tau\right)d\theta_i = \frac{1}{2}\cos\omega_i\tau$$

A comparison of Eq. (3.220) with Eq. (3.98) shows that

$$a_i = \sqrt{2\,S(\omega_i)\,\Delta\omega_i} \tag{3.221}$$

where $\Delta\omega_i = \omega_i - \omega_{i-1}$. Therefore, with the knowledge of the spectral density the stochastic process $H(X, t)$, Eq. (3.214), is completely defined. It is seen from Eq. (3.221) that $S(\omega)$ must be an non-negativ function.

For each set of θ_i, $i = 1, 2, \ldots, n$ Eq. (3.214) represents a sample of the process H in X or t. The number n of terms should be taken so that $\left[\omega_1, \omega_n\right]$ covers the range where the spectral density is (nearly) different from zero. Note that the sample repeats itself after a period of time of $2\pi/\omega_1$ or in space after a length of $2\pi/k_1$, if an equidistant step length $\Delta\omega$ (or Δk) is chosen.

3.3.1.4 *Random Second Order Surface Waves*

Inclusion of the second order terms Eq. (3.208), in the stochastic model, does not introduce new parameters. The stochastic process is still stationary due to the periodicity in its variation with X, t.

In order to calculate the statistical moments of the wave elevation H, it is appropriate again to take $X = t = 0$ and write H as

$$\begin{aligned} H &= \sum_{i=1}^{n} a_i\cos\theta_i + \frac{1}{4}\sum_{i=1}^{n}\sum_{j=1}^{n}\Big[\left(k_i + k_j\right)\left(\cos\theta_i\,\cos\theta_j - \sin\theta_i\,\sin\theta_i\right) \\ &\qquad - \left|k_i - k_j\right|\left(\cos\theta_i\,\cos\theta_j + \sin\theta_i\,\sin\theta_j\right)\Big] \\ &= \sum_{i=1}^{n} a_i\cos\theta_i + \frac{1}{2}\sum_{i=1}^{n}\sum_{j=1}^{n} a_i\,a_j\Big[\min\left(k_i, k_j\right)\,\cos\theta_i\,\cos\theta_j \\ &\qquad - \max\left(k_i, k_j\right)\,\sin\theta_i\,\sin\theta_j\Big] \\ &\equiv H^{(1)} + H^{(2)} \end{aligned} \tag{3.222}$$

where $H^{(1)}$, $H^{(2)}$ are the linear and the second order terms, respectively.

The mean value μ_H of H becomes

$$\mu_H = E[H] = E\left[H^{(1)}\right] + E\left[H^{(2)}\right] = 0 \tag{3.223}$$

and the variance

$$s_H^2 = E\left[H^2\right] = E\left[(H^{(1)})^2\right] + 2E\left[H^{(1)}H^{(2)}\right] + E\left[(H^{(2)})^2\right] \simeq \frac{1}{2}\sum_{i=1}^{n} a_i^2 \tag{3.224}$$

neglecting the last term, which is of the order a_i^4. Thus, the mean value and the standard deviation are the same as in the linear model. However, as the second order part of H contains products of statistically independent variables the central limit theorem no longer holds. Therefore, the probability distribution is not necessarily a normal distribution. For moderate values of the amplitudes a_i, or actually the wave slopes $a_i k_i$, the second order terms may be expected only to change the distribution slightly so that the Gram-Charlier series, Eq. (3.31), can be applied. To lowest order the probability density function becomes

$$p(h) = \frac{1}{\sqrt{2\pi}\, s_H} e^{-\frac{1}{2}(h/s_H)^2}\left[1 + \frac{1}{6}\lambda_3\left[\left(\frac{h}{s_H}\right)^3 - 3\frac{h}{s_H}\right]\right] \tag{3.225}$$

where the skewness

$$\lambda_3 = \gamma_1 = \frac{\zeta_3}{\zeta_2^{3/2}} = \frac{E\left[H^3\right]}{s_H^3} \tag{3.226}$$

From Eq. (3.222)

$$E\left[H^3\right] = E\left[(H^{(1)})^3\right] + 3E\left[(H^{(1)})^2 H^{(2)}\right] + 3E\left[H^{(1)}(H^{(2)})^2\right] + E\left[(H^{(2)})^3\right]$$

The first and third terms are of odd order in $\cos\theta_i$ and therefore vanish in evaluations of the expected value. In the following, the last term, which is of the order a_i^6, is neglected leaving

$$\begin{aligned} E\left[H^3\right] \simeq 3E\left[(H^{(1)})^2 H^{(2)}\right] &= 6\sum_{I=1}^{n}\sum_{j=1}^{n} a_i^2 a_j^2 \frac{1}{2}\min\left(k_i, k_j\right) E\left[\cos^2\theta_i \cos^2\theta_j\right] \\ &= 3\sum_{i=1}^{n}\sum_{j=1}^{n}\left(\frac{1}{2}a_i^2\right)\left(\frac{1}{2}a_j^2\right)\min\left(k_i, k_j\right) \end{aligned} \tag{3.227}$$

In the limit $\Delta\omega \to 0$ and $n \to \infty$, Eq. (3.227) becomes

$$E\left[H^3\right] = \frac{3}{g}\int_0^{\infty} S(\omega)\int_0^{\infty}\min\left(\omega^2, \tilde{\omega}^2\right) S(\tilde{\omega})\, d\tilde{\omega}\, d\omega$$

by use of Eq. (3.202). Integration by parts shows that

$$\int_0^\infty S(\omega) \int_0^\omega \tilde{\omega}^2\, S(\tilde{\omega})\, d\tilde{\omega}\, d\omega = \int_0^\infty \omega^2\, S(\omega) \int_\omega^\infty S(\tilde{\omega})\, d\tilde{\omega}\, d\omega$$

so that $E\left[H^3\right]$ can be written

$$E\left[H^3\right] = \frac{6}{g} \int_0^\infty \omega^2\, S(\omega) \int_\omega^\infty S(\tilde{\omega})\, d\tilde{\omega}\, d\omega \qquad (3.228)$$

and, finally, the skewness γ_1 is obtained by Eq. (3.226). It is seen that the second order stochastic modelling of the surface elevation is described completely by the spectral density $S(\omega)$ of the first order elevation. Accurate higher order moments (kurtosis etc.) require inclusion of third and higher order terms in the expansion, Eq. (3.197) which is not feasible.

3.3.2 Spectral Density of Ocean Waves

Basically, the wave spectral density $S(\omega)$ at a given location can be obtained by a Fourier analysis of a measured time history (sample) by use of Eq. (3.221), and by assuming stationarity and neglecting all non-linearities. Such spectral densities may be appropriate for structures located in the same position throughout their lifetime. Otherwise, more general spectral density formulations applicable to world-wide use are desirable. These are mostly smooth functional relations which involve a limited number of free parameters. Thereby, different stationary sea states at different locations taking account of average wind speed, direction, duration and fetch can be modelled.

For the analysis of ocean-going ships, fully developed sea states are usually assumed. Only one free parameter, the average wind speed, is left, but often an additional parameter is included to add some flexibility to the model. Only spectral densities valid for fully developed sea state will be treated here. For a treatment of other spectral models, see e.g. Chakrabarti (1987).

An upper boundary on the spectral density $S(\omega)$ was developed by Phillips (1958). In this solution it is assumed that a kind of equilibrium exists in the sense that the waves have reached their maximum size independent of the wind speed. Hence, as seen from the governing equations, (3.201) - (3.203), the only parameters on which $S(\omega)$ may depend are the acceleration of gravity g and the frequency ω. From Eq. (3.221) $S(\omega)$ has the dimension *length*$^2 \cdot$ *time* $\left(L^2T\right)$ and a dimensional analysis yields

$$S(\omega) = \alpha\, g^n\, \omega^m$$

$$L^2T = \alpha \left(LT^{-2}\right)^n \left(T^{-1}\right)^m$$

implying $n = 2$, $m = -5$. Hence

$$S(\omega) = \alpha\, g^2\, \omega^{-5} \qquad (3.229)$$

The constant α is denoted Phillips' constant and takes the value $\alpha = 0.0081$ (from measurements). This spectrum is an upper limit independent of the severity of the sea state. The assumption of equilibrium is in practice only valid for the high-frequency range. In the low-frequency range modifications are needed. Based on a large number of measurements Pierson and Moskowitz (1964) proposed a spectral density in the form of

$$S(\omega) = \alpha\, g^2\, \omega^{-5} \exp\left[-\,0.74\left(\frac{\omega U_w}{g}\right)^{-4}\right] \tag{3.230}$$

where U_w is the average wind speed. It is seen that for $\omega \to \infty$ this spectral density approaches Eq. (3.229). Today the explicit dependence on wind speed U_w is generally of less practical use as stationary sea states are normally characterised by two parameters related to the spectral moments m_0, m_1 or m_2. To obtain such formulations, Eq. (3.230) is written

$$S(\omega) = A\,\omega^{-5}\, e^{-B\,\omega^{-4}} \tag{3.231}$$

Eq. (3.231) allows for an explicit calculation of all spectral moments, Eq. (3.102):

$$\begin{aligned} m_n &= \int_0^\infty \omega^n\, S(\omega)\, d\omega = A \int_0^\infty \omega^n\, \omega^{-5}\, e^{-B\,\omega^{-4}} d\omega \\ &= \frac{A}{4B}\, B^{n/4}\, \Gamma\left(1 - \frac{n}{4}\right) \end{aligned} \tag{3.232}$$

where $\Gamma(\,)$ is the Gamma function, Eq. (3.40). Thus

$$m_0 = \frac{A}{4B}$$

$$\bar{\omega}_1 = \frac{m_1}{m_0} = B^{1/4}\, \Gamma\left(\frac{3}{4}\right) \simeq 1.2254\, B^{1/4}$$

$$\bar{\omega}_2 = \sqrt{\frac{m_2}{m_0}} = B^{1/4} \sqrt{\Gamma\left(\frac{1}{2}\right)} = (\pi\, B)^{1/4}$$

so that the *Pierson-Moskowitz (P-M) spectrum* can be written

$$S(\omega) \simeq 1.774\, \frac{m_0}{\bar{\omega}_1} \left(\frac{\omega}{\bar{\omega}_1}\right)^{-5} e^{-0.4435\,(\omega/\bar{\omega}_1)^{-4}} \tag{3.233}$$

or

$$S(\omega) = \frac{4}{\pi}\, \frac{m_0}{\bar{\omega}_2} \left(\frac{\omega}{\bar{\omega}_2}\right)^{-5} e^{-(\omega/\bar{\omega}_2)^{-4}/\pi} \tag{3.234}$$

Most often periods rather than characteristic frequencies are used:

$$\textit{Mean period:} \qquad T_s = \frac{2\pi}{\overline{\omega}_1} \tag{3.235}$$

$$\textit{Zero-upcrossing period:} \qquad T_z = \frac{2\pi}{\overline{\omega}_2} \tag{3.236}$$

Hence

$$S(\omega) \simeq 2765\, m_0\, T_s\, (\omega\, T_s)^{-5} e^{-691\, (\omega\, T_s)^{-4}} \tag{3.237}$$

and

$$S(\omega) = 64\, \pi^3\, m_0\, T_z\, (\omega\, T_z)^{-5} e^{-\pi^3\, (\omega\, T_z/2)^{-4}} \tag{3.238}$$

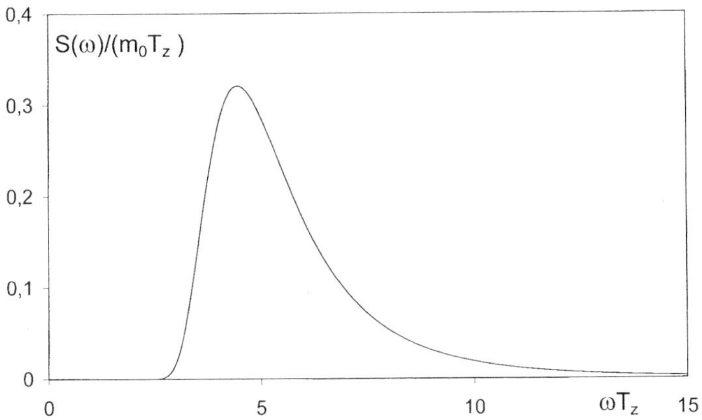

Figure 3.10 *Pierson-Moskowitz spectrum.*

Figure 3.10 shows the Pierson-Moskowitz (P-M) spectrum. It is seen that it has a steep (exponential) increase towards its peak value and a more modest decrease, following Eq. (3.229), in the high-frequency tail. The *peak frequency* ω_p follows from

$$\left.\frac{dS}{d\omega}\right|_{\omega=\omega_p} = 0$$

which implies

$$\omega_p = \left(\frac{4}{5}\, B\right)^{1/4} = \left(\frac{4}{5\pi}\right)^{1/4} \overline{\omega}_2 = 0.710\, \overline{\omega}_2$$

and the P-M spectrum can thus also be written

$$S(\omega) = 80\,\pi^4\, m_0\, T_P \left(\omega T_P\right)^{-5} \exp\left[-20\,\pi^4 \left(\omega T_P\right)^{-4} \right] \tag{3.239}$$

where T_p is the spectral peak period:

$$T_p = \frac{2\pi}{\omega_p} = 1.408\, \frac{2\pi}{\overline{\omega}_2} \tag{3.240}$$

All formulations, Eqs. (3.231), (3.233), (3.234), (3.237), (3.238) and (3.239) for $S(\omega)$, are of course identical. The actual choice only depends on the characteristic period or frequency available to characterise the stationary sea state. Most oceanographic data uses the zero-crossing period T_z and, therefore, the formulation Eq. (3.238) is mostly applied. Table 3.3 yields the relations between the different characteristic periods.

Table 3.3 *Relations between different characteristic periods for the P-M spectrum.*

	T_s	T_z	T_p
T_s	T_s	$1.086\ T_z$	$0.772\ T_p$
T_z	$0.920\ T_s$	T_z	$0.710\ T_p$
T_p	$1.296\ T_s$	$1.408\ T_z$	T_p

The variance m_0 of the wave elevation can either be derived from Eq. (3.230):

$$m_0 = \frac{A}{4B} = \frac{\alpha g^2}{4B} = \frac{\alpha g^2}{64\pi^3}\, T_z^4 \tag{3.241}$$

for one-parameter spectra or be allowed to vary independently of T_z in two-parameter spectra. As m_0 has a dimension length^2 it is obvious to relate $\sqrt{m_0}$ to some characteristic wave elevation. The measure chosen is the mean value of the one-third largest wave heights, measured from a wave trough to the next wave crest. This value is denoted $2\overline{h}_{1/3}$ and in its derivation a narrow-banded spectral density is assumed. Thus, the probability density of a peak follows closely the Rayleigh distribution, Eq. (3.137) with $\varepsilon = 0$:

$$p_P(h) = \frac{h}{m_0}\, e^{-\frac{1}{2} h^2/m_0} \quad ; \quad h \geq 0 \tag{3.242}$$

The probability density function of a minimum is exactly the same in this linear wave model, which consists of a sum of sinusoidal wave components. Therefore, the mean value $2\overline{h}_{1/3}$ of the one-third largest wave heights is given by

$$\bar{h}_{1/3} = \frac{\int_{h_{1/3}}^{\infty} h\, p_p(h)\, dh}{\int_{h_{1/3}}^{\infty} p_p(h)\, dh} = \frac{\int_{h_{1/3}}^{\infty} h\, p_p(h)\, dh}{1 - F_p(h_{1/3})}$$

where $h_{1/3}$ is the lower limit for the one-third largest wave peaks. From

$$P\left(H_p > h_{1/3}\right) = 1 - F_p\left(h_{1/3}\right) = \frac{1}{3}$$

it follows that

$$h_{1/3} = \sqrt{2m_0 \ln 3} \tag{3.243}$$

and thus

$$\begin{aligned}
\bar{h}_{1/3} &= 3 \int_{h_{1/3}}^{\infty} h\, p_p(h)\, dh \\
&= 3\sqrt{m_0} \left[-u\, e^{-\frac{1}{2}u^2} + \int e^{-\frac{1}{2}u^2}\, du \right]_{\sqrt{2\ln 3}}^{\infty} \\
&= 3\sqrt{m_0} \left[\frac{1}{3}\sqrt{2\ln 3} + \sqrt{2\pi}\left(1 - \Phi\left(\sqrt{2\ln 3}\right)\right) \right] \\
&\simeq 2.00\sqrt{m_0}
\end{aligned} \tag{3.244}$$

or

$$\sqrt{m_0} = s_H = \frac{1}{4}\left(2\,\bar{h}_{1/3}\right) \tag{3.245}$$

which gives the standard deviation s_H of the wave elevation a precise statistical interpretation. For simplicity $2\bar{h}_{1/3}$ is usually called the *significant wave height* H_s:

$$H_s \equiv 2\,\bar{h}_{1/3} \tag{3.246}$$

so that

$$s_H = \sqrt{m_0} = \frac{1}{4} H_s \tag{3.247}$$

EAST DELLWOOD BUOY GIANT WAVE

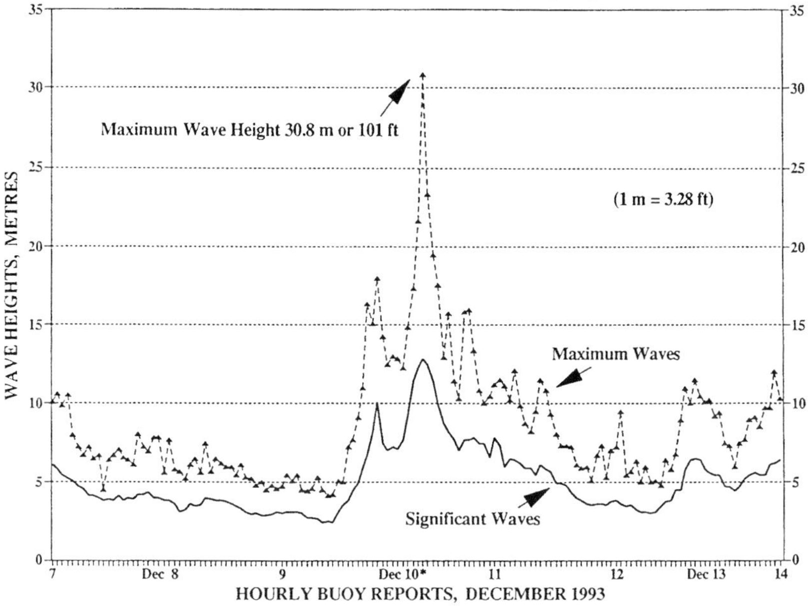

The chart above compares the variation of the maximum wave heights with the significant wave heights from December 7 to December 14, 1993. The chart below compares the peak wind speed to the maximum waves. The maximum wave height occurred well after the peak gust but during the period of the strongest winds. This region is the very same where the USCG icebreaker Polar Sea *ran into a series of "three sister" waves back in October of 1985 (*Mariners Weather Log, *Volume 30, Number 4) resulting in one death and two injuries.*

EAST DELLWOOD BUOY GIANT WAVE

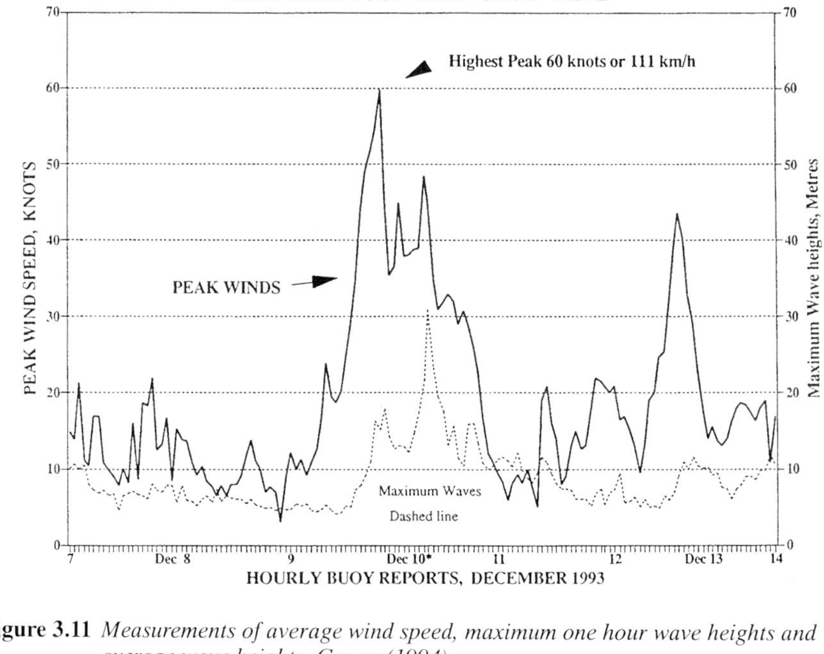

Figure 3.11 *Measurements of average wind speed, maximum one hour wave heights and average wave heights. Gower (1994).*

For one-parameter spectra of the form Eq. (3.231), H_s and T_z are related through Eq. (3.241):

$$T_z = \sqrt[4]{\frac{4\pi^3}{\alpha}} \sqrt{\frac{H_s}{g}} = 11.12 \sqrt{\frac{H_s}{g}} \tag{3.248}$$

using $\alpha = 0.0081$. Furthermore, from Eq. (3.230) H_s can be related to the average wind speed U_W:

$$H_s = 0.209 \frac{U_W^2}{g} \tag{3.249}$$

Example 3.3.1

In Figure 3.11 simultaneous measurements of significant wave heights H_s and wind speed are shown for a period of 7 days, divided into stationary sea conditions which are each of the duration of one hour. There is clearly a relation between H_s and U_w but it is not as well-defined as Eq. (3.249). The maximum wave height H_m measured in each one-hour period is also shown. Ideally, it should be related to H_s through Eq. (3.156). By application of

$$\nu(0) = \frac{1}{T_z} \simeq \frac{1}{11.12} \sqrt{\frac{g}{H_s}}$$

this relation becomes*

$$H_m \simeq 2\,\frac{1}{4}\,H_s\,\sqrt{2\ln\nu(0)T}$$

with T equal to one hour in the present case. As an example, the highest $H_s \simeq 13$ m yields $H_m \simeq 22$ m in reasonable agreement with the measurements except for the single largest value. Eq. (3.249) also yields quite reasonable results as a wind speed of 50 knots ($\simeq 25.7$ m/s) corresponds to $H_s \simeq 14$ m.

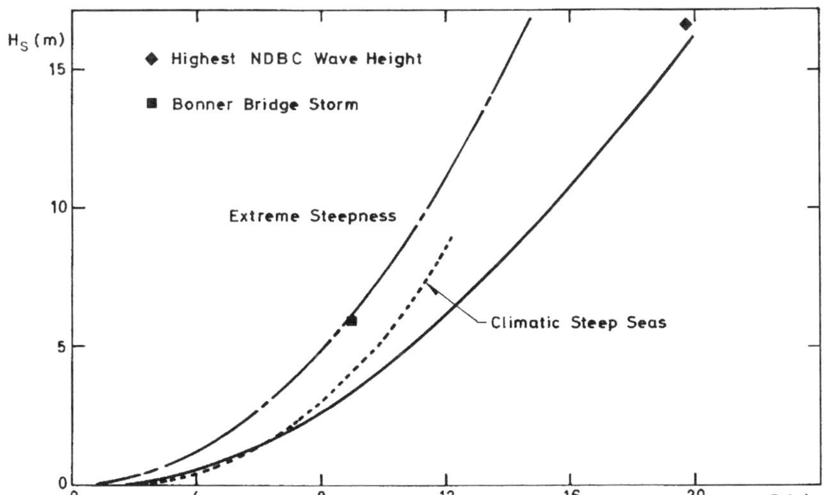

Figure 3.12 *Significant wave height H_s as a function of T_p, Buckley (1988), with inclusion of Eq.(3.248) (full line) and Eq. (3.250) (dash-dot line).*

* The factor 2 appears because H_m is measured from trough to crest.

Very steep waves will break. Therefore, the significant wave height H_s and the zero-upcrossing period cannot be chosen completely freely in two-parameter spectra. A limitation is indicated in Figure 3.12:

$$T_p > 11.4 \sqrt{\frac{H_s}{g}} \tag{3.250}$$

or, if a P-M spectrum is representative of the sea states:

$$T_z > 8.1 \sqrt{\frac{H_s}{g}} \tag{3.251}$$

The figure shows that Eq. (3.248) represents waves with moderate steepness.

In the derivation of $\bar{h}_{1/3}$ it was assumed that the spectral density was narrow-banded. However, it follows from Eq. (3.232) that for a P-M spectrum $m_4 = \Gamma(0)$ is infinite. Thus, the bandwidth $\varepsilon = 1$ and the spectrum are actually broad banded. This behaviour is mainly due to improper inclusion of the high frequency tail in the spectrum. Usually, a cut-off of the tail is applied making m_4 finite and with a value which makes $\varepsilon \simeq 0.6$ as representative of bandwidths derived from measured wave records. Of course, this inconsistency is somewhat disappointing but it does not pose any difficulties in analyses of wave loads on ships, as will be discussed in the next chapter.

The autocorrelation $R(\tau)$, Eq. (3.98), is shown in Figure 3.13 for the P-M spectrum. It is seen that two adjacent peaks are slightly correlated, violating the assumption of statistically independent peaks in the extreme value predictions. However, the results derived in Section 3.2.5 are still very accurate.

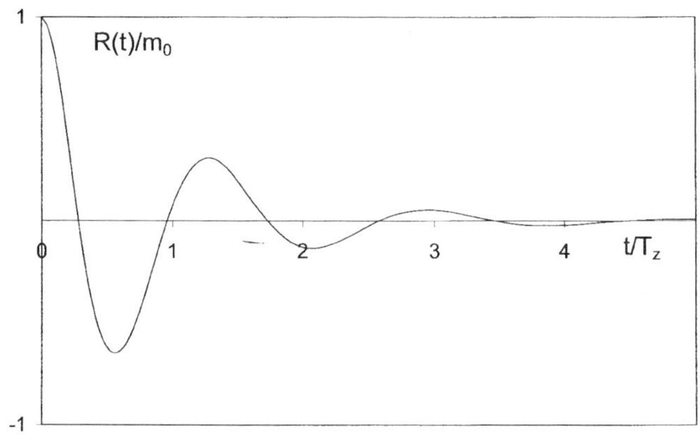

Figure 3.13 *Autocorrelation R(τ) for the P-M spectrum.*

The formulations of spectral densities given above are in the frequency domain. In some applications it may be preferred to use the wave number k or the wave length λ:

$$\lambda = \frac{2\pi}{k} \tag{3.252}$$

as independent variable. By use of Eq. (3.202) and

$$S(\omega)\,d\omega = S(k)\,dk = S(\lambda)\,d\lambda \tag{3.253}$$

spectral densities as functions of k or λ may be obtained from (3.231) or equivalent formulations in ω. Likewise, characteristic wave numbers or wave lengths may be defined. For example, for deep water waves the *mean wave number* $\bar{k}$ becomes

$$\bar{k} = \frac{\int_0^\infty k\,S(k)\,dk}{\int_0^\infty S(k)\,dk} = \frac{\int_0^\infty \omega^2\,S(\omega)\,d\omega}{g\int_0^\infty S(\omega)\,d\omega} = \frac{m_2}{gm_0} = \frac{\overline{\omega}_2^2}{g} \tag{3.254}$$

Alternatively, a *mean wave length* $\bar{\lambda}$:

$$\bar{\lambda} = \frac{\int_0^\infty \lambda\,S(\lambda)\,d\lambda}{\int_0^\infty S(\lambda)\,d\lambda} = 2\pi g\,\frac{m_{-2}}{m_0} = \frac{\pi^2 g}{\overline{\omega}_2^2}$$

is found by use of Eq. (3.232) with n = -2. It should, however, be noted that often the mean wave length is defined as $2\pi/\bar{k}$, which is different from $\bar{\lambda}$.

Example 3.3.2

A critical wave episode can be defined as a deterministic wave profile $<H>$ which in some average sense represents the most severe wave a ship or offshore structure might encounter during its operational lifetime. The photograph, Figure 3.14, might be such a severe wave.

The simplest form for such a critical wave episode is a linear, uni-directional regular wave:

$$< H(X,t) > = h_0 \cos(kX - \omega t)$$

where the wave parameters h_0, k and ω must be determined so as to maximize the load, keeping the wave parameters within reasonable limits.

The wave amplitude h_0 is usually defined as the most probable largest value within a 3 hours operation in an extreme sea state, characterized by a suitable wave spectral density. For ships the *P-M* spectrum is normally assumed and from Eqs. (3.156) and (3.247) it follows that

$$h_0 = \tilde{\mu}_N = s_H\sqrt{2\ln N_z}$$

$$= \frac{H_s}{4}\sqrt{2\ln\,(3\text{hours}/T_z)} \simeq H_s$$

where the last approximation is slightly on the conservative side as T_z is around 8-12 seconds in extreme sea states.

Figure 3.14 *An extreme wave! Reproduced from Buckley (1983)*

The most obvious choice for the two other parameters k, ω in $< H(X,t) >$ is

$$k = \frac{1}{g}\left(\frac{2\pi}{T_z}\right)^2 = \bar{k}$$

$$\omega = \frac{2\pi}{T_z} = \bar{\omega}_2$$

satisfying the dispersion relation, Eq. (3.202) and representing spectral mean values of these quantities. However, the wave loads on ships are usually largest for wave lengths close to the length L of the ship. Hence,

$$k = \frac{2\pi}{\lambda} \simeq \frac{2\pi}{L}$$

is a better choice for such applications than $k = \bar{k}$. The corresponding value of ω follows from Eq. (3.202).

Before applying any of these proposals as a design wave it must be verified that the wave $< H(x,t) >$ is not too steep. For regular deep water waves the following criteria must be fulfilled to avoid wave breaking

$$\frac{h_0}{\lambda} < \frac{1}{14}$$

which for $h_0 = H_s$, $k = \bar{k}$ yields

$$T_z > 9.4\sqrt{\frac{H_s}{g}}$$

i.e. slightly more restrictive than Eq. (3.251).

The second choice requires

$$L > 14\,H_s$$

Usually $H_s = 12 - 15$ m for the extreme waves and therefore the choice $\lambda \simeq L$ is preferable for ships with length greater that about 160 m. For shorter vessels the wave length λ should be taken as

$$\lambda = 14\,H_s$$

The periodicity of the regular waves does not represent the real behaviour of ocean waves very well, see e.g. Figure 3.5. In addition the spectral density $S(\omega)$ of the waves does not influence the wave profile, except for H_s and, possibly, $\bar{k}$.

An alternative critical wave episode has therefore been proposed ("New Wave", Tromans et al. (1991)), based on conditional mean processes. The expected wave elevation $< H(X,t) >$ around a peak (crest) of size h_0 is given by Eq. (3.188):

$$< H(X,t) > = r(X,t)\,h_0$$

where the normalized autocorrelation $r(X,t)$ can be written

$$\begin{aligned} r(X,t) &= \frac{1}{2m_0}\sum_{i=1}^{n} a_i^2 \cos\left(k_i X - \omega_i t\right) \\ &= \frac{1}{m_0}\sum_{i=1}^{n} S(\omega_i)\,\Delta\omega_i \cos\left(k_i X - \omega_i t\right) \end{aligned}$$

using Eqs. (3.220) - (3.221) and a straightforward extension to include $k_i X$ in the trigonometric arguments. Substituting e.g. $S(\omega)$ from Eq. (3.234) yields

$$< H(X,t) > = \frac{4}{\pi}\frac{h_0}{\overline{\omega}_2}\sum_{i=1}^{n}\left(\frac{\omega}{\overline{\omega}_2}\right)^{-5} \exp\left[-\frac{1}{\pi}\left(\frac{\omega}{\overline{\omega}_2}\right)^{-4}\right] \cos\left(k_i X - \omega_i t\right)\Delta\omega_i$$

which clearly shows that this deterministic wave depends on the complete shape of the wave spectral density. Furthermore, $<H(X,t)>$ depends linearly on h_0 and is independent of H_S. With the dimensionless time τ, frequency $\tilde{\omega}$ and length ξ scales

$$\begin{aligned} \tau &\equiv \overline{\omega}_2\, t \\ \tilde{\omega} &\equiv \frac{\omega}{\overline{\omega}_2} \\ \xi &= \bar{k}X \end{aligned}$$

the result can be written

$$< H(X,t) > \;=\; < H(\xi,\tau) > \;=\; \frac{4}{\pi}h_0\sum_{i=1}^{n}\tilde{\omega}_i^{-5}\exp\left(-\frac{1}{\pi}\tilde{\omega}_i^{-4}\right)\cos\left(\tilde{\omega}_i^2\xi - \tilde{\omega}_i\tau\right)\Delta\tilde{\omega}_i$$

This result is shown in Figure 3.15 for the two limiting cases: $\xi = 0$ (the variation in time at a fixed point in space) and $\tau = 0$ (the variation in space at a fixed point in time). The difference is solely due to the different factors on ξ and τ in the trigonometric function.

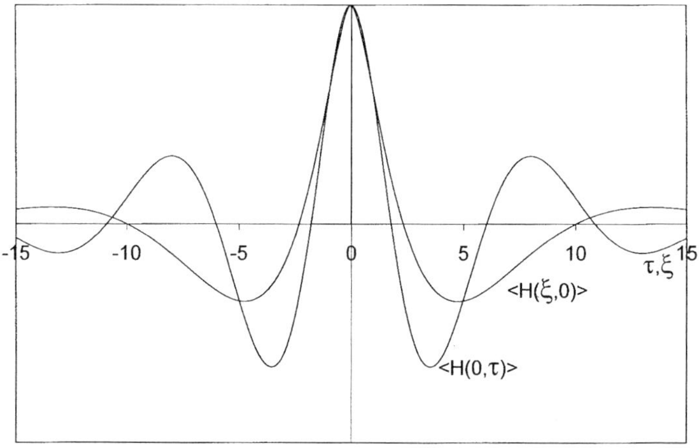

Figure 3.15 *Most probable extreme surface displacement for Pierson-Moskowitz spectrum.*

As discussed in Section 3.2.6 the conditional process is normal distributed around it mean $<H(X,t)>$ with a standard deviation depending of X,t. Close to the crest $(X=0, t=0)$, the standard deviation is much smaller than the average value and thus the process is nearly deterministic here. However, already at the first crossing of the zero wave elevation level, the standard deviation becomes close to the value $\sqrt{m_0}$ of the unconditional process. If h_0 is taken as H_s as discussed above for the regular wave case $h_0/\sqrt{m_0} = 4$, implying a rather deterministic form of the extreme crest profile. For further discussions, see e.g. Tromans et al. 1991.

3.3.2.1 *Skewness of Deep Water Waves*

For slightly non-linear stationary stochastic waves, the skewness γ_1 of the probability distribution of the wave elevation is obtained from Eq. (3.228). Substitution of Eq. (3.231) into Eq. (3.228) yields

$$E\left[H^3\right] = \frac{6}{g}\int_0^{\infty} \omega^2 A\,\omega^{-5} e^{-B\omega^{-4}} \frac{A}{4B}\left(1 - e^{-B\omega^{-4}}\right) d\omega$$

$$= \frac{6}{g} m_0 \left[m_2 - A\int_0^{\infty} \omega^2\,\omega^{-5} e^{-2B\omega^{-4}}\, d\omega\right]$$

$$= \frac{6}{g} m_0 \left[m_2 - \frac{1}{\sqrt{2}} A\int_0^{\infty} u^2\,u^{-5} e^{-Bu^{-4}}\, du\right]$$

$$= \frac{6}{g} m_0\, m_2\left(1 - \frac{1}{\sqrt{2}}\right)$$

and the skewness

$$\gamma_1 = \frac{E\left[H^3\right]}{m_0^{3/2}} = \frac{6}{g} \frac{m_2}{\sqrt{m_0}} \left(1 - \frac{1}{\sqrt{2}}\right)$$

Expressed in terms of the significant wave H_s, Eq. (3.247), and the peak period T_p, Eq. (3.240), the skewness becomes

$$\gamma_1 = \frac{6}{g} \left(\frac{5\pi}{4}\right)^{1/2} \left(\frac{2\pi}{T_p}\right)^2 \frac{1}{4} H_s \left(1 - \frac{1}{\sqrt{2}}\right)$$
$$= 3\,\pi^2 \sqrt{5\pi} \left(1 - \frac{1}{\sqrt{2}}\right) \frac{H_s}{g\,T_p^2} \simeq 34.4 \frac{H_s}{g\,T_p^2} \tag{3.255}$$

In Figure 3.16 this theoretical result is compared with measurements made during some storm events in the northern North Sea, Vinje and Haver (1994). The narrow-banded approximation given by

$$\left(\gamma_1\right)_{nb} = \frac{\frac{6}{g}\, m_0^2\, \omega_p^2}{m_0^{3/2}} = 6\,\pi^2 \frac{H_s}{g\,T_p^2} \simeq 59.2 \frac{H_s}{g\,T_p^2} \tag{3.256}$$

is also included in the figure. It is clear that the P-M spectrum models the measured skewness better than the narrow-banded approximation, although the spread of the measurements is significant.

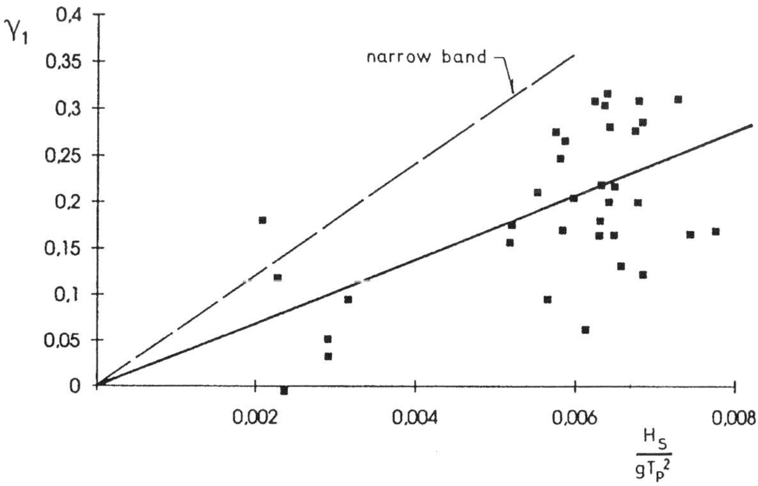

Figure 3.16 *Skewness γ_1 as a function of the average wave steepness for the P-M spectrum and a narrow-banded spectrum compared to measurements in the northern North Sea, Vinje and Haver (1994).*

The skewness is proportional to the average wave steepness H_s/gT_p^2. From Eq. (3.250) the steepness is, due to the breaking of the waves, bounded by the following (Buckley (1988)):

$$\frac{H_s}{gT_p^2} < 0.00776$$

which, for a P-M spectrum, implies

$$\gamma_1 < 0.27$$

Example 3.3.3

A first approximation for the probability distribution for the non-linear wave elevation process is found by integration of the probability density function, Eq. (3.225):

$$F_H(u) = \Phi(u) - \frac{\gamma_1}{6\sqrt{2\pi}}\left(u^2 - 1\right)e^{-\frac{1}{2}u^2} = \Phi(u) - \frac{\gamma_1}{6}(u^2 - 1)\,\varphi(u)$$

where $u = h/s_H$ and where $\Phi(u)$ and $\varphi(u)$ are the probability distribution and density function for the standard normal distribution, Eqs. (3.17) and (3.16). The probability that the wave elevation is greater than, say, $3s_H$ becomes

$$P(h > 3s_H) = 1 - F_H(3) = 1 - \Phi(3) + 0.0016$$

assuming that $\gamma_1 = 0.27$, i.e. very steep waves. From Table 3.1, $1 - \Phi(3) = 0.0013$ and the second order term then double the probability of exceeding the level $3s_H$.

From Eq. (3.225) it is seen that $p(h) < 0$ if

$$1 + \frac{1}{6}\lambda_3\left(u^3 - 3u\right) < 0$$

or,

$$h < -3.165\, s_H$$

which shows the limited applicability of the Gram-Charlier series expansion in the tail of the distribution.

Provided the non-linear waves are narrow-banded, the individual peak distribution is given by Eq. (3.119) with the upcrossing rate $\nu(x)$ determined from Eq. (3.107). The calculations make use of the joint probability density function $p(h, \dot{h})$, which can be obtained by a procedure similar to the one used for deriving Eq. (3.31). The final result for the probability distribution function $F_p(h)$ of the individual peaks becomes (Longuet-Higgins (1963)):

$$F_p(h) \simeq 1 - e^{-\frac{1}{2}u^2}\left(1 + \frac{1}{6}\lambda_3\,(u^3 - 3u) + \frac{1}{2}\lambda_{12}\,u\right)$$

again only retaining the dominant non-linear terms. The coupling term λ_{12} between the wave elevation h and its derivative $\dot{h}$ is given by

$$\lambda_{12} = \frac{E\left[H\,\dot{H}^2\right]}{s_H\; s_{\dot{H}}^2}$$

Substitution of Eq. (3.222) into this expression yields λ_{12} in a form similar to Eq. (3.228), but with the fourth spectral moment m_4 involved. Thus, for a Pierson-Moskowitz wave spectrum, Eq. (3.231), λ_{12} will be infinite without introduction of a cut-off frequency, which is of course also needed to ensure a narrow-banded spectral density.

Higher order terms may also be included in a consistent way, Jensen and Pedersen (1978), which, unfortunately, only results in a modest increase in the range of applicability of the Gram-Charlier series expansion. A generalisation to broad-banded spectral densities is given by Longuet-Higgins (1964).

An alternative to the Gram-Charlier series expansion is the transformation procedure Eq. (3.34). By use of only the three lowest statistical moments of the normalized wave elevation

$$U = \left(H^{(1)} + H^{(2)}\right)/s_H$$

the series expansion becomes

$$U = c_0 + c_1\,V + c_2\,V^2 \equiv g(V)$$

where $V \equiv H^{(1)}/s_H$ is standard normal distributed. The coefficients c_j follows from Eq. (3.36) with $c_3 = 0$ and can be reordered as

$$4c_2^3 - 6c_2 + \gamma_1 = 0$$

$$c_1^2 = 1 - 2c_2^2$$

$$c_0 = -\,c_2$$

The analytical solution to the cubic equation in c_2 has one real root. For small values of γ_1, the solution becomes $c_2 \simeq \gamma_1/6$ and, hence, $c_1 \simeq 1$ and $c_0 \simeq -\gamma_1/6$. Also because γ_1 is small, the reverse relation is

$$V \equiv g^{-1}(U) \simeq U - \frac{\gamma_1}{6}\left(U^2 - 1\right)$$

Thereby,

$$F_H(u) = P(U < u) = P\left(V < g^{-1}(u)\right) = \Phi\left(g^{-1}(u)\right)$$

Clearly $F_H(u)$ is well-behaved for all value of u. Now

$$P\left(h > 3s_H\right) = 1 - \Phi\left(g^{-1}(3)\right) = 1 - \Phi\,(2.64) = 0.0041$$

with $\gamma_1 = 0.27$. This value differs from the corresponding Gram-Charlier result: 0.0029, calculated above. However, assymptotically, for $\gamma_1 \rightarrow 0$, both procedures yield the same result as

$$\Phi\left(g^{-1}(u)\right) = \Phi(u) + \frac{1}{\sqrt{2\pi}} \int_u^{u-\frac{\gamma_1}{6}(u^2-1)} \exp\left(-\frac{1}{2}t^2\right)dt \simeq \Phi(u) - \frac{\gamma_1}{6}\left(u^2 - 1\right)\varphi(u)$$

3.3.2.2 Directional Spreading

The P-M spectrum is a unidirectional spectrum describing waves travelling in one direction in accordance with the wave elevation process, Eq. (3.214). However, wind driven waves will be generated in all directions relative to the wind. For each angle φ relative to the wind direction, a spectral density $S(\omega, \varphi)$ for the wave elevation can be defined. The spectral density must satisfy

$$s_H^2 = \int_{-\pi}^{\pi} \int_{0}^{\infty} S(\omega, \varphi)\, d\omega\, d\varphi \tag{3.257}$$

Due to lack of more accurate formulations, $S(\omega, \varphi)$ is usually taken in the form:

$$S(\omega, \varphi) = S(\omega)\, f(\varphi) \tag{3.258}$$

As

$$\int_{0}^{\infty} S(\omega)\, d\omega = s_H^2$$

the *spreading function* $f(\varphi)$ must satisfy

$$\int_{-\pi}^{\pi} f(\varphi)\, d\varphi = 1 \tag{3.259}$$

and obviously also $f(\varphi) = f(-\varphi)$. Furthermore, only very small waves are generated opposite to the wind direction, which implies that $f(\varphi) \simeq 0$ for $|\varphi| > \pi/2$. Finally, it must be expected that $f(\varphi)$ is a decreasing function of φ. The most commonly used spreading function is

$$f(\varphi) = \begin{cases} A_n \cos^n \varphi & \varphi < \left|\frac{\pi}{2}\right| \\ 0 & \text{otherwise} \end{cases} \tag{3.260}$$

with $n = 2$ or 4. From Eq. (3.259) it follows that $A_2 = 2/\pi$ and $A_4 = 8/(3\pi)$.

3.3.3 Long-Term Predictions

The analysis presented so far has been limited to stationary processes. The ocean waves only behave as stationary processes over a period of time measured in hours c.f. Figure 3.11, so over a time scale measured in years the wave process is clearly a non-stationary process which covers everything between nearly calm sea and extreme storm events.

To overcome this problem, two approaches may be used for estimating the highest waves over a period of years. The first is to assume that the largest waves appear in the severest stationary sea state encountered in that period. However, because the maximum wave height depends on the number of peaks in this period, see Eq. (3.156) as well as on the standard deviation $s_H = H_s/4$, the maximum waves may be found in a lower sea state which occurs much more frequently. Therefore, in the second method all the sea states encountered at a given location are weighted according to their probability of occurrence. The probability density function $p_p^{\ell}(h)$ for the individual peak values in the non-stationary process is derived by taking the probability density function $p_p(h)$ in each stationary sea state to be a probability, which is conditional on the parameters describing the stationary condition. For waves, these parameters are usually taken to be the significant wave height H_s and the zero upcrossing period T_z. From Eq. (3.60) it follows that

$$\begin{aligned} p_p^{\ell}(h) &= \frac{1}{\nu(0)} \int_{H_s}\int_{T_z} \nu(0|H_s, T_z)\, p_p(h, H_s, T_z)\, dT_z\, dH_s \\ &= \frac{1}{\nu(0)} \int_{H_s}\int_{T_z} \nu(0|H_s, T_z)\, p_p(h|H_s, T_z)\, p(H_s, T_z)\, dT_z\, dH_s \end{aligned} \tag{3.261}$$

where $p_p(h|H_s, T_z)$ is given by Eq. (3.137) or, for a narrow-banded process, by Eq. (3.242).

The use of zero-crossing rate $\nu(0)$ rather than the peak rate ν_p is justified by the derivations of Eqs. (3.154) - (3.155). For a Gaussian wave process, Eq. (3.118) yields

$$\nu(0|H_s, T_z) = \frac{1}{T_z} \tag{3.262}$$

so that

$$\nu(0) = \int_{T_z} \frac{1}{T_z}\, p(T_z)\, dT_z \tag{3.263}$$

where $p(T_z)$ is the marginal probability density function of T_z :

$$p(T_z) = \int_{H_s} p(H_s, T_z)\, dH_s \tag{3.264}$$

The peak distribution $p_p(h|H_s, T_z)$ only depends on $m_0 = H_s^2/16$ and the bandwidth parameter ε, but as ε normally is less than 0.9, the approximation, Eq. (3.153), with a fixed value of ε or even Eq. (3.242) may be applied. Thereby,

$$
\begin{aligned}
p_p^{\ell}(h) &= \frac{1}{\nu(0)} \int_{H_s} \int_{T_z} \frac{1}{T_z} p_P\,(h|H_s)\; p\,(H_s, T_z)\; dT_z\, dH_s \\
&= \int_{H_s} p_P\,(h|H_s)\; \hat{p}\,(H_s)\; dH_s
\end{aligned}
\tag{3.265}
$$

where $\hat{p}(H_{s})$ is the marginal distribution of H_S, weighted by the zero-crossing rate:

$$
\hat{p}(H_s) \equiv \frac{1}{\nu(0)} \int_{T_z} \frac{1}{T_z}\, p(H_s, T_z)\; dT_z \tag{3.266}
$$

The integration should cover all possible vaues of H_S and T_Z. The distribution $p(H_S, T_Z)$ cannot be derived by any theoretical method, but must be obtained from measurements. The most comprehensive measurements are those published by Hogben et al. (1986). This book contains tabulated values of $p(H_S, T_Z)$ for 104 ocean areas, denoted Marsden areas, covering all the major ship trading routes. The data are given in increments of 1 m for H_S and 1 sec for T_Z and are furthermore given for 8 global directions. An example is shown in Figure 3.17. By use of these data and the condition probability density $p_P(h|H_s)$ the *long-term probability density function* $p_p^{\ell}(h)$ of the individual peak values for the wave elevation can be carried out numerically.

SIGNIFICANT WAVE HEIGHT (m) \ ZERO CROSSING PERIOD (s)	<4	4–5	5–6	6–7	7–8	8–9	9–10	10–11	11–12	12–13	>13	TOTAL
>14	–	–	–	–	–	–	–	–	–	–	–	1
13–14	–	–	–	–	–	–	–	–	–	–	–	1
12–13	–	–	–	–	–	–	–	–	–	–	–	1
11–12	–	–	–	–	–	–	–	1	–	–	–	2
10–11	–	–	–	–	–	–	1	1	1	–	–	3
9–10	–	–	–	–	–	1	2	1	1	–	–	6
8–9	–	–	–	–	1	2	3	3	2	1	–	11
7–8	–	–	–	–	1	4	6	5	3	1	–	20
6–7	–	–	–	1	3	9	11	8	4	1	–	37
5–6	–	–	–	1	9	19	20	13	6	2	–	70
4–5	–	–	–	4	21	37	34	18	7	2	–	124
3–4	–	–	1	13	45	64	47	21	6	2	–	197
2–3	–	–	4	31	77	80	44	15	4	1	–	255
1–2	–	–	11	53	78	51	18	4	1	–	–	216
0–1	–	2	13	22	14	5	1	–	–	–	–	57
TOTAL	–	3	28	124	249	271	186	90	34	10	3	1000

Figure 3.17 *Scatter diagram for the North Atlantic, Hogben et al. (1986).*

To derive the most probable largest peak $\tilde{\mu}_N^{\ell}$ over a period of several years, Eq. (3.148) is applied

$$F_p^\ell(\tilde{\mu}_N^\ell) = \int_{-\infty}^{\tilde{\mu}_N^\ell} p_p^\ell(u)\, du = 1 - \frac{1}{N} \tag{3.267}$$

The solution to this implicit equation in $\tilde{\mu}_N^\ell$ can be carried out numerically when the total number of peaks is specified. An estimate of this number N is the average number

$$N - \nu(0)T \tag{3.268}$$

where T is the total time period considered. Alternatively, the number N of wave peaks during a 20 year period is often taken to be simply 10^8.

A closed-form expression of $\tilde{\mu}_N^\ell$ can be obtained from Eq. (3.267) if $p_p^\ell(h)$ is fitted in with an analytical distribution. Often the Weibull distribution, Eq. (3.38), gives a very good fit. Hence, Eq. (3.159) reads (see Example 3.2.2):

$$\tilde{\mu}_N^\ell = \alpha\, (\ln N)^{1/\beta} \tag{3.269}$$

where α, β are the two parameters in the Weibull distribution.

The tabulated values for $p(H_S, T_Z)$ given in Hogben et al. (1986) can also be fitted in with marginal, conditional or joint distributions, see for example Friis Hansen (1994), where Weibull distributions are found to be most accurate for the marginal distributions of H_S and the conditional distribution of T_Z given H_S. As the largest wave usually occurs in the extreme and very rare sea states, care must be taken in applications of such analytical formulas as they may be quite inaccurate in the tails of the distributions, especially in extrapolations to values of (H_S, T_Z) which are not measured in practice. On the other hand, the tabulated values in Hogben et al. (1986) only cover sea states which appear with a frequency greater than or equal to 0.001. This corresponds to at least 7 days over a period of 20 years, so these tabulated values may also have filtered out some very rare but physically possible extreme events. This may result in an underestimation of the extreme loads on ships by about 30 %. As a final comment, it may be mentioned that at present the most uncertain part of the derivation of the extreme wave loads on ships appears to be the choice of the *scatter diagram* $p(H_S, T_Z)$. Therefore, the classification societies usually specify the scatter diagram to be used in the analysis.

Example 3.3.4

Assume that the marginal distribution of the significant wave height H_s, as defined by Eq. (3.266) is a one-sided normal distribution:

$$\hat{p}(H_s) = \frac{2}{\sqrt{2\pi}\, H_0}\, e^{-\frac{1}{2}(H_s/H_0)^2} \quad ; \quad H_s \geq 0 \tag{3.270}$$

where H_0 is a constant depending on the ocean area considered. Substitution of Eqs. (3.270) and (3.242) into Eq. (3.265) yields

$$
\begin{aligned}
F_p^\ell(h) &= \int_0^\infty F_p(h|H_z)\, p(H_s)\, dH_s \\
&= \int_0^\infty \left(1 - e^{-8\,(h/H_s)^2}\right) \frac{2}{\sqrt{2\pi}\, H_0}\, e^{-\frac{1}{2}(H_s/H_0)^2} dH_s \\
&= 1 - \frac{2}{\sqrt{\pi}} \int_0^\infty e^{-(2h/H_0)^2 u^{-2} - u^2} du \\
&= 1 - e^{-4h/H_0}
\end{aligned}
\tag{3.271}
$$

which is an exponential distribution.

By use of Eq. (3.269) with $\alpha = H_0/4$ and $\beta = 1$ the most probable largest peak during a period of 20 years becomes using

$$
\tilde{\mu}_N^\ell = \frac{H_0}{4} \ln N \tag{3.272}
$$

As the troughs are distributed like the peaks in a linear wave analysis, the most probable largest trough-to-crest height becomes $1/2\, H_0 \ln N$. For the North Atlantic $H_0 \simeq 3.5$ m, so that the maximum wave height is about 36 m in 20 years.

The probability distribution function $F_N^\ell(h)$ for the largest peak becomes

$$
\begin{aligned}
F_N^\ell(h) &= \left[F_p^\ell(h)\right]^N = \left(1 - e^{-4h/H_0}\right)^N \\
&\simeq 1 - Ne^{-4h/H_0} \simeq exp\left(- Ne^{-4h/H_0}\right)
\end{aligned}
\tag{3.273}
$$

provided

$$
\frac{4h}{H_0} >> lnN \text{ or } h > \tilde{\mu}_N^\ell
$$

If this result is compared with Eq. (3.170) it is seen that the long-term distribution of the largest peak is Gumbel distributed in this example as expected.

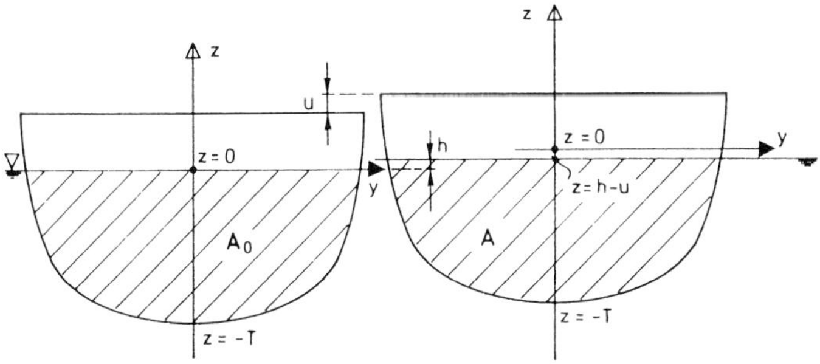

Figure 4.2 *Sectional contour of X=x. Left: section at rest. Right: section displaced vertically in waves.*

Figure 4.2 show a sketch of a section. The water elevation is $h(x, t)$ and the ship motion is assumed to be vertical with an unknown displacement $u = u(x, t)$. The integral in Eq. (4.8) is conveniently carried out in the local yz-system fixed in the section. Thus,

$$q_{FKz} = - \int_{-T}^{h-u} \frac{\partial p}{\partial z} B(x, z)\, dz \tag{4.10}$$

where $T = T(x)$ is the draft of the section measured with respect to z=0 and where $B(x, z)$ is the sectional breadth as function of z. The derivative

$$\frac{\partial p}{\partial z} = \left. \frac{\partial p}{\partial Z} \right|_{Z=z+u} \tag{4.11}$$

as

$$Z = z + u \tag{4.12}$$

has replaced the last equation in (4.6). Hence, the final expression for the vertical force q_{FKz} per unit length on a prismatic section becomes

$$q_{FKz}(x; h, u, t) = - \int_{-T}^{h-u} \left. \frac{\partial p}{\partial Z} \right|_{Z=z+u} B(x, z)\, dz \tag{4.13}$$

If the waves can be represented by the linear wave solution, Eqs. (3.201)-(3.204), then

$$\left. \frac{\partial p}{\partial Z} \right|_{Z=z+u} = \left. \frac{\partial \left(p_s + p^{(1)} \right)}{\partial z} \right|_{Z=z+u} = -\rho g + \rho g k h^{(1)} e^{k(z+u)}$$

such that

$$q_{FKz} = \rho g \int_{-T}^{h-u} B(x,z)dz - \rho g k h \int_{-T}^{h-u} e^{k(z+u)} B(x,z)dz \tag{4.14}$$

omitting the index on h for the sake of brevity.

Provided the wave elevation h and the vertical motion u of the section are small compared to the wave length and, furthermore, that the breadth does not vary too much within $0 \leq z \leq h - u$, then Eq. (4.14) can be approximated by

$$q_{FKz} = \rho g A_0 + \rho g B_0 (h - u) - \rho g k h \int_{-T}^{0} e^{kz} B(x,z)dz \tag{4.15}$$

neglecting quadratic and higher order terms in h, u. In Eq. (4.15)

$$A_0 = A_0(x) \equiv \int_{-T}^{0} B(x,z)dz \tag{4.16}$$

and

$$B_0 = B_0(x) \equiv B(x, z = 0) \tag{4.17}$$

The first term on the right hand side of Eq. (4.15) is the *hydrostatic load* in calm water q_s per unit length. The remaining part due to the dynamic undisturbed pressure in the waves is called the *Froude-Krylov force* and can be written

$$q_{FKz} - q_s = -\rho g B_0 (u - \kappa h) \tag{4.18}$$

where the *Smith correction factor*

$$\kappa = \kappa(x) = 1 - k \int_{-T}^{0} e^{kz} \frac{B(x,z)}{B_0(x)} dz \tag{4.19}$$

has been introduced. It is seen that the force $q_{FK} - q_s$ is proportional to the measure

$$\tilde{z} = \tilde{z}(x,t) \equiv u(x,t) - \kappa(x)\, h(x,t) \tag{4.20}$$

of the relative motion between the wave surface and the section. The Smith correction factor arises because the dynamic pressure decays exponentially with respect to the vertical distance from the free surface contrary to the linear increase of the hydrostatic pressure.

For many sections, see Example 4.1.1, the Smith correction factor can be approximated by

$$\kappa(x) \simeq e^{-k\,A_0(x)/B_0(x)} \tag{4.21}$$

Example 4.1.1

Rectangular section:

Eq. (4.19): $\kappa = 1 - k \int_{-T}^{0} e^{kz}\, dz = e^{-kT}$

Eq. (4.21): $\kappa = e^{-kT}$

Wedge section:

Eq. (4.19): $\kappa = 1 - k \int_{-T}^{0} e^{kz} \left(1 + \frac{z}{T}\right) dz = \frac{1}{kT}\left(1 - e^{-kT}\right)$

Eq. (4.21): $\kappa = e^{-\frac{1}{2}kT}$

As shown in Figure 4.3 the difference between the two results is small.

Circular section:

Eq. (4.19): $\kappa = 1 - kR \int_{0}^{\pi/2} \sin^2\alpha \; e^{-kR\cos\alpha}\, d\alpha$

$$= \int_{0}^{\pi/2} \cos\alpha \; e^{-kR\cos\alpha}\, d\alpha$$

using $z = -R\cos\alpha$ and $B = 2R\sin\alpha$; $\alpha\epsilon[0, \pi/2]$.

Eq. (4.21): $\kappa = e^{-\frac{\pi}{4}kR}$

Again, see Figure 4.3, the difference between the two expressions is small.

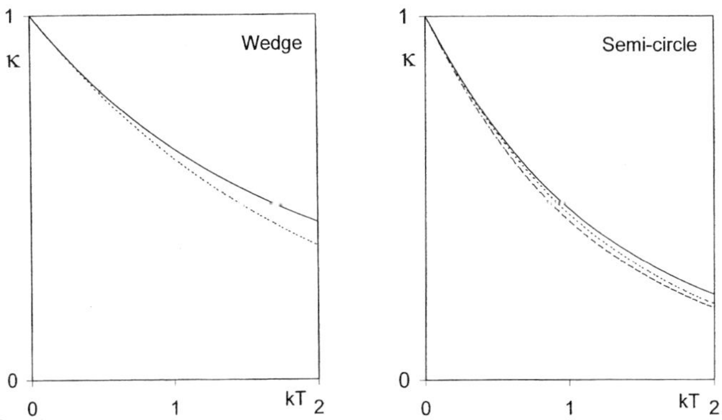

Figure 4.3 *Comparison between exact (full lines) and approximate (dashed lines) values of the Smith correction factor for two sections. For $\bar{\kappa}$ (dashed-dot line), see Example 4.2.4.*

4.2 LINEAR HYDRODYNAMIC FORCES

The Froude-Krylov force, Eq. (4.13), must be supplemented with forces accounting for the modification of the wave potential required to satisfy the interaction condition (4.3). In the general non-linear case this is, as discussed in the beginning of this chapter, very difficult. Therefore, a linear analysis is considered first.

In the linear analysis the wave elevation and the ship motion are assumed so small that the total fluid potential can be represented as a sum of different potentials each representing specific parts of the solution.

The Froude-Krylov force, Eq. (4.18), yields a quasi-static force in the sense that it only depends on the instantaneous position of the section in the waves. Hydrodynamic forces proportional to the velocity and acceleration of the section and the wave particles must be added. Contrary to the Froude-Krylov force these forces not only depend on instantaneous value of the motion, but also on its past history. This may be illustrated by the observation that if a floating body is given an impulse, then the body will oscillate for a long period. The system behaves as having a memory in the sense that what happens at one instant of time will affect the system at all later times. The free surface is responsible for this behaviour, as it allows for a transmission of energy away from the body by generating surface waves.

4.2.1 Impulse Response Functions

Consider a time-invariant linear system. The response $r(t)$ at time t due to a disturbance (impulse) $f(\tau)\Delta\tau$ at time $\tau < t$ can then be written

$$r(t) = f(\tau)\Delta\tau\, k(t-\tau) \tag{4.22}$$

where the unit *impulse response function* or *kernel function* $k(t-\tau)$ only depends on the time difference $t-\tau$. For a continuous excitation, Eq. (4.22) can be integrated to

$$r(t) = \int_{-\infty}^{t} f(\tau)\, k(t-\tau)\, d\tau = \int_{0}^{\infty} f(t-\tau)\, k(\tau)\, d\tau \tag{4.23}$$

Dealing with stationary stochastic processes, a transformation from time to frequency domain is very useful as illustrated in Chapter 3. Fourier transformation of the unit impulse response function $k(t)$ yields

$$K(\omega) = \int_{-\infty}^{\infty} k(t) e^{-i\omega t}\, dt = \int_{0}^{\infty} k(t)\, e^{-i\omega t}\, dt \tag{4.24}$$

Notice that the lower limit can be replaced by zero as $k(t)=0$ for $t<0$. The inverse relation becomes

$$k(t) = \frac{1}{2\pi} \int_{-\infty}^{\infty} K(\omega)\, e^{i\omega t}\, d\omega \tag{4.25}$$

The factor $1/2\pi$ has for convenience here been applied to the integral over the frequency domain, opposite to the convention used when defining spectral densities in Chapter 3. As *k(t)* is a real function it follows from Eq. (4.24) - (4.25) that the real *(Re)* and imaginary (*Im*) part of the *transfer function* $K(\omega)$ satisfy

$$Re(K(\omega)) = Re(K(-\omega)) = \int_0^{\infty} k(t) \cos \omega t\, dt \tag{4.26}$$

$$Im(K(\omega)) = -Im(K(-\omega)) = -\int_0^{\infty} k(t) \sin \omega t\, dt \tag{4.27}$$

Provided

$$\lim_{\omega \to \infty} K(\omega) = 0 \tag{4.28}$$

Fourier cosine and sine transformations of Eq. (4.26) and (4.27) can be performed, yielding

$$k(t) = \frac{2}{\pi} \int_0^{\infty} Re\,(K(\omega))\, \cos \omega t\, d\omega \tag{4.29}$$

and

$$k(t) = -\frac{2}{\pi} \int_0^{\infty} Im\,(K(\omega)) \sin \omega t\, d\omega \tag{4.30}$$

Hence, *k(t)* can be determined from either $Re(K(\omega))$ or $Im(K(\omega))$. The implicit relation between the real and imaginary part of $K(\omega)$ is due to the property that *k(t)* is real. The relations (4.29)-(4.30) are a special case of more general relations known as the Kramers-Konig relations.

4.2.2 Sinusoidal Excitation

If the excitation *f(t)* is a sinusoidal function

$$f(t) = f_0\,(\omega) \cos\,(\omega t + \epsilon) \tag{4.31}$$

the response $r(t)$ is obtained from Eq. (4.23):

$$
\begin{aligned}
r(t) &= f_0 \int_0^\infty \cos(\omega(t-\tau)+\epsilon)\, k(\tau)\, d\tau \\
&= f_0 \cos(\omega t+\epsilon) \int_0^\infty \cos(\omega\tau)\, k(\tau) d\tau \\
&\quad + f_0 \sin(\omega t+\epsilon) \int_0^\infty \sin(\omega\tau)\, k(\tau) d\tau \\
&= f_0 \left[Re\,(K(\omega)) \cos(\omega t+\epsilon) - Im\,(K(\omega)) \sin(\omega t+\epsilon) \right] \\
&= f_0 \sqrt{K^*(\omega)\, K(\omega)}\, \cos(\omega t+\epsilon_r(\omega)) \\
&\equiv r_0 \cos(\omega t+\epsilon_r(\omega))
\end{aligned}
\tag{4.32}
$$

using Eqs. (4.26) - (4.27). Here $K^*(\omega)$ is the complex conjugate of $K(\omega)$ and

$$
\epsilon_r = \epsilon + Arctan \left(\frac{Im\,(K(\omega))}{Re\,(K(\omega))} \right) \tag{4.33}
$$

The significance of the transfer function is then that it is the amplitude ratio of the sinusoidal response to a sinusoidal excitation:

$$
\frac{r_0}{f_0} = \sqrt{K^*(\omega)\, K(\omega)} = |K(\omega)| \tag{4.34}
$$

In general, the transfer function $K(\omega)$ is obtained by substitution of

$$
f(t) = f_0\, e^{i\omega t}
$$

into Eq. (4.23):

$$
\begin{aligned}
r(t) &= \int_0^\infty f_0\, e^{i\omega(t-\tau)}\, k(\tau)\, d\tau \\
&= f_0\, e^{i\omega t} \int_0^\infty e^{-i\omega\tau}\, k(\tau)\, d\tau \\
&= f(t)\, K(\omega)
\end{aligned}
$$

Example 4.2.1

Consider a single degree of freedom linear mass-spring-dashpot system. The equation of motion can be written

$$\ddot{r}(t) + 2\zeta\omega_0 \dot{r}(t) + \omega_0^2 r(t) = f(t) \quad ; \quad t \geq 0$$

where ω_0 and ζ are the eigenfrequency and damping ratio, respectively. A overdot denotes as usual differentiation with respect to time t.

The solution $r(t)$ is given by the sum of a particular $r_p(t)$ and the homogeneous $r_h(t)$ solution. The particular solution is taken as the Duhamel integral:

$$r_p(t) = \frac{1}{\omega_d} \int_0^t f(\tau) \sin(\omega_d (t - \tau)) e^{-\zeta\omega_0 (t-\tau)} d\tau$$

where

$$\omega_d = \omega_0 \sqrt{1 - \zeta^2}$$

The homogeneous solution is

$$r_h(t) = (A \cos\omega_d t + B \sin\omega_d t)\, e^{-\zeta\omega_0 t}$$

where the constants A and B follow from the initial conditions. As $r(0) = \dot{r}(0) = 0$ the homogeneous solution vanishes. Hence, $r(t) = r_p(t)$ and it is seen that $r(t)$ is written in the form (4.23) with the kernel

$$k(\tau) = \frac{\sin\omega_d \tau}{\omega_d} e^{-\zeta\omega_0 \tau}$$

which then is the unit impulse response function for this system. Hence, usual dynamic problems within rational mechanics also exhibit a memory.

The transfer function $K(\omega)$ can be obtained by substitution of $k(\tau)$ into Eq. (4.24). However, as the transfer function is the solution for a sinusoidal excitation it is much easier to insert

$$f(t) = f_0\, e^{i\omega t}$$

into the governing equation. Thereby,

$$r(t) = r_0\, e^{i\omega t}$$

with

$$r_0 \left[(i\omega)^2 + 2\zeta\omega_0 i\omega + \omega_0^2 \right] = f_0$$

such that

$$K(\omega) = \frac{r(t)}{f(t)} = \frac{r_0}{f_0} = \left[\omega_0^2 - \omega^2 + 2i\zeta\omega_0\omega \right]^{-1}$$

and

$$|K(\omega)| = \left[\left(\omega_0^2 - \omega^2\right)^2 + \left(2\zeta\omega_0\omega\right)^2 \right]^{-1/2}$$

The result Eq. (4.34) makes response predictions of linear systems subjected to stationary stochastic excitations very easy as will be discussed in Section 4.4. First, however, the determination of $k(t)$ and $K(\omega)$ for the hydrodynamic loads must be carried out.

4.2.3 Added Mass and Hydrodynamic Damping

For a section of the hull executing a vertical motion $u = u\,(t)$ in calm sea, the hydrodynamic vertical force $q_H\,(t)$ can in general be written

$$q_H\,(t) = -\,c_2\ddot{u}(t) - c_1\dot{u}(t) - c_0 u(t) - \int_{-\infty}^{t} \ddot{u}(\tau)\,k_2\,(t-\tau)\,d\tau$$
$$-\int_{-\infty}^{t} \dot{u}(\tau)\,k_1\,(t-\tau)\,d\tau - \int_{-\infty}^{t} u(\tau)\,k_0\,(t-\tau)\,d\tau \qquad (4.35)$$

as the fluid motion around the section is completely characterized by the motion $u(t)$, velocity $\dot{u}(t)$ and acceleration $\ddot{u}(t)$ of the section, provided these variables are small enough to neglect non-linear terms.

The three coefficients c_2, c_1 and c_0 represent the parts of the force proportional to the instantaneous values of $\ddot{u}(t)$, $\dot{u}(t)$ and $u(t)$, respectively. They could of course be included in the kernel functions k_2, k_1 and k_0 using Dirac's delta function. The kernel functions as well as the coefficients c_2, c_1 and c_0 must be determined by solving the governing equations for the motion of the fluid, which as stated previously is very complicated. A thorough discussion is given by Ogilvie (1964). It is found that a Fourier transformation to the frequency domain provides an effective procedure.

First, however, Eq. (4.35) can be simplified by integration by parts of the first and last integral on the right hand side:

$$q_H\,(t) = -\,c_2\ddot{u}(t) - c_1\dot{u}(t) - c_0 u(t)$$
$$-\left[\dot{u}(\tau)\,k_2\,(t-\tau)\right]_{-\infty}^{t} - \int_{-\infty}^{t} \dot{u}(\tau)\,\dot{k}_2\,(t-\tau)\,d\tau - \int_{-\infty}^{t} \dot{u}(\tau)\,k_1\,(t-\tau)\,d\tau$$
$$-\left[u(\tau)\,\bar{k}_0\,(t-\tau\right]_{-\infty}^{t} + \int_{-\infty}^{t} \dot{u}(\tau)\,\bar{k}_0(t-\tau)\,d\tau$$

where

$$\bar{k}_0\,(t-\tau) = \int_{t-\tau}^{\infty} k_0\,(\xi)\,d\xi$$

Hence,

$$q_H(t) = -\,c_2\ddot{u}(t) - b'\dot{u}(t) - c'u(t) - \int_{-\infty}^{t} \dot{u}(\tau)k'(t-\tau)d\tau \tag{4.36}$$

where the coefficients

$$b' = c_1 + k_2(0)$$
$$c' = c_0 + \int_0^{\infty} k_0(\tau)d\tau \tag{4.37}$$

and the kernel function

$$k'(\tau) = \dot{k}_2(\tau) + k_1(\tau) + \int_{\tau}^{\infty} k_0(\xi)d\xi \tag{4.38}$$

The transfer function $K(\omega)$ relating the force q_H to the amplitude u_0 in a sinusoidal excitation is obtained by substitution of

$$u(t) = u_0\,e^{i\omega t} \tag{4.39}$$

into Eq. (4.36):

$$q_H(t) = \left[c_2\omega^2 - i\omega b' - c' - i\omega K'(\omega)\right]u_0 e^{i\omega t} \tag{4.40}$$

using

$$\int_{-\infty}^{t} \dot{u}(\tau)k'(t-\tau)d\tau = \int_{0}^{\infty} \dot{u}(t-\tau)\,k'(\tau)\,d\tau$$

$$= i\omega u_0 e^{i\omega t}\int_{0}^{\infty} e^{-i\omega\tau}\,k'(\tau)\,d\tau = i\omega K'(\omega)u_0\,e^{i\omega t}$$

Hence

$$K(\omega) = \frac{q_H(t)}{u_0\,e^{i\omega t}} = c_2\omega^2 - i\omega b' - c' - i\omega\,K'(\omega) \tag{4.41}$$

The transfer function $K'(\omega)$ corresponding to the kernel function $k'(\tau)$ will in general by a complex function. The transfer function $K(\omega)$ can then be written

$$\begin{aligned} K(\omega) &= c_2\omega^2 - c' + \omega Im(K'(\omega)) \\ &- i\left[\omega b' + \omega Re(K'(\omega))\right] \end{aligned} \tag{4.42}$$

Rather than finding the coefficients and kernel functions in the time domain formulation (4.36), it is normally much easier to determine the transfer function for different frequencies ω. This may for instance be done by calculations or measurements of q_H, assuming a sinusoidal motion with a given frequency. Usually, q_H is then given as a differential equation

$$q_H(t) = -\,m_w(\omega)\ddot{u}(t) - N(\omega)\,\dot{u}(t) - Cu(t) \tag{4.43}$$

where the real coefficients $m_w(\omega)$ and $N(\omega)$ depend on the frequency ω of the motion. Substitution of Eq. (4.39) into Eq. (4.43) yields

$$K(\omega) = m_w(\omega)\omega^2 - i\omega N(\omega) - C \tag{4.44}$$

The coefficients m_w, N and C can then be expressed in terms of the coefficients derived from the time domain formulation Eq. (4.42). Taking C as independent of ω ensure a unique definition of $m_w(\omega)$ and C

$$m_w(\omega)\omega^2 = c_2\omega^2 + \omega Im(K'(\omega)) \tag{4.45}$$

$$N(\omega) = b' + Re\,(K'(\omega)) \tag{4.46}$$

$$C = c' \tag{4.47}$$

It should be emphasised that whereas Eq. (4.36) is a true time domain formulation, Eq. (4.43) is not as it implicitly assumes a sinusoidal motion.

Provided the transfer function $K'(\omega)$ satisfy Eq. (4.28) the corresponding impulse response function $k'(t)$ can be written

$$k'(t) = \frac{2}{\pi} \int_0^\infty (N(\omega) - b') \cos \omega t \, d\omega \tag{4.48}$$

or

$$k'(t) = \frac{2}{\pi} \int_0^\infty \omega \big(c_2 - m_w(\omega) \big) \sin \omega t \, d\omega \tag{4.49}$$

using Eqs. (4.29), (4.30) and (4.45), (4.46). As $K'(\omega)$ is assumed to vanish for $\omega \to \infty$, a and b' can be interpretated as

$$c_2 = \lim_{\omega \to \infty} m_w(\omega) = m_w^\infty \tag{4.50}$$

$$b' = \lim_{\omega \to \infty} N(\omega) \tag{4.51}$$

For surface waves generated by a point source $b' = 0$. The sectional hydrodynamic load $q_H(t)$ can now finally be written

$$\begin{aligned} q_H(t) &= - m_w^\infty \, \ddot{u}(t) - C\, u(t) - \int_{-\infty}^{t} \dot{u}(\tau)\, k'(t - \tau)\, d\tau \\ &= - m_w^\infty \, \ddot{u}(t) - C\, u(t) - \int_0^\infty \dot{u}(t - \tau)\, k'(\tau)\, d\tau \end{aligned} \tag{4.52}$$

where the second expression follows from Eq. (4.23).

Whether to use Eq. (4.48) or (4.49) to calculate $k'(t)$ depends on the procedures available for determination of $m_w(\omega)$ and $N(\omega)$. However, as both equations should give the same result, they can also be used to check the consistency of the values for $m_w(\omega)$ and $N(\omega)$.

The coefficients $m_w(\omega)$ and $N(\omega)$ are denoted the *added mass (of water)* and the *damping coefficient*, respectively. This notation follows from the fact that $-m_w \ddot{u}$ is of the form of an inertia force whereas $-N\dot{u}$ represent a damping term responsible for the decay in the motion due to generation of surface waves. Both $m_w(\omega)$ and $N(\omega)$ depend on the sectional shape as well as on the frequency ω. Several procedures exist to determine $m_w(\omega)$ and $N(\omega)$, ranging from simple Lewis transformation, Example 4.2.3, to very accurate boundary element methods, Andersen and He (1985).

Finally, the term C proportional to the motion is the *restoring term* from the linearized Froude-Krylov force, Eq. (4.18):

$$C = \rho g B_0 \tag{4.53}$$

Example 4.2.2

An approximate formula for the sectional damping coefficient $N(\omega)$ is given by Yamamoto et al (1986):

$$N(\omega) = \frac{\rho g^2}{\omega^3}\left[2\sin\left(\frac{\omega^2 B}{2g}\right)\exp\left(-\frac{\omega^2 T}{g}\right)\right]^2 \quad ; \quad \omega \geq 0$$

where B and T are the water line breadth and the draught, respectively. It is clear, that with only two parameters to describe the sectional contour, the formula cannot describe the variation of $N(\omega)$ with different sectional shapes very well. A more accurate formula is given in Example 4.2.3.

For the draught to beam ratio

$$\frac{T}{B} = 1.2$$

Figure 4.4 shows $N(\omega)$ and Figure 4.5 $k'(t)$, the last obtained by numerical integration of Eq. (4.48).

It is seen that $k'(t) \to 0$ for relative modest values of t. Thereby, the upper boundary on the convolution integral in Eq. (4.52) can be replaced by a finite value, greatly facilitating the numerical evaluation of the integral.

Finally, the sectional added mass is obtained by numerical integration of Eq. (4.27) using Eq. (4.45):

$$m_w(\omega) - m_w^\infty = -\frac{1}{\omega}\int_0^\infty k'(t)\sin\omega t \, dt$$

This relation yields the variation of the added mass with frequency. The result is shown in Figure 4.6. To get its absolute value, the added mass a at $\omega \to \infty$ must be derived. This is rather easy, because no surface waves are generated as $\omega \to \infty$. Mapping procedures, see e.g. Bishop and Price (1979), including the simple Lewis transformation are convenient methods. It is also seen that $m_w(\omega) \to \infty$ for $\omega \to 0$. This can lead to numerical problems for ships sailing in following sea, where the frequency of encounter can be close to zero.

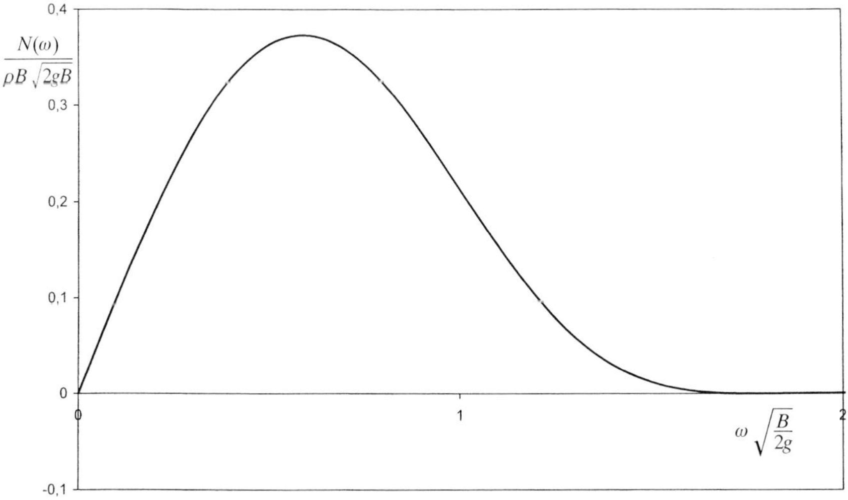

Figure 4.4 *Sectional hydrodynamic damping coefficient N(ω) for T/B = 1.2. Yamamoto et al. (1986).*

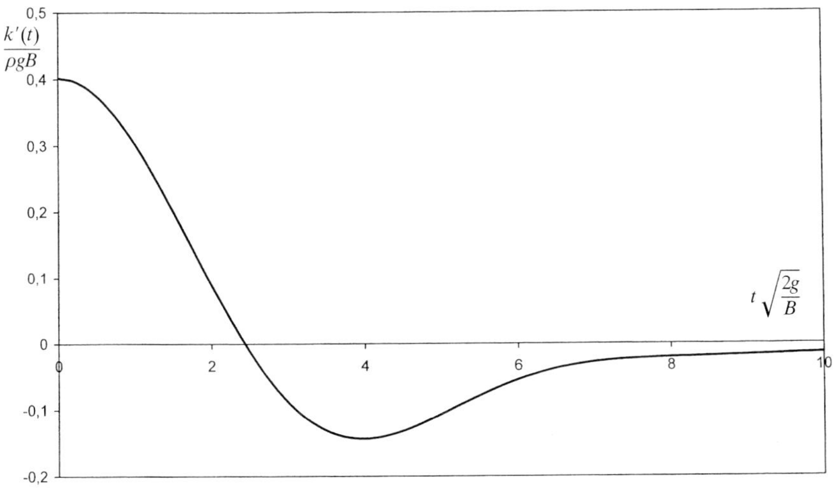

Figure 4.5 *Sectional impulse response function k'(t) for the hydrodynamic load q_H. T/B = 1.2.*

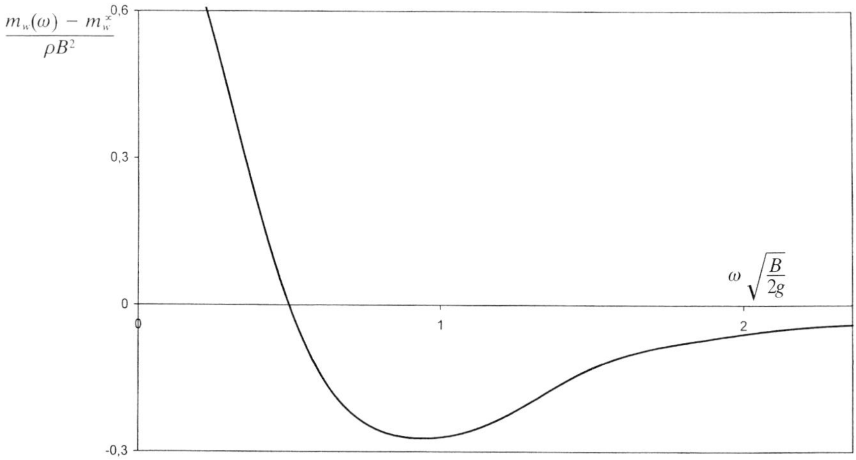

Figure 4.6 *Sectional added mass $m_w(\omega)$ relative to its value m_w^∞ at $\omega \to \infty$. T/B = 1.2.*

Example 4.2.3

A very useful transformation of a unit circle to ship-like sections was proposed by Lewis (1929):

$$y = d\left[(1 + a_1)\cos\theta + a_3\cos 3\theta\right]$$
$$z = -d\left[(1 - a_1)\sin\theta - a_3\sin 3\theta\right]$$

Here θ is the angle in the unit circle, measured from horizontal axis, see Figure 4.7.

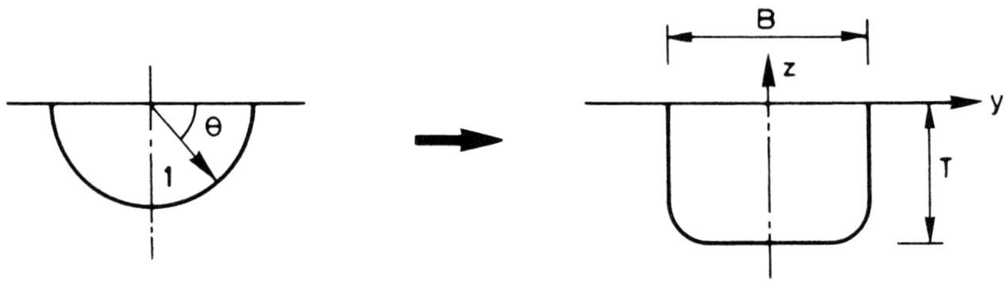

Figure 4.7 *Transformations of the unit circle to a ship-like sectional contour.*

The non-dimensional constants a_1, a_3 can be expressed in the draught to beam ratio

$$\beta \equiv \frac{2T}{B}$$

and the sectional area coefficient

$$\alpha = \frac{A}{BT}$$

where B, T and A are the water line breadth, the draught and the sectional submerged area respectively.

From

$$\beta = \frac{1 - a_1 + a_3}{1 + a_1 + a_3}$$

and

$$A = \int y dz = 2 \int_0^{\pi/2} y(\theta) \frac{dz}{d\theta} d\theta$$

the following relations are found

$$a_1 = C(1 - \beta) \quad ; \quad a_3 = C(1 + \beta) - 1$$

where

$$C = \left\{ \frac{3}{2}(1 + \beta) - \sqrt{\frac{1}{4}(1 + \beta)^2 + 2\beta\left(1 - \frac{4}{\pi}\alpha\right)} \right\}^{-1}$$

The scale parameter d with dimension length becomes

$$d = \frac{B}{4C}$$

Figure 4.8 shows sectional contours derived from this Lewis transformation. It is readily seen that ship-like sectional contours are obtained, provided α is not too small.

The usefulness of the transformation is that it can be shown that the added mass of water m_w^∞ in the high frequency limit, Eq. (4.50) becomes

$$m_w^\infty = C_m \rho A$$

where ρ is the mass density of water and

$$C_m = \frac{(1 + a_1)^2 + 3a_3^2}{1 - a_1^2 - 3a_3^2}$$

A derivation of the damping coefficient $N(\omega)$ for these sectional shapes has been given by Tasai (1959). The result is written as an infinite integral to be evaluated numerically:

$$N(\omega) = \frac{\rho g^2 \overline{A}^2}{\omega^3}$$

with

$$\overline{A} = \frac{2\tau}{1 + a_1 + a_3} \int_1^\infty \left(\frac{1 + a_1}{\gamma^2} + \frac{3a_3}{\gamma^4} \right) \cos\left\{ \tau\left(\frac{\gamma^4 + a_1\gamma^2 + a_3}{(1 + a_1 + a_3)\gamma^3} - 1 \right) \right\} d\gamma$$

and

$$\tau = \frac{\omega^2 B}{2g}$$

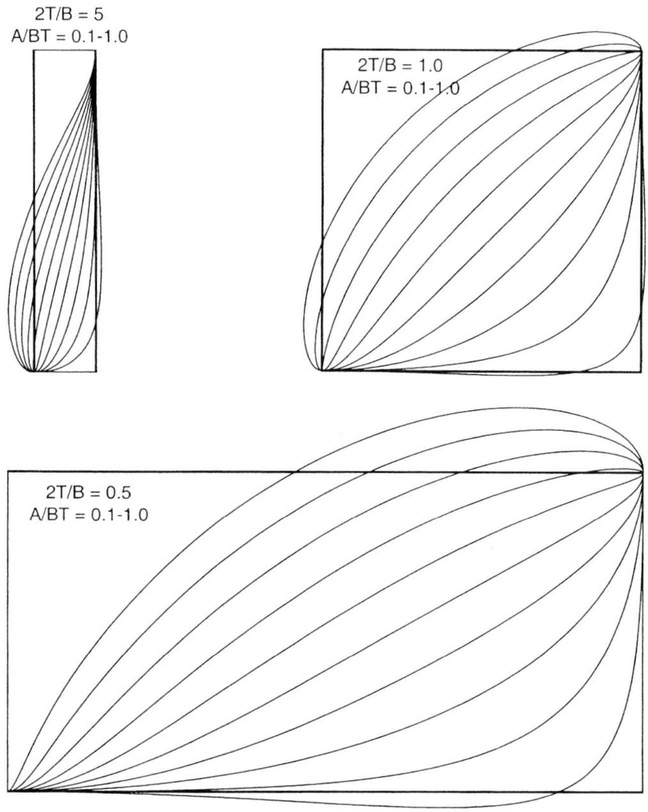

Figure 4.8 *Lewis form sections.*

With both $N(\omega)$ and m_w^∞ known, the added mass $m_w(\omega)$ can be obtained analogous to the derivation in Example 4.2.2. It is seen that $\overline{A} \to 2\tau$ irrespectively of a_1, a_3 for $\tau \to 0$ such that $N(\omega) \to \omega\rho B^2$ for $\omega \to 0$ for <u>all</u> sectional shapes.

4.2.4 Total Hydrodynamic Force

If waves are present, then a wave-induced force must be added to Eq. (4.43). Like the force due to the motion of the section, this force can be found by a Fourier transformation of the wave-induced force in the time domain solution. For the restoring term it was found in Section 4.1 that this term was equal to $C\kappa h$, see Eq. (4.18). This is the same form as the restoring term due to u, only u should be replaced by $-\kappa h$. For the damping and inertia terms substitution of u by $-\kappa h$ in Eq. (4.43) also gives fairly accurate expressions for these parts of the wave-induced load. This is illustrated in Example 4.2.4 for a circular section. Hence,

$$q_H = -\left[m_w(\omega)\ddot{\tilde{z}} + N(\omega)\dot{\tilde{z}} + C\tilde{z}\right] \tag{4.54}$$

can be used as an expression for the total hydrodynamic force per unit length for a section executing sinusoidal motion in regular waves. Here the relative vertical motion $\tilde{z}$ is given by Eq. (4.20).

A more consistent approach is to calculate the diffraction forces on the ship using the velocity potential for the waves with the boundary condition (4.3), and assuming the ship at rest. Such formulation has been used by Salvesen, Tuck and Faltinsen (1970) in their derivation of the most often sited strip theory (STF) in the literature. However, the correlation between results from both strip theory formulations and measurements are usually very good and no conclusive statements can be made whether one or the other formulation yields the most accurate results a priori. See e.g. Odabasi and Hearn (1977) for a very thorough discussion of the various strip theory formulations.

For the expression, Eq. (4.54), to be valid the frequency for the motion and the excitation frequency as experienced at the section must be the same. The linear wave elevation $h(X,t)$ is given by Eq. (3.203):

$$h(X,t) = a\cos(kX - \omega t + \epsilon)$$

such that at the section $x = x$,

$$h(x,t) = a\cos\left(kx\cos\beta - ky\sin\beta + \epsilon - \omega_e t\right) \tag{4.55}$$

where the *frequency of encounter* ω_e:

$$\omega_e = \omega - kV\cos\beta = \omega - \omega^2 \frac{V}{g}\cos\beta \tag{4.56}$$

is the apparent wave frequency seen from the section. The Doppler shift $kV\cos\beta$ depends on the wave frequency ω, forward speed V and the heading angle β.

The frequency used when evaluating the sectional coefficients m_w and N should therefore be ω_e. In addition the time derivative

$$\dot{(\ )} = \frac{d}{dt}$$

should be replaced by the *total derivative*

$$\frac{D}{Dt} = \frac{\partial}{\partial t} + \frac{\partial}{\partial x}\frac{dx}{dt} = \frac{\partial}{\partial t} - V\frac{\partial}{\partial x} \tag{4.57}$$

to account for the fact that Eq. (4.43) is formulated in the global XYZ system, but represents a load on a section moving with the speed V in this system. It is not straightforward, however, to make the replacement of d/dt by D/Dt consistently because the coefficients m_w and N depend on x. Two formulations have often been applied in the literature. The first is due to Gerritsma and Beukelman (1964) and here the *sectional hydrodynamic load* q_H is taken in the form

$$q_H(x,t) = -\left[\frac{D}{Dt}\left[m_w(x,\omega_e)\frac{D\tilde{z}}{Dt}\right] + N(x,\omega_e)\frac{D\tilde{z}}{Dt} + \rho g B_0 \tilde{z}\right] \tag{4.58}$$

in which the first term in Eq. (4.54) has been replaced by the change in momentum because m_w depends on t through the last term in Eq. (4.57).

In the time-domain the corresponding equation is

$$q_H(x,t) = -\left[\frac{D}{Dt}\left(m_w^\infty(x)\frac{D\tilde{z}}{Dt}\right) + \rho g B_0(x)\,\tilde{z} + \int_0^\infty \frac{D\tilde{z}(t-\tau)}{D\tau}k^*(x,\tau)\,d\tau\right] \tag{4.59}$$

where $k^*(x,t)$ can be found by inserting $\tilde{z}(x,t) = z_0(x)e^{i\omega_e t}$ in Eq. (4.59). The resulting equation should be equal to Eq. (4.58) and thus

$$\begin{aligned} K^*(x,\omega_e) &= \int_0^\infty k^*(x,\tau)e^{-i\omega_e\tau}d\tau \\ &= N + \frac{D}{Dt}(m_w - m_w^\infty) + i\omega\,(m_w - m_w^\infty) \\ &= K'(\omega_e) + \frac{D}{Dt}(m_w - m_w^\infty) \end{aligned}$$

where $N = N(x,\omega_e)$ and $m_w = m_w(x,\omega_e)$. From Eq. (4.25) the kernel $k^*(x,t)$ then becomes

$$\begin{aligned} k^*(x,t) &= \frac{1}{2\pi}\int_{-\infty}^{\infty} K^*(x,\omega)e^{i\omega t}d\omega \\ &= k'(x,t) + \frac{1}{\pi}\int_0^\infty \frac{D}{Dt}(m_w - m_w^\infty)\cos\omega t\, d\omega \end{aligned}$$

The somewhat pecular second term is due to the empirical nature of Eq (4.58).

The second formulation follows from the solution of the linearized potential flow problem. Thereby, it is more rational than Eq. (4.58). It comply with the STF approach, but based on the relative motion assumption. In the time domain the result becomes, Xia et al. (1998):

$$q_H(x,t) = -\frac{D}{Dt}\left[m_w^\infty(x)\frac{D\tilde{z}}{Dt} + \int_0^\infty \frac{D\tilde{z}\,(t-\tau)}{D\tau}\,k''(x,\tau)\,d\tau\right] - \rho g B_0(x)\,\tilde{z} \tag{4.60}$$

where

$$k''(x,\tau) = \int k'(x,t)\,dt$$

$$= \frac{2}{\pi}\int_0^\infty \frac{N(x)}{\omega}\sin\omega\tau\,d\omega = \frac{2}{\pi}\int_0^\infty (m_w(x,\omega) - m_w^\infty(x))\cos\omega\tau\,d\omega \tag{4.61}$$

It is easily seen that both Eq. (4.59) and (4.60) become identical to Eq. (4.52) for zero forward speed, $V = 0$. The frequency domain equation corresponding to Eq. (4.60) is obtained by inserting $\tilde{z}(x,t) = z_0(x)e^{i\omega_e t}$ in Eq. (4.60). Hence,

$$q_H(x,\omega_e) = -i\omega_e\tilde{z}\left[\frac{Da}{Dt} + i\omega_e m_w^\infty\right] + \omega_e^2\tilde{z}K''(\omega) - i\omega_e\tilde{z}\frac{DK''}{Dt} - \rho g B_0\tilde{z}$$

and with

$$K''(\omega_e) = m_w - m_w^\infty - i\frac{N}{\omega_e}$$

from Eq. (4.61), the result becomes

$$q_H(x,\omega_e) = \tilde{z}\left[\omega_e^2 m_w - i\omega_e\frac{Dm_w}{Dt} - i\omega_e N - \frac{DN}{Dt} - \rho g B_0\right]$$

or, by eliminating ω_e

$$q_H(x,\omega_e) = -\left[\frac{D}{Dt}\left(m_w(x,\omega_e)\frac{D\tilde{z}}{Dt}\right) + \frac{D}{Dt}(N(x,\omega_e)\tilde{z}) + \rho g B_0\tilde{z}\right] \tag{4.62}$$

The difference between the two frequency domain formulations Eq. (4.58) and Eq. (4.62) is seen to be in the damping term N, only. As $DN/dt = -VdN/dx$ the difference depends on the forward speed V.

The unknown is the equations (4.58)-(4.60) is the absolute vertical motion $u(x,t)$ of the ship. This function is determined from the vertical equilibrium equations for the ship using some assumptions on the variation of u along the length of the vessel. Usually, the ship is treated as a rigid body, but hull flexibility can also be taken into account. The equilibrium equations are derived in Section 4.3 and solved for the case, where q_H has a sinusoidal variation in time, Eq. (4.58). In Section 4.4 the results in the frequency domain are extended to cover the ship responses in stationary and instationary sea states, using the stochastic methods derived in Chapter 3. Section 4.5 deals with special non-linear effects and, finally Section 4.6 consider time-domain solutions based on e.g. (4.59) or (4.60). Such time-domain formulations are necessary, albert often very time consuming with respect to computer resources, for transient problem (e.g. collision or grounding) or when significant non-linearities need to be included (e.g. ship responses in severe sea).

Example 4.2.4

For the very simple case of a uniform flow past a circular cylinder embedded in an infinite fluid domain the velocity potential ϕ is known:

$$\phi = UR^2 \frac{\cos \alpha}{r}$$

Here R is the radius of the cylinder and r, α the polar coordinates for a point in the fluid, see Figure 4.9.

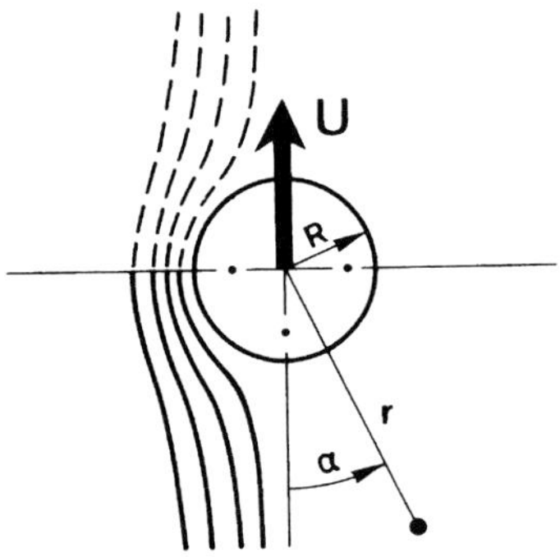

Figure 4.9 *Flow past a circular cylinder in a infinite fluid domain.*

The pressure p in the fluid is given by the linearized Bernoulli equation

$$p = -\rho \frac{d\phi}{dt} = -\frac{2}{\pi} \frac{\cos \alpha}{r} \cdot \frac{d(\rho A U)}{dt}$$

where, for later comparisons, the sectional area

$$A = \frac{\pi}{2} R^2$$

of the lower half of the circular shape is introduced. The sectional vertical force q acting on this lower half of the cylinder becomes

$$q = 2 \int_0^{\pi/2} p\Big|_{r=R} \cos \alpha R d\alpha$$

$$= -\frac{4}{\pi} \int_0^{\pi/2} \frac{d}{dt}(\rho A U) \cos^2 \alpha \, d\alpha$$

In order to relate this result to the vertical motions of a ship section, the fluid velocity is taken as

$$U = \frac{du}{dt} - v_z$$

where u is the motion of the cylinder and v_z the vertical wave particle velocity

$$v_z = \frac{\partial \Phi}{\partial z} = e^{-kR\cos\alpha}\frac{dh}{dt}$$

for a fictitious Airy wave h with a still water level in the horizontal symmetry plane of the cylinder and wave potential Φ given by Eq. (3.201). The exponential decay function is valid for deep water waves. Inserting this expression for U in the sectional force q yields

$$q = -\frac{d}{dt}\left[\rho A\left\{\frac{du}{dt} - \bar{\kappa}\frac{dh}{dt}\right\}\right]$$

where the non-dimensional parameter

$$\bar{\kappa} = \frac{4}{\pi}\int_0^{\pi/2} \exp(-kR\cos\alpha)\cos^2\alpha\, d\alpha$$

is shown in Figure 4.3. It is seen that $\bar{\kappa}$ is very close to the Smith correction factor κ for the circular cylinder, such that

$$q \simeq -\frac{d}{dt}\left[\rho A\frac{d\bar{z}}{dt}\right] \equiv -\frac{d}{dt}\left[m_w\frac{d\bar{z}}{dt}\right]$$

Hence, the relative motion assumption is seen to be a very accurate model for the added mass term for a circular cylinder supporting the use of this concept also for other sections. Finally, it is noted, that the added mass, m_w derived above, actually is

$$m_w = \lim_{\omega\to\infty} m_w(\omega) = m_w^\infty = \rho A$$

due to the neglecting of surface wave generation by use of the velocity potential for an infinite fluid domain. Thus the high frequency added mass for a circular cylinder is equal to the displaced fluid. This is also in accordance with the result in Example 4.2.3 for $a_1 = a_3 = 0$.

4.3 EQUATIONS OF MOTION

As mentioned above the sectional hydrodynamic load q_H depends on the vertical motion $u(x, t)$ of the section. This motion is determined by the equations of motion for the ship

$$\int_L m(x)\frac{d^2u}{dt^2}dx = \int_L q_H(x, u, t)dx \tag{4.63}$$

$$\int_L m(x)x\frac{d^2u}{dt^2}dx = \int_L q_H(x, u, t)xdx \tag{4.64}$$

where the integrations extend over the length L of the ship i.e.

$$\int_L = \int_{x_{min}}^{x_{max}} \quad ; \qquad L = x_{max} - x_{min}$$

The left hand sides are the inertia contributions due to the mass distribution *m(x)* of the ship whereas the right hand sides contain the total time-varying external forces on the ship. The equations express dynamic force equilibrium in the vertical direction and moment equilibrium about the y-axis, respectively. The remaining four equilibrium equations do not depend on the vertical sectional force q_H, provided the ship and its mass distribution are symmetrical about the centre-plane $y=0$. If this is not the case, coupling between the vertical and horizontal motions may occur. Here only the vertical motion *u(x)* and the associated sectional forces are considered. For coupled horizontal-torsional loads, see e.g. Bishop and Price (1979).

To perform the integrations over the length of the ship, Eqs. (4.63)-(4.64) must be supplemented with an equation specifying the variation of *u(x,t)* with the respect to *x*. If the hull girder can be considered rigid, then *u(x,t)* can be written

$$u(x, t) = w(t) - (x - x_{cog})\theta(t) \tag{4.65}$$

where the *heave w(t)* is the vertical motion of the centre of gravity x_{cog} and the *pitch* $\theta(t)$ is the angle of rotation about a axis parallel to the y-axis. The centre of gravity

$$x_{cog} = \int_L m(x)xdx \left[\int_L m(x)dx \right]^{-1} \tag{4.66}$$

For very flexible or very fast ships, the elastic global response of the ship subjected to wave loads might be of some importance, especially as regard fatigue damage. In such cases Eq. (4.65) must be replaced by a suitable beam formulation of the hull girder. This is considered in details in Chapter 6.

Substitution of Eq. (4.65) into the equilibrium equations (4.63) and (4.64) yields

$$M\ddot{w} = \int_L q_H(x, w, \theta, t)dx \tag{4.67}$$

$$J\ddot{\theta} = -\int_L \left(x - x_{cog}\right) q_H(x, w, \theta, t)dx \tag{4.68}$$

where displacement M and mass moment of inertia J are

$$M = \int_L m(x)dx \tag{4.69}$$

$$J = \int_L \left(x - x_{cog}\right)^2 m(x)dx \tag{4.70}$$

and an overdot, as before, denotes differentiation with respect to time.

The right hand sides in Eqs. (4.67) and (4.68) can be expressed explicitly in the heave *w*, pitch θ and the wave elevation *h(x,t)* using an appropriate formulation of q_H like Eq. (4.58). In this formulation the coefficients $m_w(x,\omega_e), N(x,\omega_e)$ and $B_0(x)$ must be known. Furthermore, the relative motion $\tilde{z}(x,t)$, Eq. (4.20), becomes

$$\tilde{z}(x,t) = w(t) - (x - x_{cog})\theta(t) - \kappa(x)h(x,t) \tag{4.71}$$

Inserting Eq. (4.71) and the wave elevation *h(x,t)*, Eq. (4.55), with *y=0* and $\epsilon = 0$ in Eq. (4.58) and carrying out the differentiations according to Eq. (4.57) yields

$$\begin{aligned} q_H(x,t) = & - \Big[m_w(x)(\ddot{w} + 2V\dot{\theta} - \bar{x}\ddot{\theta}) \\ & + (N(x) - Vm'_w(x))(\dot{w} + V\theta - \bar{x}\dot{\theta}) \\ & + \rho g B_0(x)(w - \bar{x}\theta)\Big] \\ & + \kappa(x)a\Big[(\rho g B_0(x) - m_w(x)\omega^2)\cos(k_e x - \omega_e t) \\ & + (N(x) - Vm'_w(x))\,\omega\sin(k_e x - \omega_e t)\Big] \end{aligned} \tag{4.72}$$

where $(\,)'$ denotes differentiation with respect to *x* and where

$$\bar{x} = x - x_{cog} \tag{4.73}$$

$$k_e = k\cos\beta \tag{4.74}$$

Substitution of Eq. (4.72) into the equilibrium equations (4.67)-(4.68) and collecting all terms depending on the heave *w* and pitch θ on the left hand sides yield two coupled second order ordinary differential equations with constant coefficients:

$$
\begin{aligned}
&\left(M + I_0\right)\ddot{w} + \left(N_0 - V[m_w]\right)\dot{w} + \rho g A_w w - I_1\ddot{\theta} \\
&+ \left(VI_0 + V\,[m_w\bar{x}] - N_1\right)\dot{\theta} + \left(VN_0 - V^2[m_w] - \rho g S_y\right)\theta \\
&= a\left(F_1 \cos\omega_e t + F_2 \sin\omega_e t\right)
\end{aligned}
\tag{4.75}
$$

and

$$
\begin{aligned}
&- I_1\ddot{w} - \left(N_1 + VI_0 - V[m_w\,\bar{x}]\right)\dot{w} - \rho g S_y w \\
&+ \left(J + I_2\right)\ddot{\theta} + \left(N_2 - V\left[m_w\,\bar{x}^2\right]\right)\dot{\theta} \\
&+ \left(- VN_1 - V^2 I_0 + V^2[m_w\,\bar{x}] + \rho g I_{yy}\right)\theta \\
&= a\left(F_3 \cos\omega_e\, t + F_4 \sin\omega_e\, t\right)
\end{aligned}
\tag{4.76}
$$

where the following constants have been introduced:

$$
I_i\,(\omega_e) = \int_L m_w(x, \omega_e)(x - x_{cog})^i dx \tag{4.77}
$$

$$
N_i\,(\omega_e) = \int_L N(x, \omega_e)(x - x_{cog})^i dx \tag{4.78}
$$

together with the water plane area A_w, Eq. (2.82)

$$
A_w = \int_L B_0(x)dx \tag{4.79}
$$

and its first and second moments with respect to the centre of gravity:

$$
S_y = \int_L (x - x_{cog})B_0(x)dx \tag{4.80}
$$

$$
I_{yy} = \int_L (x - x_{cog})^2 B_0(x)dx \tag{4.81}
$$

Furthermore, with $x_{min} \leq x \leq x_{max}$, terms of the type

$$[H(x)] \equiv H(x_{max}) - H(x_{min}) \tag{4.82}$$

appear in the differential equations (4.75) - (4.76) due to possible abrupt termination of the sectional constants at the ends of the ship. These terms are important for relative fast ships with a transom stern, as for instance container ships.

Finally, the four coefficients F_1, F_2, F_3 and F_4 follow from the terms in q_{II} depending on the wave elevation:

$$F_1(\omega_e) = \int_L f_1(x, \omega_e) dx$$

$$F_2(\omega_e) = \int_L f_2(x, \omega_e) dx$$

$$F_3(\omega_e) = -\int_L (x - x_{cog}) f_1(x, \omega_e) dx$$

$$F_4(\omega_e) = -\int_L (x - x_{cog}) f_2(x, \omega_e) dx$$

where

$$f_1 = \kappa(\rho g B_0 - m_w \omega^2) \cos k_e x + \kappa\omega(N - Vm'_w) \sin k_e x$$

$$f_2 = \kappa(\rho g B_0 - m_w \omega^2) \sin k_e x - \kappa\omega(N - Vm'_w) \cos k_e x$$

As q_{II} depends linearly on the wave amplitude a, the right hand sides in Eqs. (4.75)-(4.76) are as shown linear in a, too.

4.3.1 Response Amplitude Operator

The solution to Eqs. (4.75)-(4.76) can be written as the sum of a particular and the homogeneous solution. The homogeneous solution depends on the initial conditions and will decay exponentially with time. It is therefore omitted as the response to a steady state excitation $h(x, t)$ is requested here. Later in Section 4.6 transient motions are discussed. The particular solution will have the same frequency ω_e as the right hand sides:

$$w(t) = a\Phi_w(\omega_e)\cos(\omega_e t + \epsilon_w(\omega_e)) \tag{4.83}$$

$$\theta(t) = a\Phi_\theta(\omega_e)\cos(\omega_e t + \epsilon_\theta(\omega_e)) \tag{4.84}$$

where the amplitude functions Φ_w and Φ_θ and the phase lags ϵ_w and ϵ_θ are determined by substitution of Eqs. (4.83) - (4.84) into Eq. (4.75) - (4.76). As the excitation is given by the wave elevation,

$$h(x,t) = a\cos(k_e x - \omega_e t)$$

the *amplitude functions* Φ_w and Φ_θ are closely related to the complex transfer functions $K_w(\omega_e)$ and $K_\theta(\omega_e)$ associated with the heave and pitch motion, respectively. From Eq. (4.34) it follows that

$$\Phi_w(\omega_e) = |K_w(\omega_e)| \tag{4.85}$$

$$\Phi_\theta(\omega_e) = |K_\theta(\omega_e)| \tag{4.86}$$

Often the amplitude functions Φ are named *frequency response functions.*

The square of the amplitude function

$$\Phi^2(\omega_e) = |K(\omega_e)|^2 = K^*(\omega_e)K(\omega_e) \tag{4.87}$$

is denoted the *response amplitude operator**.

Having determined the heave *w(t)* and pitch motion $\theta(t)$ for the ship sailing in a regular wave, the hydrodynamic force $q_H(x,t)$ per unit length, Eq. (4.72), is also known and can be written on the form:

$$q_H(x,t) = a\Phi_{q_H}(x,\omega_e)\cos(\omega_e t + \epsilon_{q_H}(x,\omega_e)) \tag{4.88}$$

as seen from Eq. (4.72). The total force q per unit length becomes

$$\begin{aligned} q(x,t) &= -m(x)(\ddot{w}(t) - (x - x_{cog})\ddot{\theta}(t)) + q_H(x,t) \\ &= a\Phi_q(x,\omega_e)\cos(\omega_e t + \epsilon_q(x,\omega_e)) \end{aligned} \tag{4.89}$$

where the first term is the inertia (D'Alembert) force. Integration of $q(x,t)$ yields the *vertical wave-induced shear force* $Q_z(x,t)$

* In the literature Φ is also often denoted the response amplitude operator (RAO)! Therefore, care has to be taken when interpretating results, quoted in the literature.

$$
\begin{aligned}
Q_z(x,t) &= -\int_{x_{\min}}^{x} q(\tilde{x},t)d\tilde{x} \\
&= a\Phi_Q(x,\omega_e)\cos(\omega_e t + \epsilon_q(x,\omega_e))
\end{aligned}
\tag{4.90}
$$

and the *vertical wave bending moment*

$$
\begin{aligned}
M_y(x,t) &= -\int_{x_{\min}}^{x} (x-\tilde{x})q(\tilde{x},t)d\tilde{x} \\
&= a\Phi_M(x,\omega_e)\cos(\omega_e t + \epsilon_M(x,\omega_e))
\end{aligned}
\tag{4.91}
$$

using the same sign convention as in Chapter 2. As seen from Eqs. (4.88)-(4.91) any response R depending linearly on the wave elevation can be written on the same form, given by the wave amplitude a, an amplitude function Φ_R, a phase lag ϵ_R and the frequency of encounter ω_e:

$$
R(t) = a\Phi_R(\omega_e)\cos(\omega_e t + \epsilon_R(\omega_e)) \tag{4.92}
$$

The validity of the results can be established by experiments in model basins, where regular waves can be generated. Extensive comparisons have been performed in the past, see for instance Flokstra (1974) and generally, the various hydrodynamic strip theory formulations yield quite accurate results. As seen from Eqs (4.75)-(4.76) the forward speed V may influence the results quite significantly and usually the applications of strip theories are restricted to forward speeds V corresponding to a Froude number F_n

$$
F_n = \frac{V}{\sqrt{gL}} < 0.3 \tag{4.93}
$$

However, recent experiments, Naonosuke et al. (1993) have shown very good agreement with strip theory predictions at Froude numbers up to 0.6. Also comparisons with results from three-dimensional linear formulations have indicated that strip theory with its assumption of a two-dimensional flow pattern may be used for load predictions even for very fast slender monohull ships.

Example 4.3.1

For a homogeneous loaded prismatic barge with length L, breadth B and draft T, the equations (4.75)-(4.76) can be solved analytically. The constants in the equations become

$$I_0 = m_w L \quad ; \quad I_1 = 0 \quad ; \quad I_2 = \frac{1}{12} m_w L^3$$

$$N_0 = NL \quad ; \quad N_1 = 0 \quad ; \quad N_2 = \frac{1}{12} NL^3$$

$$A_w = BL \quad ; \quad S_y = 0 \quad ; \quad I_{yy} = \frac{1}{12} BL^3$$

$$M = mL \quad ; \quad J_y = \frac{1}{12} mL^3$$

$$[m_w] = 0 \quad ; \quad [m_w \bar{x}] = m_w L \quad ; \quad [m_w \bar{x}^2] = 0$$

Assuming head sea, $\beta = 180^{\circ}$, and a wave crest at the centre of gravity at t=0, the coefficients F_1, F_2, F_3 and F_4 become

$$F_1 = \frac{2}{k}\kappa\left(\rho g B - m_w \omega^2\right)\sin\frac{kL}{2}$$

$$F_2 = -\frac{2}{k}\kappa\omega N \sin\frac{kL}{2}$$

$$F_3 = \kappa\omega N \frac{2}{k^2}\left(\sin\frac{kL}{2} - \frac{kL}{2}\cos\frac{kL}{2}\right)$$

$$F_4 = \kappa\left(\rho g B - m_w \omega^2\right)\frac{2}{k^2}\left(\sin\frac{kl}{2} - \frac{kL}{2}\cos\frac{kL}{2}\right)$$

With these coefficients, Eqs. (4.75)-(4.76) can be written

$$(1 + \eta)M\ddot{w} + NL\dot{w} + \rho gBLw + 2\eta VM\dot{\theta} + VNL\theta = a(F_1 \cos\omega_e t + F_2 \sin\omega_e t) \tag{4.94}$$

and

$$(1 + \eta)M\ddot{\theta} + NL\dot{\theta} + \rho gBL\theta = a\frac{12}{L^2}(F_3 \cos\omega_e t + F_4 \sin\omega_e t) \tag{4.95}$$

where η is the ratio between the added mass of water and the displacement of the ship:

$$\eta = \frac{I_o}{M} = \frac{m_w}{m}$$

Substitution of Eqs. (4.83)-(4.84) into Eqs. (4.94)-(4.95) yields the amplitude functions Φ_w and Φ_θ after some algebra. The results are shown in the Figure 4.10 as function of λ/L, where the wave length $\lambda = 2\pi/k$. This is the only independent wave parameter as

$$\omega^2 = kg = \frac{2\pi}{\lambda}g$$

$$\omega_e = \omega - kV\cos\beta = \omega + kV$$

The following values have been assumed:

$$L/B = 8 \quad ; \quad B/T = 2$$

$$m_w = m \quad ; \quad N = \rho g^2 (0.45)^2 / (\omega_e)^3$$

The values for m_w and N are reasonable values, except for very long waves.

As the right hand side of Eq. (4.95) is zero for

$$\sin\frac{kL}{2} - \frac{kL}{2}\cos\frac{kL}{2} = 0$$

the amplitude function Φ_θ will be zero for the value of λ/L satisfying:

$$\tan\frac{kL}{2} = \frac{kL}{2} \Rightarrow \frac{kL}{2} = 4.4934094$$

or

$$\frac{\lambda}{L} = 0.699$$

irrespectively of the value of all other parameters (V, L, B, T, m_w etc.) for a homogeneous loaded box-shaped ship.

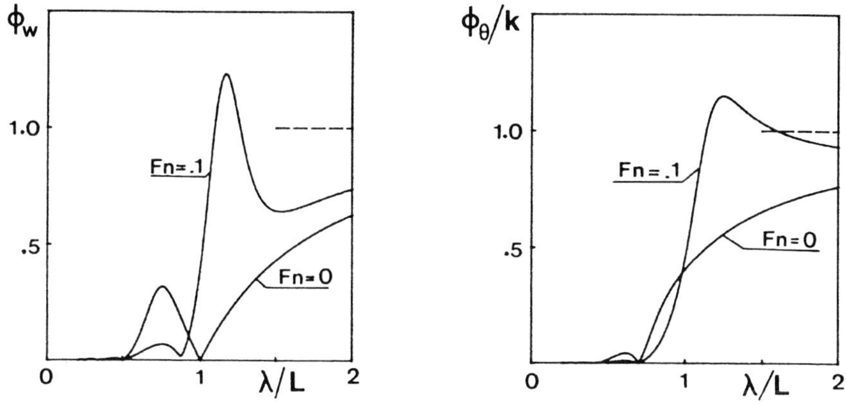

Figure 4.10 *Amplitude functions Φ_w and Φ_θ for heave and pitch of a homogeneous loaded barge.*

For the heave motion, the coupling between heave and pitch through Eq. (4.94) implies that a similar type of relation does not exist. However, for zero forward speed, the coupling terms disappear and hence $\Phi_w = 0$ for $\sin kL/2 = 0$ or $\lambda/L = 1$, see Figure 4.10.

For very long wave length, $\lambda >> L$, the ship will behave as cork moving with the wave surface. Hence, in general

$$\Phi_w \rightarrow 1 \quad \text{for} \quad \lambda/L \rightarrow \infty$$

$$\Phi_\theta \rightarrow k \quad \text{for} \quad \lambda/L \rightarrow \infty$$

as the pitch will tends towards the wave slope given by

$$\frac{dh}{dX} = -ak\sin(kX - \omega t + \epsilon)$$

For $\lambda/L \to 0$ the wave number $k \to \infty$. As the coefficients F_1, F_2, F_3 and F_4 are inversely proportional to k the right hand sides in Eq. (4.94)-(4.95) will tend to zero such that

$$\Phi_w \to 0 \quad \text{for} \quad \lambda/L \to 0$$

$$\Phi_\theta k \to 0 \quad \text{for} \quad \lambda/L \to 0$$

As seen in Figure 4.10 large variations in Φ_w and Φ_θ occur for wave lengths comparable with the length of the ship. With the solutions Φ_w and Φ_θ inserted in Eq. (4.89) the total force per unit length $q(x,t)$ is obtained. The result is shown in Figure 4.11 for three different values of λ/L. The forward speed is taken as zero, but the results only depend on the forward speed through the frequency of encounter ω_e in the expression for the damping coefficient N. This special result is due to the extreme symmetry in the present example. As usual the amplitude function Φ_q for $q(x,t)$ is shown in non-dimensional form using appropriate normalization parameters.

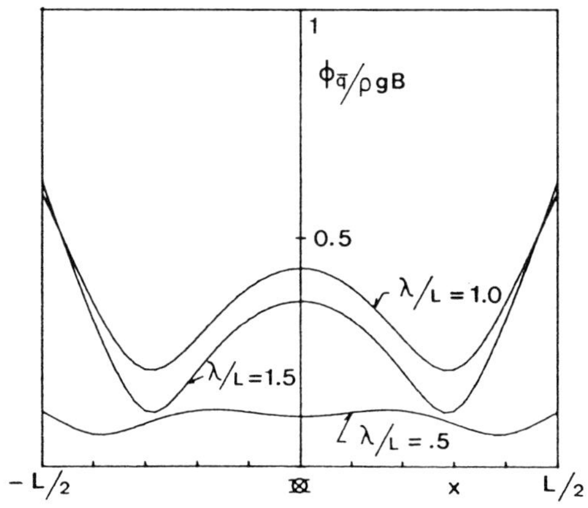

Figure 4.11 *Amplitude function* Φ_q *for the load q along the length of the hull.*

Finally, the vertical bending moment M_y, Eq. (4.91), is determined. Usually, it will be largest around amidship with a smooth decay to zero at the ends of the ship. For the present example a very simple result is obtained for the bending moment amidship, albeit after some algebra:

$$\Phi_M = \frac{\kappa}{k^2}\left(1 - \cos\frac{kL}{2} - \frac{kL}{4}\sin\frac{kL}{2}\right) \cdot (\rho g B - m_w \omega^2)^2 + (N\omega)^2 \qquad (4.96)$$

This result is shown in Figure 4.12 in non-dimensional form. For both very long or very short wave lengths the wave bending moment tends to zero, which also is obvious by physical arguments. It is interesting to compare the Froude-Krylov result ($m_w = N = 0$) with the result Eq. (4.96). The inclusion of dynamic effects through especially the added mass of water m_w significantly reduce the wave bending moment. The result shown in Figure 4.12 is for $V=0$ but as V only enters in Eq. (4.96) through the frequency of encounter ω_e in N, the variation with V is very modest in the present example. This is not generally so. Normally, the bending moment will increase quite significantly when the Froude number $F_n = V/\sqrt{gL}$ exceeds about 0.2 to again reach a stable level for F_n greater than about 0.25. The maximum value of the amplitude function Φ_M is typically in the range of $0.02\,\rho g B L^2$ to $0.035\,\rho g B L^2$.

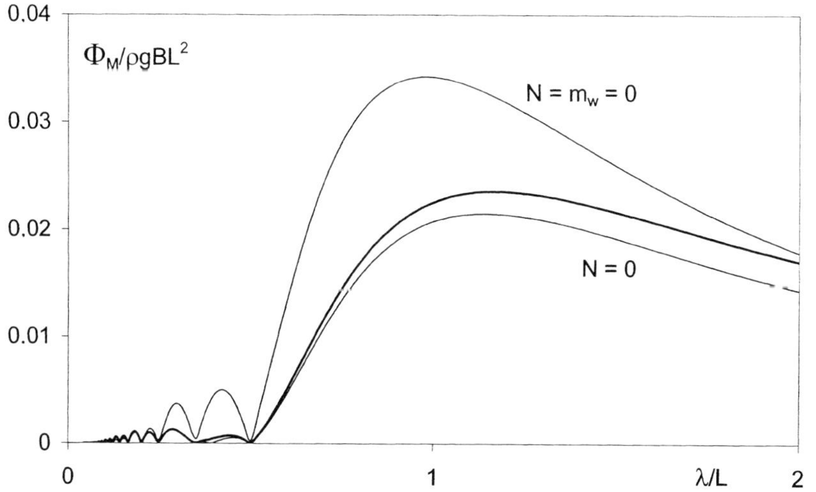

Figure 4.12 *Amplitude function* Φ_M *for the vertical bending moment amidship.*

4.4 LINEAR WAVE RESPONSES IN RANDOM SEAS

It was shown in Section 3.3.1.3 that a stationary stochastic seaway can be represented by a wave elevation $H(X,t)$ given as a sum of sinusoidal functions each with different amplitude, Eq. (3.214). For each of these sinusoidal wave components, the response is conveniently found by Fourier transformation as shown in Section 4.2. The result is given by Eq. (4.34) or more explicitly for ship motion responses by Eqs. (4.83)-(4.84) and Eqs. (4.88)-(4.92).

Due to linearity the response R to the wave elevation given by Eq. (3.214) becomes

$$R(t) = \sum_{i=1}^{n} a_i \, \Phi_R(\omega_{e,i}) \cos\left(\omega_{e,i}t + \epsilon_R(\omega_{e,i})\right) \tag{4.97}$$

where the phase angle ϵ_R includes the deterministic contribution,

$$Arctan\left(\frac{Im(K(\omega_e))}{Re(K(\omega_e))}\right)$$

derived from the transfer function $K(\omega)$, but also the random phase lag, θ_i, from Eq. (3.214). The statistical properties of the response can be derived as in Section 3.3.1.3. The only difference in that the amplitude a_i now is replaced by $a_i \, \Phi_R(\omega_{e,i})$. Hence any statistical moment $E[G(R(t)]$ will be independent of t, such that the response R becomes a stationary stochastic process. Analogous to Eqs. (3.216)-(3.220), the mean value is

$$\mu_R = 0 \tag{4.98}$$

whereas the variance becomes

$$\zeta_{R2} = s_R^2 = \frac{1}{2} \sum_{i=1}^{n} \left(a_i \, \Phi_R(\omega_{e,i})\right)^2 \tag{4.99}$$

Further moments are not needed if n is large as the response then due to the central limit theorem applied to Eq. (4.97) with statistically independent phase lags ϵ_R will be normal distributed like the wave elevation itself. Hence the probability density function p_R for the response becomes

$$p_R(r) = \frac{1}{\sqrt{2\pi}\, s_R} e^{-\frac{1}{2}(r/s_R)^2} \tag{4.100}$$

Finally, the autocorrelation $R_R(\tau)$ takes the form

$$R_R(\tau) = \frac{1}{2} \sum_{i=1}^{n} \left(a_i \, \Phi_R\left(\omega_{e,i}\right)\right)^2 \cos \omega_{e,i} \, \tau \tag{4.101}$$

The wave amplitudes a_i are given by Eq. (3.221) and in the limit $n \rightarrow \infty$, the variance and the autocorrelation then become

$$s_R^2 = \int_0^\infty \Phi_R^2(\omega_e) S(\omega) d\omega \tag{4.102}$$

and

$$R_R(\tau) = \int_0^\infty \Phi_R^2(\omega_e) S(\omega) \cos \omega_e \, \tau \, d\omega \tag{4.103}$$

respectively.

In order to perform the integrations either the frequency of encounter ω_e should be expressed in terms of the wave frequency ω or vice versa using Eq. (4.56) and Eq. (3.253). As illustrated in Example 4.3.1 the response amplitude operator Φ_R^2 is often given as function of λ/L. In such cases s_R^2 is most conveniently obtained as

$$s_R^2 = \sqrt{\frac{\pi g}{2L}} \int_0^\infty \Phi_R^2 \left(\lambda/L\right) S\left(\omega\left(\lambda/L\right)\right)\left(\lambda/L\right)^{-3/2} d\lambda/L \tag{4.104}$$

by use of Eq.(3.202).

In a short-crested seaway characterized by a wave spectral density on the form given by Eq. (3.258), the variance s_R^2 becomes

$$s_R^2 = \int_0^\infty S(\omega) \int_{-\pi}^{\pi} \Phi_R^2(\omega, \mu - \varphi) f(\varphi) d\varphi \, d\omega \qquad (4.105)$$

where μ is the angle between the ship heading and the wind direction. The response amplitude operator $\Phi_R^2(\omega, \beta)$ must then be calculated for a number of heading angles $\beta = \mu - \varphi$ covering the range of φ, see Figure 4.13.

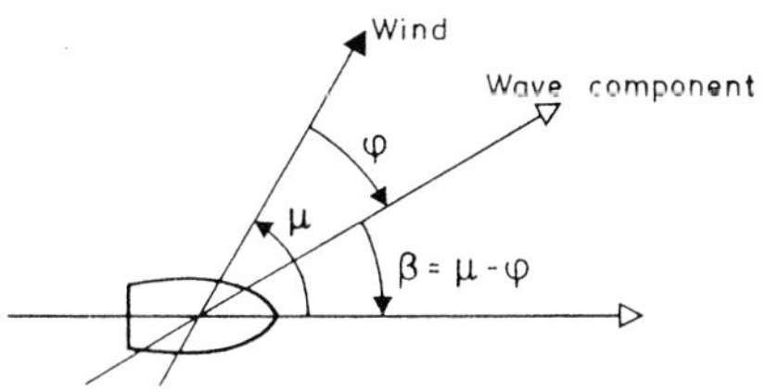

Figure 4.13 *Ship sailing in a short-crested seaway.*

Upcrossing rates and peak distributions for the response are easily obtained using the results in Section 3.2.3 and 3.2.4 for Gaussian processes as the only information needed are the spectral moments:

$$m_{R0} \equiv s_R^2 = \int_0^\infty S_R(\omega) d\omega \qquad (4.106)$$

$$m_{R2} = \int_0^\infty \omega_e^2 \Phi_R^2(\omega_e) S(\omega) d\omega = \int_0^\infty \omega_e^2 S_R(\omega) d\omega \qquad (4.107)$$

$$m_{R4} = \int_0^\infty \omega_e^4 \Phi_R^2(\omega_e) S(\omega) d\omega = \int_0^\infty \omega_e^4 S_R(\omega) d\omega \qquad (4.108)$$

where

$$S_R(\omega) = \Phi_R^2(\omega_e) S(\omega) \qquad (4.109)$$

is denoted the *response spectral density.*

The probability density function for the individual peaks is given by Eq. (3.137) with the spectral bandwidth obtained from Eq. (3.106) using Eqs. (4.106)-(4.108). As the response amplitude operator Φ_R^2 usually is narrow compared to the wave spectral density itself, the Rayleigh distribution, Eq. (3.153), is normally a very accurate approximation for the peak distribution of the response.

The spectral moments m_{Rn} are based on ω_e^n rather than on ω^n, because ω_e reflects the time-varying behavior of the response R in the ship coordinate system, where the response is measured. The zero-upcrossing frequency of the response becomes

$$\bar{\omega}_2 = \sqrt{\frac{m_{R2}}{m_{R0}}} \tag{4.110}$$

Example 4.4.1

For the homogenously loaded barge considered in Example 4.3.1, the response amplitude operator Φ_M^2 for the vertical wave bending moment amidship is given by Eq. (4.96) for the ship sailing in head sea.

The variance s_M^2 becomes

$$s_M^2 = \int_0^\infty \Phi_M^2\left(\frac{\lambda}{L}\right) S(\omega) d\omega$$

and by use of Eq. (4.104) this integral is transformed to an integral over λ/L. Numerical integration using the Pierson-Moskowitz spectrum, given by Eq. (3.238) yields the result shown in Figure 4.14.

For the present example, the forward speed does not influence the result very much, as discussed in Example 4.3.1. The result in Figure 4.14 is for zero forward speed.

Usually, the bending moment is made non-dimensional as in Figure 4.14. Only very seldom, the standard derivation s_M exceeds the value *0.01* $\rho g B L^2 H_S$ for any ship. Notice that s_M is linear in H_s because of the form of the Pierson-Moskowitz spectrum.

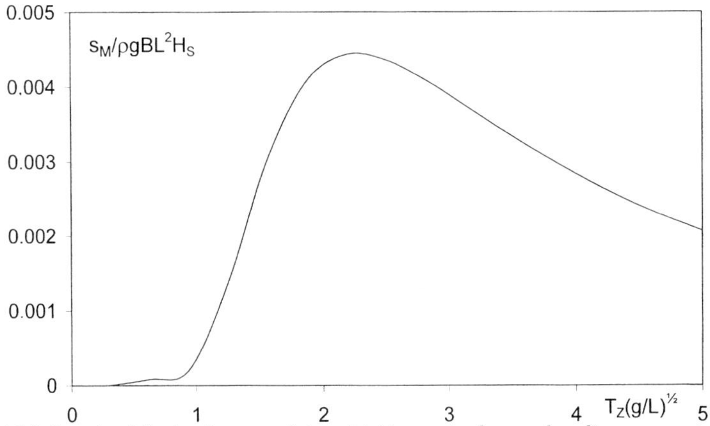

Figure 4.14 *Standard derivation* s_M *of the midship vertical wave bending moment as function of the zero-upcrossing period* T_z *for a homogeneously loaded barge.*

In a long-term analysis considering the ship during its complete operational lifetime, the statistical analysis is performed analogous to the procedure for wave peaks in Section 3.3.3.

The probability density function $p_p^\ell(r)$ for the individual peak values is then obtained as

$$p_P^{\ell}(r) = \frac{1}{\nu(0)} \int_{H_s} \int_{T_z} \int_V \int_{\mu} \nu(0|\mu, V, H_s, T_z)\, p_P(r, \mu, V, H_s, T_z)\, d\mu\, dV\, dT_z\, dH_s \tag{4.111}$$

where, compared to Eq. (3.261), the ship speed V and the ship heading μ relative to the wind direction have been included, because the standard derivation s_M and hence p_P depend on these parameter. Other parameters may also enter the expression, the most obvious is the ship's loading condition.

Writing

$$p_P(r, \mu, V, H_s, T_z) = p_P(r|\mu, V, H_s, T_z) p(\mu, V, H_s, T_z) \tag{4.112}$$

with the first term on the right hand side given by Eq. (3.137) or Eq. (3.242) using Eqs. (4.106)-(4.108) for the statistical moments, the difficult part is the joint probability density function $p(\mu, V, H_s, T_z)$. Nearly no data are available to estimate this function as it combines weather data, H_s, T_z, with ship operational data, μ, V. Usually the angle μ is taken to be uniformly distributed between 0 and 2π and statistically independent of V, H_s and T_z. Hence

$$p(\mu, V, H_s, T_z) = \frac{1}{2\pi} p(V, H_s, T_z) \quad ; \quad \mu \epsilon [0, 2\pi] \tag{4.113}$$

It would clearly be wrong to assume that the speed V is independent of H_s and T_z as the ship has to reduce its speed in heavy sea, see also Figure 4.17. A reasonable approximation could be to take V as a deterministic function of H_s:

$$V = V(H_s) \tag{4.114}$$

such that V will be between the service speed V_{ser} and a minimum steering speed V_{min}, depending on H_s. This relation may require calculation of the response amplitude operator Φ_R^2 for many values of V. To avoid this, an alternative to the relation Eq. (4.114) might be, Jensen and Dogliani (1996)

$$V = \begin{cases} V_{min} & \text{with a probability } \alpha = \alpha(H_{s0}) \\ V_{ser} & \text{with a probability } 1 - \alpha(H_{s0}) \end{cases} \tag{4.115}$$

where the probability α depends on a threshold value $H_s = H_{s0}$

Using Eq. (4.114), Eq. (4.111) becomes

$$p_P^{\ell} = \frac{1}{2\pi\, \nu(0)} \int_{H_s} \int_{T_z} \int_0^{2\pi} \nu(0\,|\,\mu, V(H_s), H_s, T_z)\, p_P(r|\,\mu, V(H_s), H_s, T_z)\, p(H_s, T_z)\, d\mu dT_z dH_s \tag{4.116}$$

whereas Eq. (4.115) leads to

$$p_p^\ell(r) = \frac{1}{2\pi\,\nu(0)} \int_{T_z}\int_{\mu} \left[\alpha_1 \int_0^{H_{s0}} f(V_{\min})dH_s + (1-\alpha_1)\int_0^{H_{s0}} f(V_{ser})dH_s \right.$$
$$\left. + \alpha_2 \int_{H_{s0}}^{H_{s\max}} f(V_{\min})dH_s + (1-\alpha_2)\int_{H_{s0}}^{H_{s\max}} f(V_{ser})dH_s \right] d\mu dT_z \qquad (4.117)$$

with

$$\alpha_1 = \alpha(H_s \le H_{s0})$$

$$\alpha_2 = \alpha(H_s > H_{s0})$$

and

$$f(V) \equiv \nu(0\,|\,\mu, V, H_s, T_z)\, p_P(r\,|\,\mu, V, H_s, T_z) p(H_s, T_z)$$

The procedure, Eqs. (3.267)-(3.269) to obtain long term extreme values for the waves is equally valid for ship responses, provided account is taken as above for the additional dependence on V and μ. The use of a Weibull fitting for $p_p^\ell(r)$ to the numerical integrated values obtained by Eq. (4.116) or (4.117) is often quite accurate. An example is given in Figure 4.15.

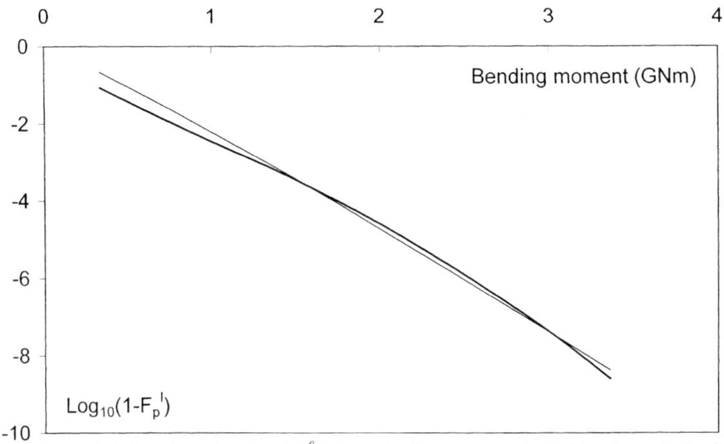

Figure 4.15 *Probability distribution $F_p^\ell(m)$ of the long-term individual peaks in the vertical wave bending moment amidship in a container ship, (L= 270m). Thick line: direct calculation, thin line: Weibull approximation.*

With $N = 10^8$ as a reasonable value for the total number of response peaks during the lifetime of the ship, the most probable value of the response occurs for

$$\log_{10}\left(1 - F_P^\ell(m)\right) = \log_{10}\frac{1}{N} = -8$$

using Eq. (3.267). This is the reason for the chosen horizontal axis in Figure 4.15.

Numerically, Eq. (4.111) is usually carried as

$$p_P^\ell(v) = \frac{1}{N_z}\sum_{i=1}^{M} N_{zi}\, p_P\,(r\,;\,i)$$

where

$$N_{zi} = v(0\,|\,\mu, V, H_s, T_z)\; Tp\,(\mu, V, H_s, T_z) d\mu dV dH_s dT_z$$

is the number of zero upcrossing in each of the M stationary conditions $(\mu, V, H_s, T_z) = (\mu, V, H_s, T_z)_i \quad ; \quad i = 1, 2, ...M$. Furthermore,

$$p_P(r; i) \equiv p_P\,(r\,|\,\mu, V, H_s, T_z)$$

is the corresponding probability density function for the individual peaks. Similarly the probability distribution becomes

$$F_P^\ell(r) = \frac{1}{N_z}\sum_{i=1}^{M} N_{zi}\, F_P\,(r; i)$$

where for Rayleigh distributed individual peaks

$$F_P\,(r; i) \simeq 1 - \exp\left[-\frac{1}{2}\left(\frac{r}{s_{Ri}}\right)^2\right]$$

with

$$s_{Ri} = s_R\,(\mu, V, H_s, T_z)$$

The probability distribution $F_N^\ell(v)$ for the largest peak amoung N peaks is finally derived from the assumption of statistically independent peaks

$$
\begin{aligned}
F_N^{\ell}(r) &\simeq \left[F_P^{\ell}(r)\right]^{N_z} \\
&= \left[\frac{1}{N_z}\sum_{i=1}^{M} N_{zi}\, F_P\,(r;i)\right]^{N_z} \\
&\simeq \left[1 - \frac{1}{N_z}\sum_{i=1}^{M} N_{zi}\exp\left[-\frac{1}{2}\left(\frac{r}{s_{Ri}}\right)^2\right]\right]^{N_z} \\
&\simeq \exp\left[-\sum_{i=1}^{M} N_{zi}\exp\left[-\frac{1}{2}\left(\frac{r}{s_{Ri}}\right)^2\right]\right]
\end{aligned}
$$

By this formulation numerical problems related to a power $N \simeq 10^8$ are avoided.

4.5 SLAMMING AND GREEN WATER ON DECK

In addition to the linear wave-induced forces considered in the previous sections, two other wave-induced loads need to be addressed. Both are non-linear functions of the relative motion of the ship and can therefore be calculated when the statistics of the relative motion is known.

Figure 4.16 *Bow flare slamming in heavy sea. Photograph by Lauritzen Group.*

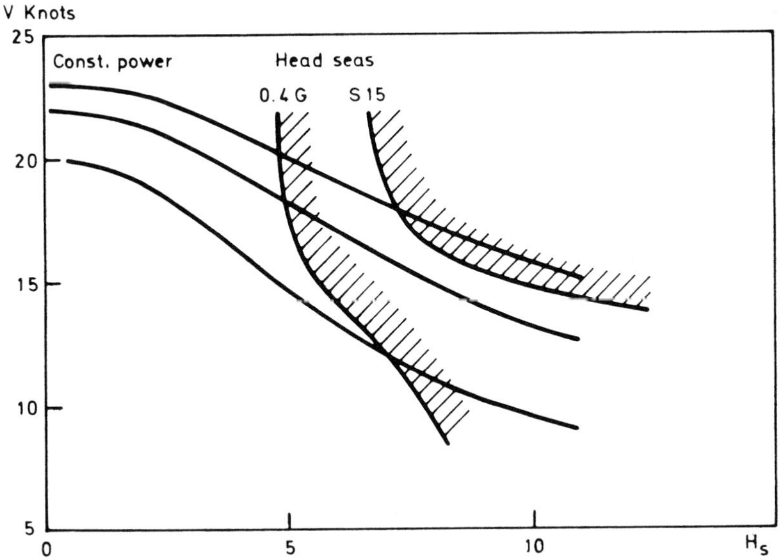

Figure 4.17 *Speed reduction scheme in stationary sea states.* 0.4G: *Vertical acceleration greather than* 0.4g *every 15 min;* S 15: *Slamming occurs every 15 min. Adapted from SSPA.*

Measurements have shown that if the forward part of the ship hits the water surface with a sufficient high relative velocity after a bow emergence, then a high impulsive loads, a slam, will be experienced. Figure 4.16 gives an impression of such event. This phenomenon can lead to structural damage and hence the captain will usually reduce the speed or change the heading if the number or the severity of the slams become too high. An example of such procedure is shown in Figure 4.17. As seen this voluntary speed reduction is usually larger in heavy sea than the automatic reduction due to added resistance in waves. The opposite to a slam is deck submergence when a wave runs over the deck. This may cause severe damage to deck cargo, deck equipment and even to the deck house.

In the following a statistical treatment of these two events is given.

4.5.1 Slamming

Two conditions must be satisfied to experience a slam during a wave cycle. First the relative vertical motion

$$z(x,t) \equiv w(t) - x\,\theta(t) - h(x,t) \tag{4.118}$$

should be larger than the draft $T(x)$ at the section considered. Secondly, the relative velocity Dz/Dt must be larger than a threshold velocity v_0:

$$z_t \equiv \frac{Dz}{Dt} = \frac{\partial z}{\partial t} - V\frac{\partial z}{\partial x} \geq v_0 \tag{4.119}$$

Due to the linear relations, Eqs. (4.118)-(4.119), the relative motion z and its time derivative z_t can be written on the form Eq. (4.92), where the amplitude functions Φ_z and Φ_{zt} are obtained by substitution of Eqs. (4.83), (4.84) and (4.55) into Eqs. (4.118)-(4.119).

In a stationary stochastic seaway both z and z_t will therefore be normal distributed with zero mean values and standard deviations s_z and s_{zt} given by Eq. (4.102):

$$s_z^2(x) = \int_0^\infty \Phi_z^2(\omega, x)\, S(\omega)\, d\omega \tag{4.120}$$

$$s_{z_t}^2(x) = \int_0^\infty \Phi_{z_t}^2(\omega, x)\, S(\omega)\, d\omega \tag{4.121}$$

or, for short-crested seaways, by Eq. (4.105).

The probability that the individual peak values z_p of the relative motion exceeds the draft $T(x)$ is

$$P\left(Z_p > T(x)\right) = 1 - F_z(T(x)) \simeq \exp\left[-\frac{1}{2}\left(\frac{T(x)}{s_z(x)}\right)^2\right] \tag{4.122}$$

provided the spectral bandwidth ε_z, Eq. (3.106) with the moments Eqs. (4.106)- (4.108) is sufficiently small ($\varepsilon_z \leq 0.6$) to apply the Rayleigh distribution, Eq. (3.153), for the individual peak values.

The second criteria, Eq. (4.119), becomes

$$P\left(\frac{DZ}{Dt} > \nu_0\right) = 1\, F_{z_t}(\nu_0) \simeq \exp\left[-\frac{1}{2}\left(\frac{\nu_0}{s_{z_t}(x)}\right)^2\right] \tag{4.123}$$

in a stationary seaway, analogous to Eq. (4.122). As $Z(t)$ and $DZ(t)/Dt$ are statistically independent processes, see Eq. (3.108), the probability to experience a slam becomes

$$P(slam) = P\left(Z_p > T(x)\right)P\left(\left.\frac{DZ}{Dt}\right|_{Z=Z_p} > \nu_0\right) = \exp\left[-\frac{1}{2}\left\{\left(\frac{T(x)}{s_z(x)}\right)^2 + \left(\frac{\nu_0}{s_{z_t}(x)}\right)^2\right\}\right] \tag{4.124}$$

The number N_s of slams within a time period t can be determined as

$$N_s = \nu(0)t\, P(slam) \tag{4.125}$$

where $\nu(0)$ is the zero-upcrossing rate for $z(t)$:

$$\nu(0) = \frac{1}{2\pi}\sqrt{\frac{m_{2z}(x)}{s_z^2(x)}} \tag{4.126}$$

The sectional force $q_{sl}(x, t)$ associated with a slam has been found to be approximately proportional to the relative velocity Dz/Dt:

$$q_{sl}(x, t) = \alpha\left(\frac{Dz}{Dt}\right)^2 \tag{4.127}$$

By comparison with Eq. (4.58), it is seen that q_{sl} has the same form as the *momentum slamming force*

$$q_{ms}(x, t) = \frac{Dm_w}{Dt}\frac{D\tilde{z}}{Dt} = \frac{\partial m_w}{\partial \tilde{z}}\left(\frac{D\tilde{z}}{Dt}\right)^2 \tag{4.128}$$

to be included in Eq. (4.58) in a non-linear generalization, in which m_w and all other coefficients depend on the instantaneous relative immersion z or, to be consistent with Eq. (4.58), $\tilde{z}$. A momentum slamming force can therefore easily be included in a non-linear time-domain solution of the equilibrium equations (4.63)-(4.64). Such a procedure is discussed in Section 4.7. It should be noticed that the force only act during downward motions ($Dz/Dt < 0$) and that the force is proportional with the change of added mass m_w with relative immersion $\tilde{z}$. This formulation covers quite well what is known as bow flare slamming, experienced in ships with a large bow flare. An other type of slamming is bottom slamming, taking place when a flat bottom hits the water surface. This is relevant for e.g. a tanker or a bulk carrier. Here the momentum slamming model cannot be used, because $\partial m_w/\partial z$ is not defined at the instant of slamming. Several empirical formulations have been proposed, see e.g. Bishop and Price (1979). Most of them apply the relative motion formulation, Eq. (4.127). More recently, Zhao and Faltinsen (1993) have developed a rational formulation for slamming loads covering all practical cases and including pressure distributions over the slamming area. Here, however, no further consideration of the slamming pressure will be given as the resulting sectional forces usually are sufficient accurately dealt with by the momentum slamming force $q_{ms}(x, t)$, defined by Eq. (4.128).

4.5.2 Green Water on Deck

Completely analogous to slamming green water will appear on the deck if the relative immersion is larger than the freeboard $F(x)$. Hence, at a section $X=x$, the probability of green water on deck becomes

$$P_{gw} \equiv P(\text{green water on deck at } X = x) = \exp\left[-\frac{1}{2}\left(\frac{F(x)}{s_z(x)}\right)^2\right] \tag{4.129}$$

and the number N_ν of deck wetnesses during a time period t is given by

$$N_v = \nu(0)t\,P_{gw} \tag{4.130}$$

The loading on the deck plates due to green water is very difficult to asses, because actually the water flow has a very complicated pattern, including spays. For applications in the present strip theory formulation, the following formula for the sectional force $q_{gw}(x,t)$ due to green water on deck has been found useful, Wang et al. (1998):

$$q_{gw}(x,t) = -\,g m_{gw}(x,t) - \frac{D}{Dt}\left[m_{gw}(x,t)\frac{Dz_e}{Dt}\right] \tag{4.131}$$

Here m_{gw} is the sectional mass of water on deck and z_e is a modified relative vertical motion, depending on z and the Smith correction factor κ.

The first term in Eq. (4.131) is simply the gravity force due to the mass m_{gw}, whereas the last term is analogous to the momentum slamming force q_{ms}, Eq. (4.128). Because m_{gw} depends on z, the green water load can only been included in a time-domain solution of the equilibrium equation (4.63)-(4.64).

4.6 TRANSIENT AND NON-LINEAR HYDRODYNAMICS

The frequency-domain formulation Eq. (4.58), considered in Section 4.3 is very powerful, because the extreme value analysis easily can be carried out as shown in Section 4.4. However, the analysis is based on two assumptions. The first is that the load depends linearly on the wave height and the second is that the wave elevation and associated wave kinematics have a sinusoidal variation. Thereby, neither non-linear or transient problems can in general be solved in the frequency-domain. In the following the application of time-domain formulation to these problems is shortly described.

4.6.1 Transient Problems

Transient hydrodynamic problems may be related to for instance collision or grounding accidents where the interest will be on the motions and sectional forces for a ship following an external load acting at, say, time $t = 0$. Hence, the motion of the vessel prior to $t = 0$ can be assumed known. In the present case where the emphasis is on the vertical motion, then a grounding at $t = 0$ of a ship in still water will imply that

$$\tilde{z}(x,t) = \begin{cases} 0 & t < 0 \\ u(x,t) & t \geq 0 \end{cases} \tag{4.132}$$

and thereby the expression $q_H(x,t)$ for the sectional hydrodynamic load, Eq. (4.60) becomes

$$q_H(x,t) = -\frac{D}{Dt}\left[m_w^{\infty}(x)\frac{Du(x,t)}{Dt} + \int_0^t \frac{Du(x,t-\tau)}{D\tau}k''(x,\tau)d\tau\right] - \rho g B_0 u(x,t) \tag{4.133}$$

with D/Dt, $m_w^\infty(x)$ and $k''(x,\tau)$ given by Eq. (4.57), (4.50) and (4.61), respectively. The range of the convolution integral is now reduced from infinity to $[0, t]$, greatly facilitating a numerical evaluation.

The unknown vertical motion $u(x,t)$ is found by vertical force and momentum equilibrium analogous to the procedure described in Section 4.3, but including the grounding force. The solution $u(x,t)$; $t > 0$ can be found numerically either by a time integration algorithm or by model superposition. Short descriptions of these methods are given in Chapter 6 dealing with hull girder vibrations, whereas an extensive discussion of the transient grounding scenario can be found in Simonsen (1997). Usually, the convolution integral poses no problem in these cases, because the integration is restricted to a small range $[0,t]$. In other cases, $\tilde{z} \neq 0$ for $t < 0$ and care has then to be taken when the upper (infinity) boundary on the convolution integral is replaced by a finite time T. If T is too small then the convolution integral becomes inaccurate whereas a too large T will increase the computational effort. It has here to be recognized, that the integral must be evaluated in full at any new time step in the solution procedure, because of the two different arguments $t - \tau$ and τ in the integral.

This problem is encountered when dealing with transient loads like slamming and green water on deck for a ship sailing in a random seaway.

4.6.2 Non-linear Hydrodynamic Analysis

Generally, the vertical motion characteristics of ships are predicted quite well by the linear strip theory but significant non-linearities are observed in measurements of the sectional forces in the hull girder.

Two examples are shown i Figure 4.18 and Figure 4.19. Both of them concern the vertical bending moment amidships which is the most important wave load parameter in the design of ships larger than 100 m (i.e. dominated by beam bending rather than local pressure on the hull platings).

Figure 4.18 shows the probability distribution functions for the peak (sag) and trough (hog) values of the bending strain obtained from measurements in a container vessel during an operational period of about 20 minutes. A sea state can normally be assumed stationary within such a short period and the theoretical linear distribution function will be the same for both the peak and trough values and close to a Rayleigh distribution. However, as seen from the figure, non-linearities tend to increase the sagging bending moment significantly and slightly decrease the hogging bending moment.

The same trend is noted in Figure 4.19, showing bending stresses obtained from measurements in a specific class of frigates. Altogether measurements covering an operational period of 9 years are included. The measured results have been obtained in different sea states, yielding a distribution function far from a Rayleigh distribution, and more resembling extreme value distributions of the Gumbel or Weibull type. However, also in this case a linear analysis will lead to exactly the same probability distribution for the sagging and hogging stresses.

In design procedures the significant difference observed between the magnitude of the sagging and hogging bending moment at the same probability level must be accounted for and in the following an outline is given of two non-linear strip theories capable of predicting at least some of these non-linearities. The methods are illustrated by recent calculations for the wave bending moment in different vessels both in severe stationary seaways and covering the whole operational life of the ship.

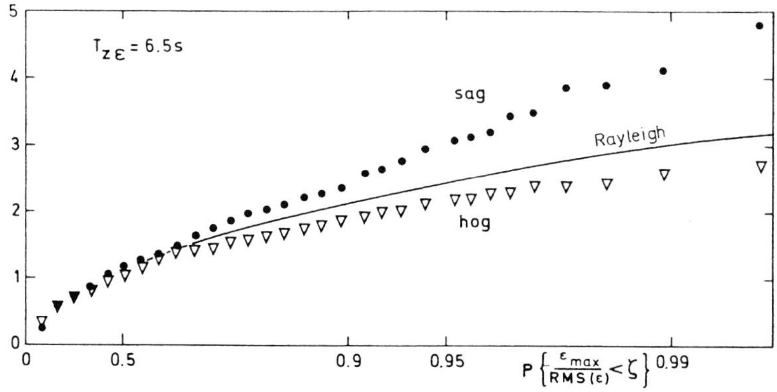

Figure 4.18 *Short term statistical representation of the peaks ε_{max} in the wave induced bending strain ε derived from Northern Atlantic measurements on CTS TOKYO EXPRESS (1018 GMT Dec 27, 1973). A low pass filter was applied to remove contributions from the 2-node vibration taking place around 5 rad/s. After Hackmann (1979).*

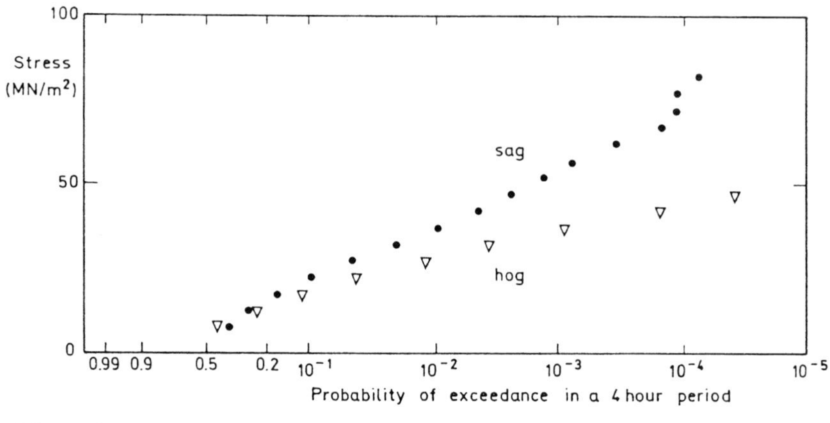

Figure 4.19 *Gumbel plot of long-term deck stress measurements in a narrow beam LEANDER Class frigate. A probability of exceedance of $2 \cdot 10^{-5}$ corresponds to a 20 years' return period. After Clarke (1986).*

Only the vertical wave-induced loads are considered. As two-dimensional strip theories usually yield predictions close to measurement in small amplitude waves, a non-linear generalisation of the linear strip theory therefore seems to be an appropriate starting point for the derivation of a non-linear wave loads analysis.

4.6.2.1 *Second-order, Frequency Domain Ship Theory*

A straightforward non-linear, but empirical generalization of Eq. (4.58) for the sectional hydrodynamic load q_H is simply to assume that the added m_w, the hydrodynamic damping N and the restoring force all depend on the instantaneous sectional draught $T(x) + \tilde{z}(x,t)$, where $T(x)$ is the still water draught used in the linear formulation, Eq. (4.58), when calculating $m_w^{(x)}$, $N(x)$ and $B_0(x)$. Hence,

$$q_H(x,t) = -\frac{D}{Dt}\left[\frac{D}{Dt}\left(m_w(x,\tilde{z},\omega)\frac{D\tilde{z}}{Dt}\right) + N(x,\tilde{z},\omega)\frac{D\tilde{z}}{Dt}\right] + q_{Fkz} \qquad (4.134)$$

where q_{FKz} is the non-linear Froude-Krylov force, Eq. (4.13).

The non-linearities responsible for the difference between the hogging and sagging responses shown in Figure 4.18 and Figure 4.19 are mainly the variations in added mass, hydrodynamic damping and Froude-Krylov force with the relative immersion of the hull. In addition, non-linearities in the wave profile and associated wave kinematics can be important, too. All these non-linearities have asymmetric terms and thus a second order model might be appropriate for moderate non-linearities. This is the basic idea behind the quadratic strip theory, proposed by Jensen and Pedersen (1979).

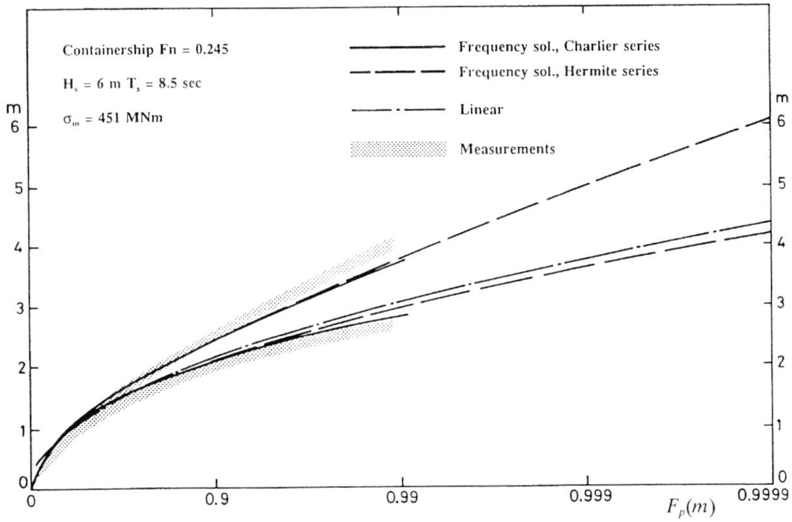

Figure 4.20 *Probability distribution function for peak values of the non-dimensional midship bending moment $(M = M_{mean})/\sigma_M$ where σ_M is the standard deviation of the first-order (linear) contribution. The ship is sailing in head sea.*

A second order formulation of q_H is obtained by a Taylor series expansion in $\tilde{z}$, retaining only terms proportional to $\tilde{z}$ and $\tilde{z}^2$. The derivatives of m_w and N with respect to $\tilde{z}$ are calculated from m_w and N determined at two water lines displaced $\pm\ \delta$ from the still water line:

$$\left.\frac{\partial m_w(x,\omega)}{\partial \tilde{z}}\right|_{\tilde{z}=0} \simeq \frac{m_w(x,\delta,\omega) - m_w(x,-\delta,\omega)}{2\delta}$$

and analogously for N. This procedure has given reasonable results although it does not comply with a consistent second order solution of the ship-wave problem. The restoring term in Eq. (4.58) is replaced by the first two terms in a Taylor series expansion in $\tilde{z}$ of $q_{Fkz}(x,\tilde{z})$, Eq. (4.13), using the second order pressure distribution Eqs. (3.207) and (3.210). Finally, second order Stokes waves, Eq. (3.208), are used as the incident waves. From the equilibrium equations, Eqs. (4.61)-(4.63) a set of first and second order wave-induced responses R can be found in the same form as for instance the wave elevation, Eq. (3.205) and (3.208). Hence, precisely the same statistical methods as discussed in Section 3.3.1.4 can applied. The result can be presented as a probability density function $p(r)$, analogous to the Gram-Charlier expansion, Eq. (3.225), or as statistical moments including the skewness, Eq. (3.226) and the kurtosis. In the second case an approximate probability density function can be defined as described in Section 3.1.3 and illustrated in Example 3.3.3. The agreement with the measured response in Figure 4.18 is very good as shown in Figure 4.20. It is seen that the Gram-Charlier distribution gives results closest to the measurements, but also that this series expansion only is valid up to a probability of exceedance equal to about 1 per cent for the individual peak values. For lower probabilities the series expansion diverges, yielding negative probability densities as discussed in Example 3.3.3. The Hermite transformation, Eq. (3.34), based on the four lowest statistical moments is seen to give nearly as accurate predictions as the Gram-Charlier series, but without numerical problems in the tail of the distribution. Of course, as no measurements are available in the tail, one cannot make conclusions regarding the validity of the transformation method for very small probabilities of exceedance, but generally is is expected to be rather good, Winterstein (1988).

4.6.2.2 *Non-linear Time-domain Strip Theory*

The second order formulation can capture some of the non-linearities, including bow flare slamming, but it cannot be applied for severe non-linearities or when transient load effects become important. For these cases a time-domain formulation must be used. Again taking the starting point from a linear strip theory, either Eq. (4.59) or (4.60), a non-linear empirical formula for $q_H(x,t)$ can be obtained by taking $k'(t)$ and thereby m_w and N as functions of the instaneous draught as before. The equilibrium equations can then be solved by a time-step procedure, but because $m_w(x,\omega,\tilde{z})$, $N(x,\omega,\tilde{z})$ and $k'(x,\tilde{z},t)$ have to be calculated at each time step together with the convolution integral, this procedure is computationally very expensive. One possible simplification is to neglect the non-linearities in m_w and N and hence also in $k'(t)$ assuming the non-linearities in the Froude-Krylov force q_{Fkz} to be the dominating non-linear term. In general reasonable results are obtained, but usually the non-linearities in m_w and N account for 20-30 per cent of the total non-linearities and should therefore not be neglected.

In order to avoid the convolution integral an approximation of m_w and N by rational polynomina in the frequency ω can be applied, Söding (1982), Xia et al. (1998). Thereby, all non-linearities can still be included, albeit in a somewhat empirical way.

In a time-domain solution, the input is a wave sequence, either regular or stochastic. An example of the results from the first type is shown in Figure 4.21. Here the amplitude of the wave-induced vertical bending moment distribution along the length of a vessel (S175, which is a container ship of length L= 175 m) is shown. The wave is sinusoidal with an amplitude a=L/60 and a wave length λ= 1.2 L. The ship is sailing in head sea with a Froude No. F_n=0.25. Results from the second-order frequency-domain and a non-linear time-domain strip theory are shown together with measured values obtained from model test. The agreement is very good and the large difference between the sagging (>0)* and hogging (<0) bending moment is clearly visible. Thus, a linear formulation will be insufficient in this case.

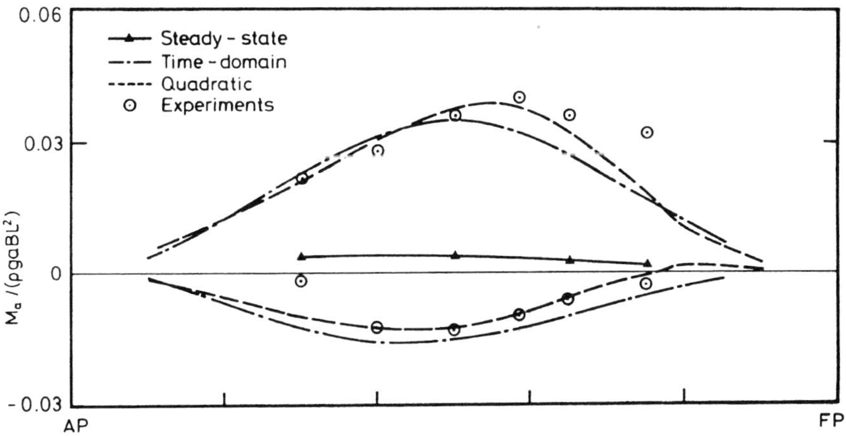

Figure 4.21 *Non-linear sagging (positive) and hogging (negative) bending moments of the S175 Containership sailing in the regular wave,* $\lambda = 1.2\,L, a = L/60$, $Fn = 0.25$. *The comparison is made for the experiments, Watanabe et al. (1989), a time-domain non-linear strip theory, Xia et al. (1998) and the second-order strip theory, Jensen and Pedersen, (1979).*

The steady-state contribution shown in the figure is due to the change of trim and draught when the ship is sailing as compared to the values at rest. This difference cannot be calculated by the strip theories given here but requires a full three dimensional hydrodynamic analysis. However, as the contribution to the load is small it is usually neglected.

An example with a stochastic wave elevation as input is shown in Figure 4.22.

The calculated response is the midship vertical wave bending moment. Two different ships (S175(O) and S175 (M)) are considered, the difference being that S175(M) has a larger bow flare than the other, original hull form. This is seen that the larger bow flare results in larger slamming forces leading to larger bending moments and to significant transistent, so-called whipping, vibrations of the hull girder. These vibrations will be discussed in Chapter 6.

The importance of the non-linearities due to bow flare slamming and green water on deck is illustrated in Figure 4.23.

For this rather extreme realization, where both slamming and green water on deck occur for two adjacent wave peaks, the green water load, calculated from Eq. (4.131), is very large and increases the hogging bending moment significantly.

Having performed a large number of realizations of the wave-induced responses needed, the extreme value statistics discussed in Section 3.2.5 can be applied.

* In Figure 4.21-Figure 4.23 sagging is positive opposite the definition in Chapter 2.

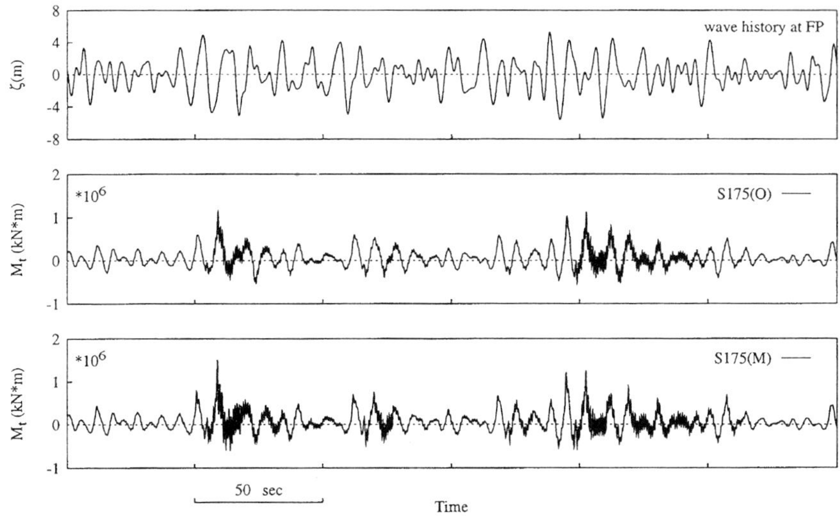

Figure 4.22 *Calculated time histories of the midship bending moment of the original and the modified S175 Containership moving in a particular realization of a moderate head sea with significant wave height H_s = 8.3 m, mean period T_s = 11.1 sec; Fn =0.25, Xia et al. (1998).*

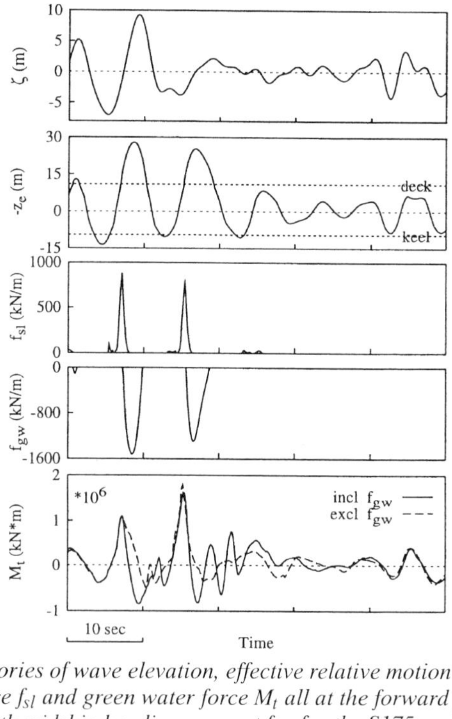

Figure 4.23 *Time histories of wave elevation, effective relative motion z_e, momentum slamming force f_{sl} and green water force M_t all at the forward perpendicular, together with midship bending moment f_{gw} for the S175 container ship in head sea. The sea state and forward speed are the same as in Figure 4.22. Wang et al. (1998).*

4.7 DESIGN LOADS

In the structural analysis of the hull girder, design loads are needed. As discussed in Chapter 3 two fundamental different approaches can be applied. The first makes use of design waves, either regular or conditional, see Example 3.3.2 and the result could be as shown in Figure 4.21 (although the wave in that figure is too small to represent a design wave). The other approach is to apply the long-term statistical method, Section 3.3.3 and Section 4.4, but taking into account the most significant non-linear effects. Such analysis is described below.

The quadratic strip theory has been used to obtain design values for the midship bending moment in various types of vessels. Two different approaches are used. In the first, Mansour and Jensen (1995), the design value is taken as the most probable largest value during a 3 hours operation in a short-term severe sea state characterised by a Pierson-Moskowitz wave spectrum with a significance wave height $H_s = 15$ m and a zero crossing period $T_z = 12$ s. This rather unrealistic sea state is often used as an equivalent to a complete long-term analysis. A low forward speed and a heading angle equal to 135 degrees are assumed.

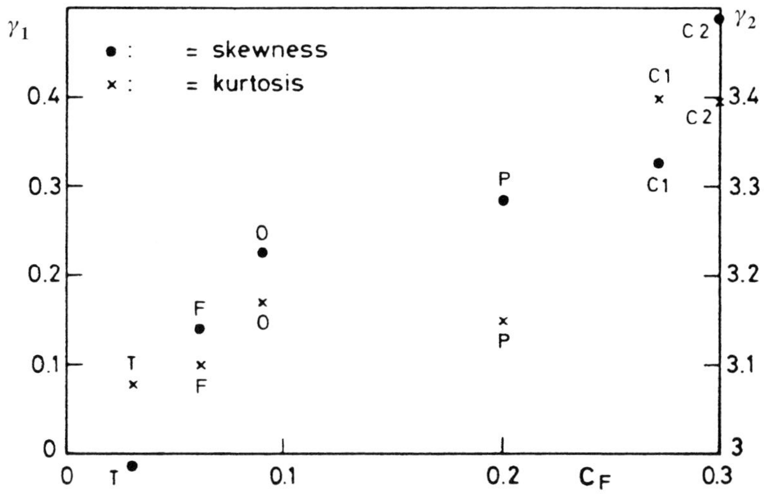

Figure 4.24 *Variation of skewness and kurtosis with the bow flare coefficient C_F for six ships. Results are for the midship sagging wave bending moment. For hog the sign of the skewness should be changed.*

The vessels considered are a tanker (T), a frigate (F), an oil-bulk-ore carrier (O), a floating oil production vessel (P) and two containerships (C1, C2). The degree of non-linearities in the bending moment is depicted in Figure 4.24 showing the skewness γ_1 and kurtosis γ_2 of the response. For comparison, a linear response to a Gaussian sea will have γ_1=0 and γ_2=3. The different ships are here characterized by one single parameter, the bow flare coefficient C_F, measuring the flare of the forward 20 per cent of the hull. Of course, this is a rather crude measure, which cannot completely describe the difference in hull forms seen among these ships. However, a clear trend towards increasing skewness and kurtosis with flare coefficient is noted. The skewness is responsible for the difference between the sagging and hogging moments, whereas a kurtosis greater than 3 implies a larger probability of obtaining larger peak and trough values than in the Gaussian case.

From the polynomial (or Hermite) series approximation, Eq. (3.34) the ratio between non-linear and linear design values of the bending moment is obtained. The results are shown in Figure 4.25.

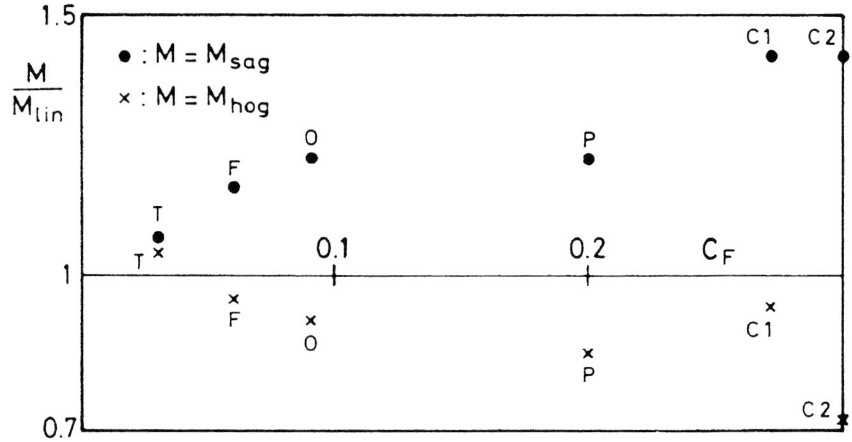

Figure 4.25 *Variation of the ratio of hogging and sagging wave bending moment to the linear wave bending moment with the bow flare coefficient C_F for six ships.*

Again some spreading in the results is seen, but the trend is an increase in the sagging and a decrease in the hogging bending moment with bow flare.

A drawback in this analysis is the rather arbitrary choice of equivalent sea state and that neither the service speed nor different heading angles are taken into account. This can only be done through a full long-term analysis in which the wave statistics in the different ocean areas covered by the ship's operational profile is weighted and combined with a speed reduction procedure in severe sea states. Such an analysis has been performed by Jensen, Banke and Dogliani (1994) for 7 ships, some of them identical to those used in the above-mentioned short-term analysis. The main results are given in Figure 4.26. The abscissa is now a combined bow flare and forward speed coefficient C_{FV}

$$C_{FV} = C_F + 0.5\, F_n \tag{4.135}$$

where the Froude No. F_n is based on the nominal service speed. Clearly the same trend and the same range for the ratio between the non-linear and linear predictions for the design values of the wave-bending moments are found in the long-term analysis as in the stationary, short-term case. It should be noted, that because of the very large extreme bow motions yielding both bow emergence and deck wetness, a proper definition of the flare must be used, see the last cited reference.

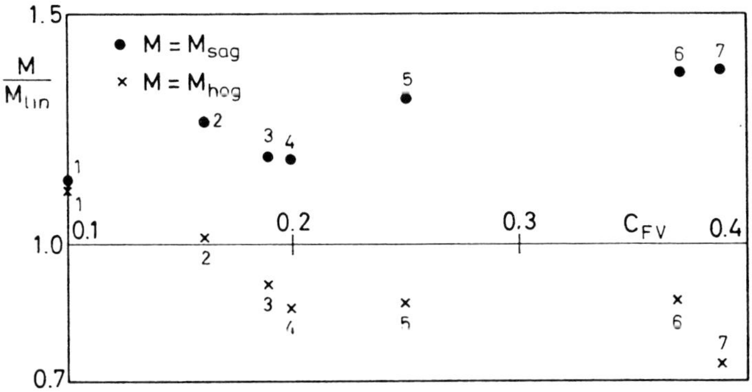

Figure 4.26 *Ratio between the median (i.e 50 per cent fractile) values of the long-term extreme values of the wave-induced sagging and hogging bending moments amidships and the corresponding results from linear strip theory for 7 ships (1: tanker, 2: dry cargo ship, 3+5 frigates, 4: floating production vessel, 6+7: container ships).*

Finally, it should be mentioned, that the classification societies provide empirical values for the design loads depending solely the main particulars of the ships. These formulas are based on a large in-house knowledge accumulated over many years. In general, these formulas predict design values in reasonable agreement with direct calculations performed as described in this chapter. However, for novel ship design care has to be taken not to rely on purely empirical formulas, derived for other kind of ships. Therefore the classification societies also recommend or even request direct calculations for novel ship types.

4.7.1 Rule Values for the Hull Girder Loads

The Classification Societies issue very detailed information on how to estimate the design loads on ship. It is outside the scope of the present treatment to discuss the rules but an example is given below, using the rules from Det Norske Veritas, 1999 edition:

Example 4.7.1

The container ship analyzed in Figure 4.15 has the main dimensions L= 270 m, B= 32.2 m and $C_B = 0.6$. The service speed is 24.5 knots = 12. 6 m/s, corresponding to a Froude No. $Fn = 0.245$.

From Figure 4.15 it follows that linear wave-induced bending moment with a returned period of 20 years ($N = 10^8$) is approximately 3.2 10^9 Nm. The non-linear corrections factors for hog and sag can be obtained from Figure 4.26 as the ship is the same as ship No. 7. Thus the design wave-bending moments become

$$M_{sag} = 1.4 \cdot 3.2\ 10^9\ \text{Nm} = \underline{4.5\ 10^9\ \text{Nm}}$$
$$M_{hog} = 0.75 \cdot 3.2\ 10^9\ \text{Nm} = \underline{2.4\ 10^9\ \text{Nm}}$$

whereas the rule values are

$$M_{sag}^{rule} = 0.11 C_w L^2 B (C_B + 0.7)\ \text{kNm} = 3.6\ 10^9 \text{Nm}$$
$$M_{hog}^{rule} = 0.19\ C_w L^2 B C_B\ \text{kNm} = \underline{2.8\ 10^9 \text{Nm}}$$

However, the rule based sagging bending moment must be corrected for speed and flare as the coefficient C_{AF} in the rules becomes

$$C_{AF} = 0.2\frac{24.5}{\sqrt{270}} + 0.3 = 0.6 > 0.5$$

The correction factor is thus 1.2 according to the rules, such that

$$M_{sag}^{rule} = 1.2 \cdot 3.6\ 10^9 Nm = \underline{4.3\ 10^9 \mathrm{Nm}}$$

It is seen that there is a very good agreement between the calculated design values and the rule values, the differences being about 10 per cent.

Finally, it should be mentioned, that the rule based design still water bending moments amidships become

$$M_{sill\ water,\ sag}^{rule} = 2.1\ 10^9\ Nm$$
$$M_{still\ water,\ hog}^{rule} = 2.8\ 10^9\ Nm$$

These values or, better direct calculated still water bending moments should be added to the wave bendings to obtain the total bending moments.

5 Hull Girder Response

When the design loads on the ship hull have been identified, a stress and deformation analysis can be carried out. In the main large parts of these loads are only dependent on the outer geometry of the ship. However, the stress calculation requires knowledge of the structural layout of the complete ship. As a rule, this geometry is determined from functional and economic requirements. It is then the purpose of the strength analysis to provide sufficient scantlings, so that the hull structure with reasonable certainty can be expected not to suffer damages during normal operational conditions, which impede its ability to function.

Figure 5.1- Figure 5.5 adapted from Guldhammer and Meldahl (1971) show a number of characteristic structural elements of ships. Figure 5.1 shows a section of a vessel with longitudinal stiffening in the bottom and the deck and transverse stiffening in the sides. A double bottom with transverse stiffening is shown in Figure 5.2. As seen from these two double-bottom structures, the transversely stiffened design is the best as regards production as a result of far less intersecting plate elements. Therefore, transverse stiffening is normally used in small ships, where the necessary longitudinal strength, i.e. the section modulus of the hull girder, is obtained automatically by the plate thicknesses for bottom and deck needed for the local strength.

For large ships (more than approximately 100 m long) longitudinal stiffening of especially deck and bottom will lead to smaller plate thicknesses of the outer plating of the ship. This saving of weight compensates for the disadvantages of a more complicated production process. Figure 5.3 shows a longitudinally stiffened section of an older medium-sized tanker. It is seen that the ship has no double bottom, now mandatory for tankers entering US ports, and that there are two longitudinal bulkheads. The division of the tanker into a number of tank compartments both longitudinally and transversely is mainly due to hydrostatic stability requirements. When the longitudinal bulkheads are designed, it must be taken into account that e.g. the central tank may be empty and the side tank full. This necessitates in Figure 5.3 the two strong horizontal beams between the side shell and the longitudinal bulkhead.

While longitudinal stiffening is thus used in the midship sections of large ships, the forepart and the stern are normally transversely stiffened, see Figure 5.4 and Figure 5.5. The shown hull cross-sections are only examples of hull design. Other examples can be found especially in shipbuilding magazines.

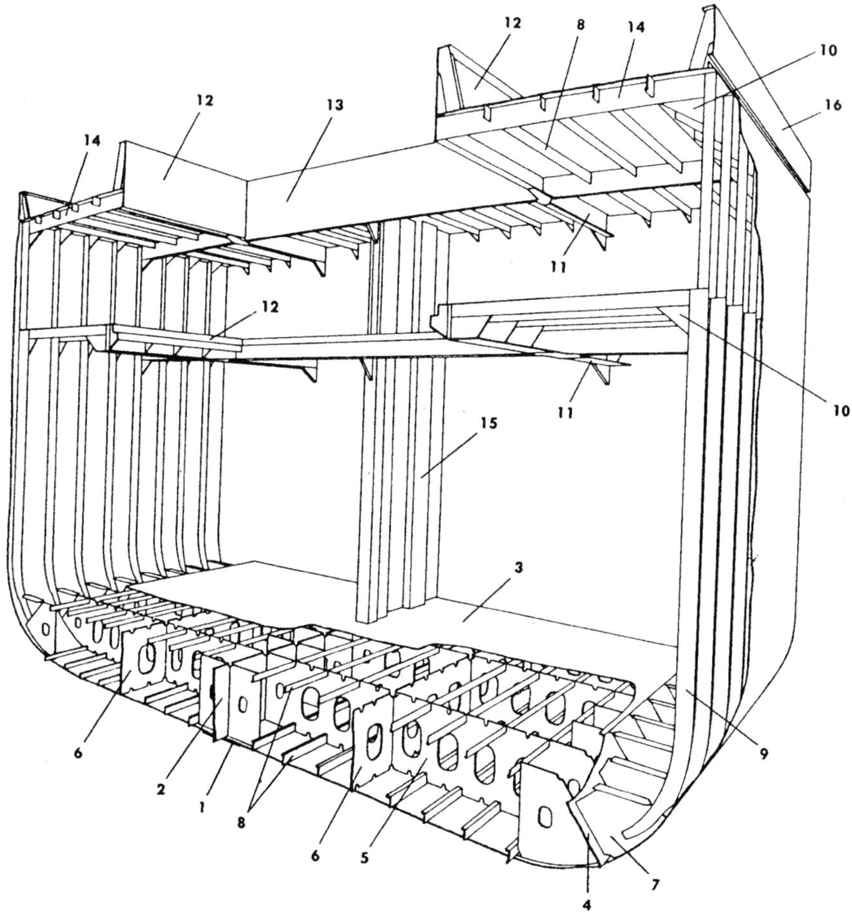

1. plate keel
2. centre girder
3. tank top
4. margin plate
5. floor
6. bilge keel
7. bilge bracket
8. longitudinal
9. frame
10. beam knee
11. deck girder
12. hatch coaming
13. hatch end coaming
14. deck transverse
15. centre bulkhead
16. gunwale

Figure 5.1 *Sectional view of a vessel with longitudinal stiffening in the deck and bottom and transverse stiffening in the sides.*

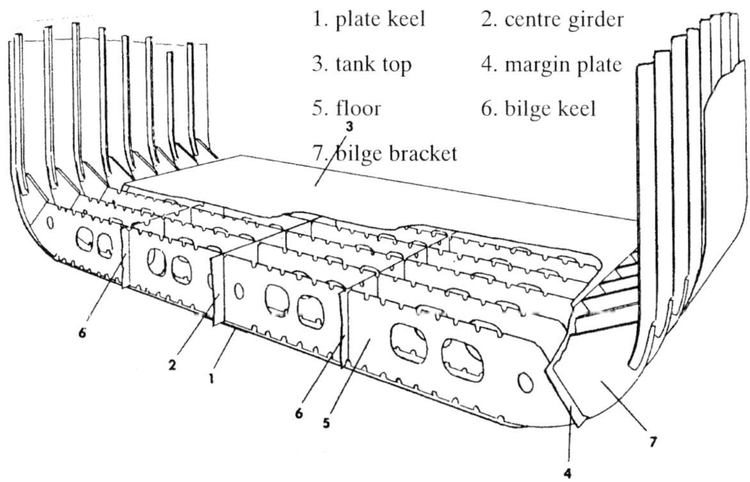

Figure 5.2 *Transversely stiffened double-bottom construction.*

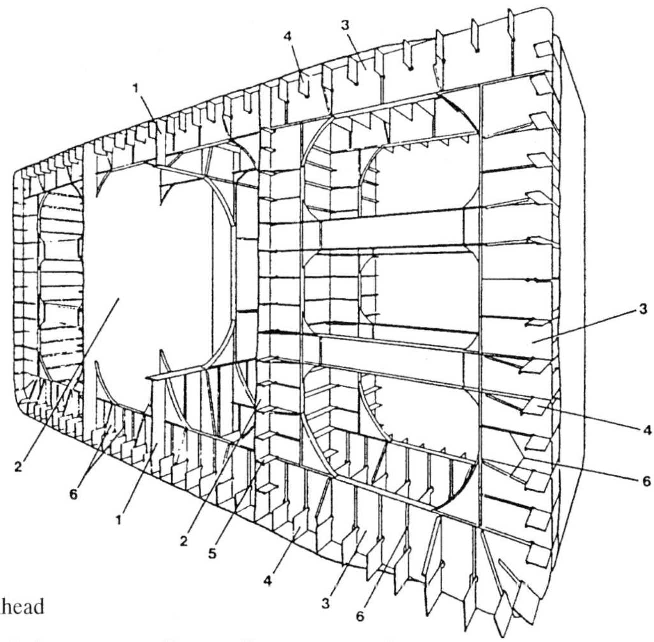

1. centre girder
2. longitudinal bulkhead
3. web (web frame, deck transverse, floor or floor transverse)
4. longitudinal
5. bulkhead stiffener
6. stiffening of web plate with stiffener or bracket

Figure 5.3 *Longitudinally stiffened tanker.*

1. forepeak bulkhead (collision bulkhead)
2. wash bulkhead
3. floor
4. stringer
5. bow plate

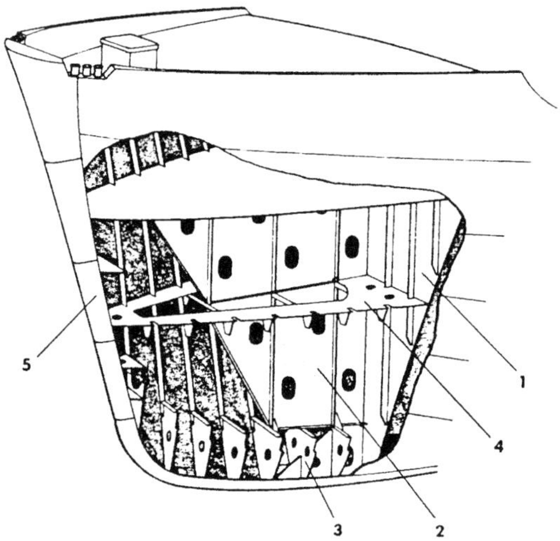

Figure 5.4 *Structural layout of a fore part.*

1. afterpeak bulkhead
2. floor
3. stern tube
4. boss
5. sole piece
6. stern

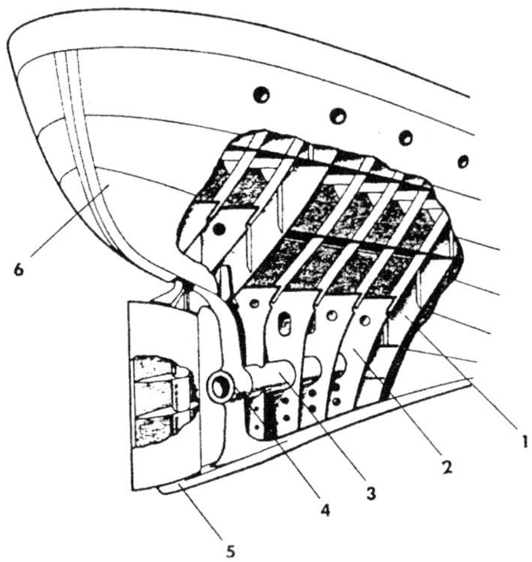

Figure 5.5 *Structural layout of a stern.*

With the tanker cross-section in Figure 5.3 in mind, the transmission of forces can in principle be described as follows. The water pressure on a plate element is transferred to the longitudinals, which then "pass" their load to the strong transverse web frames. The frames are connected to side, deck and bottom platings as well as to possible longitudinal bulkheads. These elements together form the hull girder. The transmission of force effects will thus be

plate → longitudinal → web frame → panel → hull girder

where panel designates the side plating and the longitudinal bulkheads with their stiffening. Because of the large differences in stiffness between the five levels, each level can approximately be analysed separately. Here only the hull girder analysis is considered. It should, however, be mentioned that a division into partly independent levels is not absolutely necessary, because the whole ship may be analysed by the finite element method (FEM). However, this is a time-consuming procedure, especially in preparation of input data and in verification of the results. Therefore, the present analysis is based on thin-walled beam theory, which accurately model the global stress distribution in the hull girder.

The hull girder means all the structural elements (plating, longitudinals, longitudinal bulkheads, decks and the like) with a certain longitudinal extent. As an example Figure 5.6 shows (half) of a midship section in a container ship. The stress analysis assumes that the hull girder is prismatic, which means that the cross-section does not vary much in the neighbourhood of the considered section. Moreover, it is presupposed that the deformation of the hull girder is linearly elastic.

In Chapter 2 and 4, the loads on the hull girder are treated. The results were given as sectional forces $Q_x(x)$, $Q_y(x)$, $Q_z(x)$ and sectional moments $M_x(x)$, $M_y(x)$, $M_z(x)$, c.f. Figure 5.7, referring to an xyz-coordinate system fixed in the ship, where the x-axis is the longitudinal coordinate and the z-axis lies in the centre line plane. These sectional forces and moments are, as described in Chapter 2 and 4, due to hydrostatic and hydrodynamic forces as well as inertia forces from the mass distribution of the ship. The sectional loads depend solely on the outer hull geometry and the mass distribution of the ship and are therefore assumed given here.

The stresses found in the hull girder at a section $x = x$ are, because of the assumption of a prismatic hull girder, only dependent on the sectional quantities $Q_x(x)$, $Q_y(x)$, $Q_z(x)$, $M_x(x)$, $M_y(x)$ and $M_z(x)$ and the sectional geometry of the hull girder*. The stresses from each sectional load components can be treated separately. They may be divided into four main groups:

- Bending stresses due to M_y and M_z
- Shear stresses due to Q_y and Q_z
- Torsional stresses due to M_x
- Axial stresses due to Q_x

Finally, the stresses which result from uneven heating / cooling of the cross-section as subjected to the ambient air and water temperature will also be considered:

- Temperature-induced stresses

The cross-section shown in Figure 5.6 is used as a common example.

* Except for the warping torsional stresses considered in Section 5.3.4.

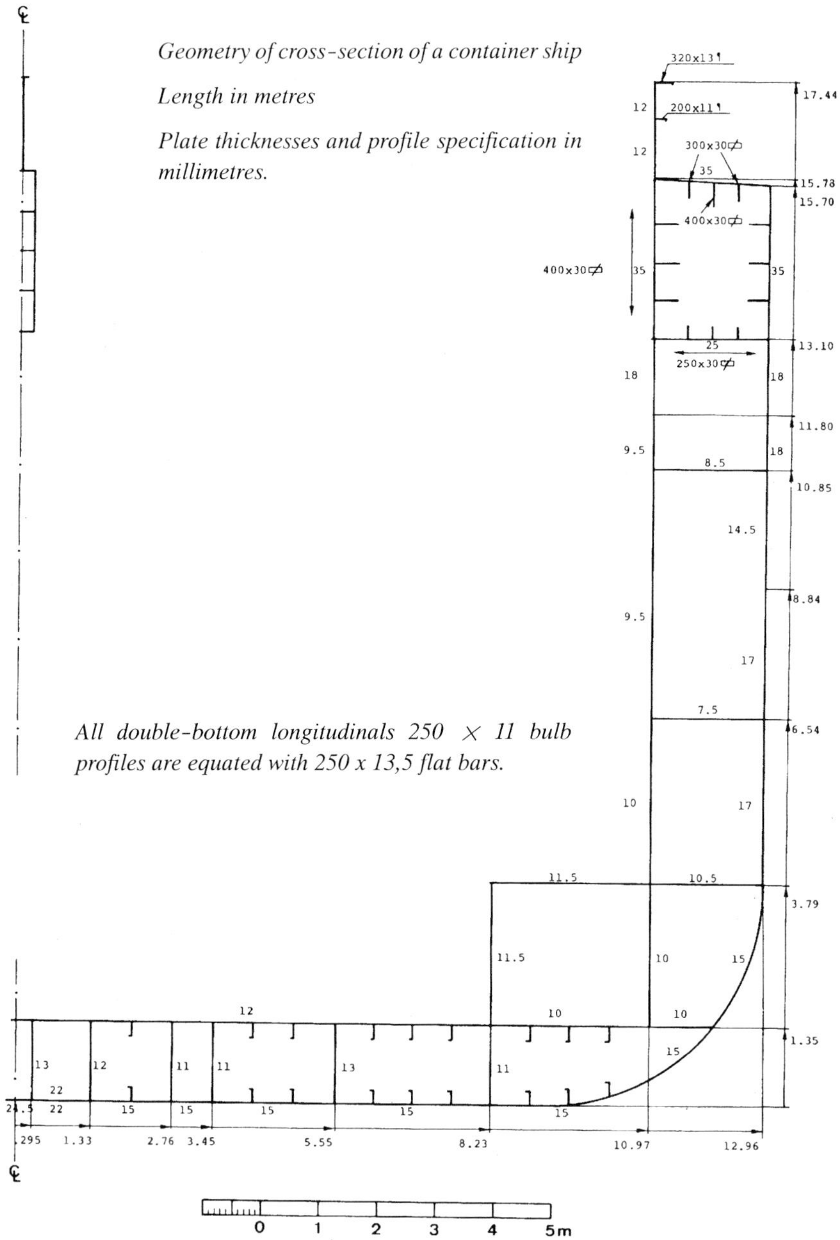

Figure 5.6 *Geometry of a cross-section of a container ship.*

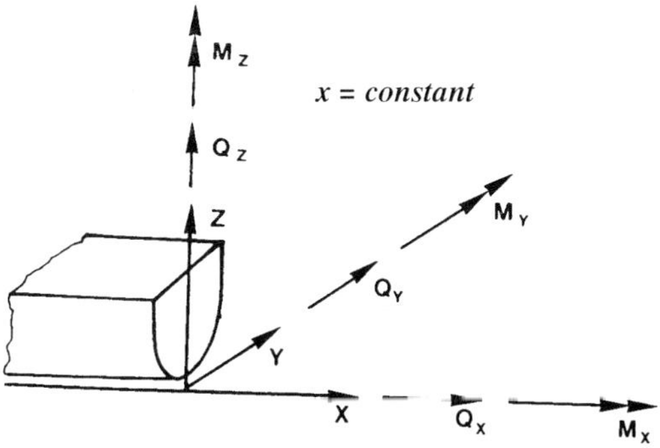

Figure 5.7 *Definition of sectional forces and sectional moments in relation to the xyz-coordinate system fixed in the ship.*

5.1 BENDING RESPONSE

If an infinitely long, prismatic beam is exposed to bending, the cross-section of the deformed beam will remain plane (the so-called *Navier's hypothesis*). Thus, the elastic normal strain ε in an arbitrary section $x = x$ becomes a linear function of the coordinates (y,z) of the cross-section:

$$\varepsilon = \varepsilon(y,x) = a + by + cz \tag{5.1}$$

The corresponding normal stress σ, according to Hooke's law for uniaxial tension will be:

$$\sigma(y,x) = E(y,z)\,(a + by + cz) \tag{5.2}$$

where E is the *modulus of elasticity* of the cross-sectional material. The value of E may vary over the cross-section, if e.g. some parts of the cross-section are made of aluminium and the rest consists of steel plates.

The constants a, b and c shall be determined so that the resultant of the stress σ over the cross-section is equal to the sectional moments M_y and M_z. Thus:

$$\int_A \sigma dA = a\int_A EdA + b\int_A EydA + c\int_A EzdA = Q_x = 0$$

$$\int_A y\sigma dA = a\int_A EydA + b\int_A Ey^2dA + c\int_A EyzdA = -M_z$$

$$\int_A z\sigma dA = a\int_A EzdA + b\int_A EyzdA + c\int_A Ez^2dA = M_y$$

where A is the cross sectional area of the longitudinal structural members (not to be confused with the immersed sectional area!).

If, as usual for ship cross-sections, it is assumed that the cross-section is symmetric about the z-axis, the terms with a integrand linear in y become zero (as $E(y,z) = E(-y,z)$), which gives

$$\sigma(y,z) = E(y,z)\left[\frac{M_y}{E_y}(z - z_0) - \frac{M_z}{E_z}y\right] \tag{5.3}$$

where

$$\begin{aligned} z_0 &\equiv \int_A EzdA \Big/ \int_A EdA \\ E_y &\equiv \int_A Ez^2dA - z_0^2 \int_A EdA \\ E_z &\equiv \int_A Ey^2dA \end{aligned} \tag{5.4a}$$

For cross-sections consisting of N rectilinear, evenly thick, thin plates the integrations can be replaced by summations

$$\begin{aligned} \int_A EdA &= \sum_{i=1}^{N} E_i h_i \ell_i \\ \int_A EzdA &= \sum_{i=1}^{N} E_i h_i \ell_i \tilde{z}_i \\ \int_A Ez^2dA &\simeq \sum_{i=1}^{N} E_i h_i \ell_i \left(\tilde{z}_i^2 + \frac{1}{12}\hat{z}_i^2\right) \\ \int_A Ey^2dA &\simeq \sum_{i=1}^{N} E_i h_i \ell_i \left(\tilde{y}_i^2 + \frac{1}{12}\hat{y}_i^2\right) \end{aligned} \tag{5.4b}$$

where, according to Figure 5.8, the modulus of elasticity, the thickness and the length of the ith plate element are respectively E_i, h_i og ℓ_i and where

$$\begin{aligned} \tilde{z}_i &= \frac{1}{2}(z_{1i} + z_{2i})\,; \quad \hat{z}_i = z_{2i} - z_{1i} \\ \tilde{y}_i &= \frac{1}{2}(y_{1i} + y_{2i})\,; \quad \hat{y} = y_{2i} - y_{1i} \end{aligned}$$

with (y_{1i}, z_{1i}) and (y_{2i}, z_{2i}) as the coordinates of the endpoints for the ith plate element. The line $z = z_0$ is normally denoted the neutral axis of the cross-section, as the bending stresses due to the moment M_y are here zero. The horizontal sectional moment M_z will usually be of much lesser importance that M_y since both $M_z < M_y$ and $E_z > E_y$.

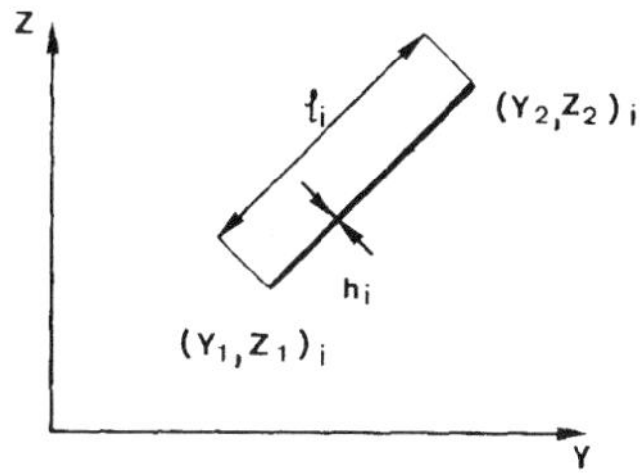

Figure 5.8 *A rectilinear, evenly thick plate element.*

For most ships the modulus of elasticity will be constant for the whole cross-section, so that Eq. (5.3) reduces to

$$\sigma(y,z) = \frac{M_y}{I_y}(z - z_0) - \frac{M_z}{I_z}y \tag{5.5a}$$

where the *moments of inertia* I_y and I_z are

$$\begin{aligned} I_y &= \int_A z^2 dA - z_0^2 A\,; \quad z_0 = \int_A z dA/A \\ I_z &= \int_A y^2 dA \end{aligned} \tag{5.5b}$$

If the modulus of elasticity E is constant, the largest stress $\sigma_{\max}$ will be

$$\sigma_{\max} = \frac{|M_y|}{I_y} z_{\max} + \frac{|M_z|}{I_z} y_{\max} \tag{5.6}$$

where

$$\begin{aligned} z_{\max} &= \max|z - z_0| \\ y_{\max} &= \max|y| \end{aligned}$$

provided $z = z_{\max}$, $y = y_{\max}$ occur at the same material point, e.g. a deck corner. The quantities W_z and W_y defined by

$$W_y = \frac{I_y}{|z - z_0|} \qquad W_z = \frac{I_z}{|y|} \tag{5.7}$$

are called the *section modulus* and are usually calculated for the outer contour of the cross-section.

When the bending stiffness of the cross-section, given by E_y, E_z or I_y, I_z, are to be calculated, the continuity of the individual longitudinal elements must be assessed. Normally hatch coamings and superstructures cannot be completely included in the stiffness of the hull girder, see the left figure in Figure 5.9. Also, possible longitudinal hatch beams cannot be effectively included as they are only connected to the hull girder in their endpoints and therefore may be vertically displaced relative to the deformed cross-section of the hull girder, see Figure 5.9 right.

Figure 5.9 *Structural ineffective parts of the ship during bending.*

For calculation of the moment of inertia of a cross-section it is common practice to use computers. For hand calculations it is advantageous to use a table as Table 5.1 to determine z_0 and I_y.

Table 5.1 *Calculation of sectional properties*

ℓ (m)	h (m)	ℓh (m^2)	$\bar{z}$ (m)	$\hat{z}$ (m)	$\ell h \bar{z}$ (m^3)	$\ell h(\bar{z}^2 + \hat{z}^2/12)$ (m^4)
		S_1			S_2	S_3

If only half the cross-section $y \geq 0$ is considered, S_1, S_2 and S_3 shall be multiplied by 2 !

Cross-sectional area $A = S_1$ (m^2)

The vertical distance of the neutral axis above keel $z_0 = S_2/S_1$ (m)

Moment of inertia $I_y = S_3 - S_2^2/S_1$ (m^4)

Section modulus at keel $W_{keel} = I_y/z_0$ (m^3)

Section modulus at deck $W_{deck} = I_y/(z_{deck} - z_0)$ (m^3)

Finally, it should be stressed that use of the beam theory implies that plane cross-sections remain plane during bending, Eq. (5.1). This assumption is not quite fulfilled for the deck and bottom plating. By use of a so-called "effective breadth" this can be corrected for, but this is not discussed here on since the effect is normally of no practical importance for the global loads on the hull girder.

Example 5.1.1

Determination of the bending stiffness for the cross-section shown in Figure 5.6. The material is steel everywhere.

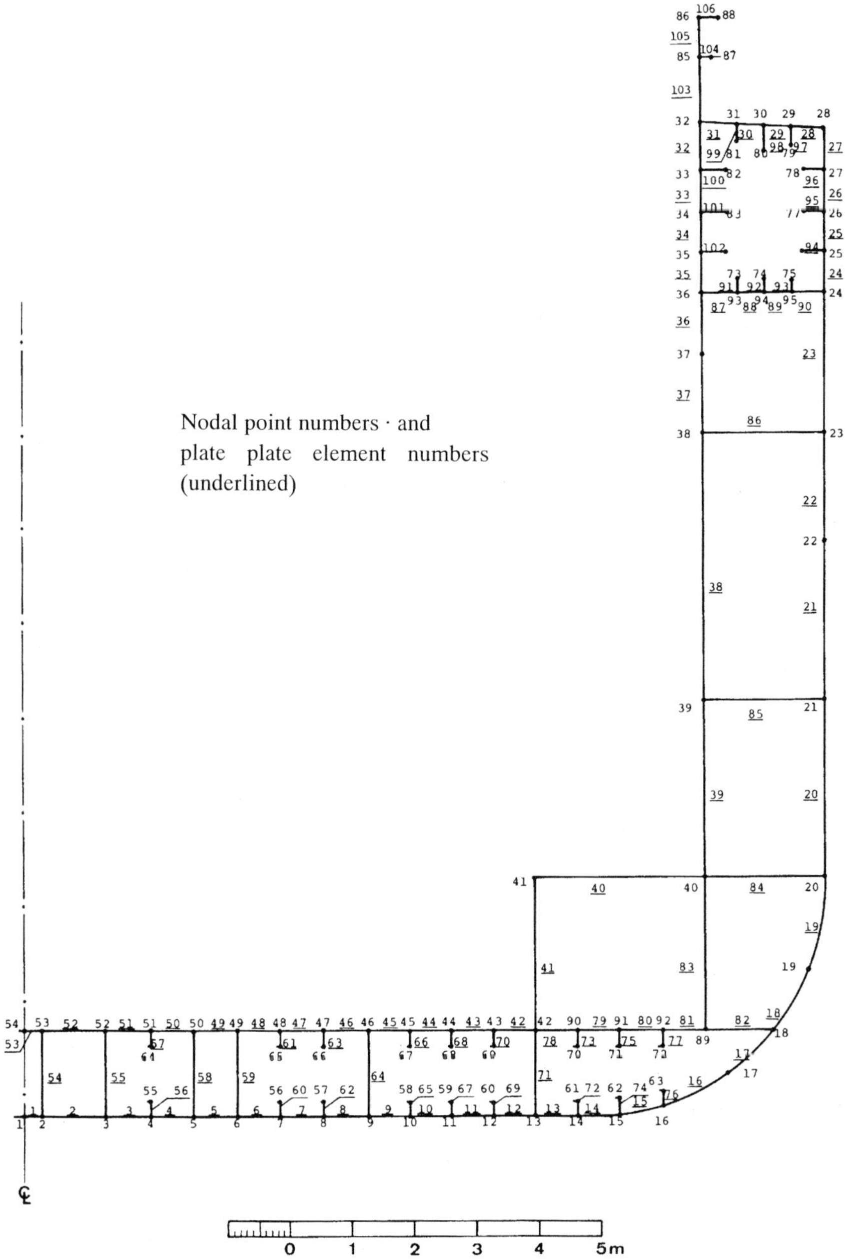

Figure 5.10 *Modelling of hull cross-section for determination of cross-sectional constants. The modelling is based on the cross-section shown in Figure 5.6.*

To be able to use Eqs. (5.4b) the cross-section must be approximated as a system of rectilinear plates of uniform thickness. This approximation is shown in Figure 5.10. It should particularly be noted that two nodal points (Nos 17 and 19) are placed in the bilge to obtain reasonable modelling of the rounding and that nodal points (Nos 22 and 37) are placed where the plate thicknesses change. Besides, the bulb profile of the longitudinals is approximated by an evenly thick plate element of the same height and area as the bulb profile. Finally, the hatch beam has not been included, while the hatch coaming is assumed to be fully effective.

The following result is obtained:

$$z_0 = 7.238 \text{ m}$$

$$I_y = 106.1 \text{ m}^4$$

$$I_z = 294.1 \text{ m}^4$$

in the (y,z)-coordinate system with origo in the keel at the centre line (that is at nodal point No 1 in Figure 5.10). The neutral axis is thus 7.238 m above the keel.

The main particulars of the ship are:

The perpendicular length of the ship	L	= 185.93 m
The breadth moulded midship	B	= 25.92 m
The block coefficient	C_B	= 0.6

A container ship will due to its slender hull form always be in a hogging condition. Therefore no longitudinal compressive stresses occur in the deck, making it possible to use flat bars rather than T-profiles or bulb profiles as longitudinals. The maximum tensile stresses in the deck can be estimated using $M_y = 2\ 10^9\ Nm$, c.f. Chapter 4, Section 4.7.1.

With this bending moment the maximum tensile stress becomes

$$\sigma_{max} = \frac{2\ 10^9 \text{ Nm}}{106.1\text{m}^4} \cdot (17.44 - 7.238) \text{ m} = 192 \text{ MN/m}^2$$

and it occurs in the upper plate element of the hatch coaming.

As in this example the neutral axis is usually closer to the keel of the ship than the deck because of the double bottom. Thus, the largest bending stresses occur in the deck, and therefore high-tensile steel is often used in here. The very heavy longitudinals (400 × 30 mm flat bars) just below the deck have been introduced to raise the neutral axis in order to lower the deck stresses.

<u>Example 5.1.2</u>

Consider the simplified cross-section of a hull with a superstructure shown in Figure 5.11.

The hull girder is made of steel and the superstructure of aluminium. The cross-sectional dimensions are shown in Figure 5.11. Only the stresses as a result of a vertical moment M_y are considered.

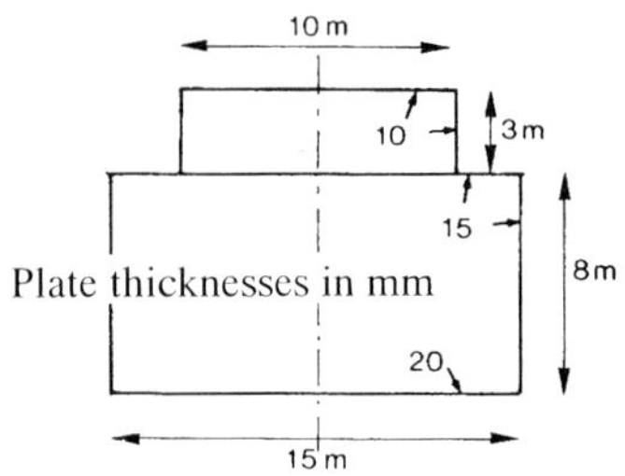

Figure 5.11 *Simplified hull cross-section with a superstructure.*

If the hull girder alone is considered, Eq. (5.5a) gives

$$\sigma = \frac{M_y}{I_y}(z - z_0) \equiv \frac{M_y}{W_y}$$

with

$$z_0 = 3.61 \text{ m} ; \qquad I_y = 9.56 \text{ m}^4$$

If the superstructure is included, calculated as fully effectively then

$$\sigma = \frac{M_y}{E_y/E}(z - z_0^*) \equiv \frac{M_y}{W_y^*}$$

is obtained where

$$E = E_{steel} = 2.1 \ 10^{11} \text{ N/m}^2 \text{ for the hull girder}$$

$$E = E_{alu} = 0.7 \ 10^{11} \text{ N/m}^2 \text{ for the superstructure}$$

and where

$$z_0^* = (2.76 + \tfrac{1}{3}\ 1.67)/(0.765 + \tfrac{1}{3}0.16) = 4.05 \text{ m}$$

$$E_y = \left[\left(19.52 + \tfrac{1}{3}\ 17.56\right) - 4.05^2\left(0.765 + \tfrac{1}{3}0.16\right)\right] \cdot E_{steel}$$

$$= 11.95\text{m}^4 \cdot E_{steel}$$

The result is given in Table 5.2 by the section modulus W_y and W_y^*, without and with the superstructure included:

$$W_y \equiv \frac{9.56 \text{ m}^4}{|z - 3.61 \text{ m}|}; \qquad W_y^* \equiv \frac{11.95 \text{ m}^4}{|z - 4.05 \text{ m}|} \cdot \frac{E_{steel}}{E}$$

Table 5.2

	z(m)	W_y(m³)	W_y^*(m³)
top deck house	11.0	–	5.15
bottom deck house	8.0	–	9.08
top hull girder	8.0	2.18	3.03
bottom hull girder	0	2.65	2.95

As the normal stress σ is inversely proportional to the section modulus, it is seen from the results that the stresses in the superstructure will be smaller than in the hull girder. However, the permissible stress for aluminium is smaller than for steel and hence the stresses in the superstructure may be the critical ones. This result applies generally to passenger ships with relatively high and long superstructures. To reduce the stress level in the superstructure, expansion joints are therefore often introduced in the upper decks of the superstructure. Alternatively, the coupling between the superstructure and the hull girder can be made flexible by e.g. letting the superstructure rest on relatively weak transverse frames.

Example 5.1.3

The vertical deflection of the hull girder can be determined accurately as discussed in Chapter 6, taking into account inertia effects as well as bending and shear deformations. However, a simple estimate of the vertical deflection δ amidships, defined relative to a straight line through AP and FP can be obtained assuming a prismatic hull with a constant bending moment M_y. With these assumptions it follows from Figure 5.12 that

$$\delta = R\left(1 - \cos\frac{L}{2R}\right) \simeq \frac{L^2}{8R} = \frac{L^2}{8}\kappa$$

where L is the length of the ship, R the radius of curvature and κ the curvature:

$$\kappa = \frac{1}{R} = \frac{d\epsilon}{dz} = \frac{M_y}{EI_y}$$

Hence,

$$\delta \simeq \frac{M_y L^2}{8EI_y}$$

As an example, consider the ship also used in Example 5.1.1:

$$\delta = \frac{2\ 10^9 \cdot 185.93^2}{8 \cdot 2.1\ 10^{11} \cdot 106.1}\text{m} = 0.39\text{ m}$$

This is the maximum upwards deflection experienced only if the vessel is subjected to its maximum design (still water + wave) bending moment. In still water typically only half or less of this value is obtained.

In the above analysis both the moment of inertia I_y and the bending moment M_y are assumed constant along the length of the vessel. This is of course not so for a real ship, but as long as the ratio M_y/I_y is constant the above analysis is rather accurate. Deflections due to shear is ignored as it only gives a small contribution.

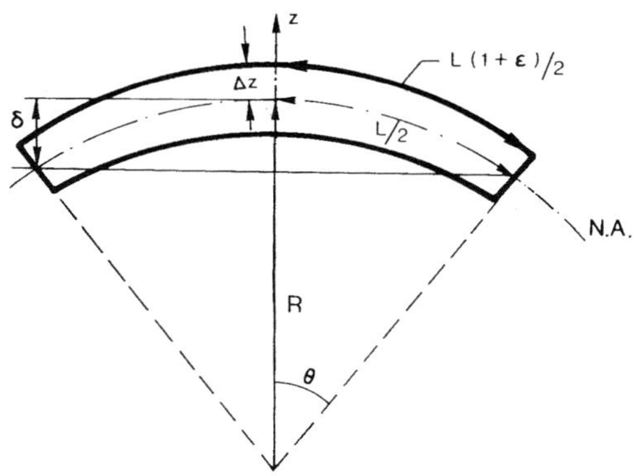

Figure 5.12 *Definition of the radius of curvature R and the height δ.*

5.2 SHEAR RESPONSE

If a section dx of a beam element is considered, Figure 5.13, with a sectional bending moment $M(x)$, a sectional shearing force $Q(x)$ and a distributed load $q(x)$ the following is obtained by moment equilibrium:

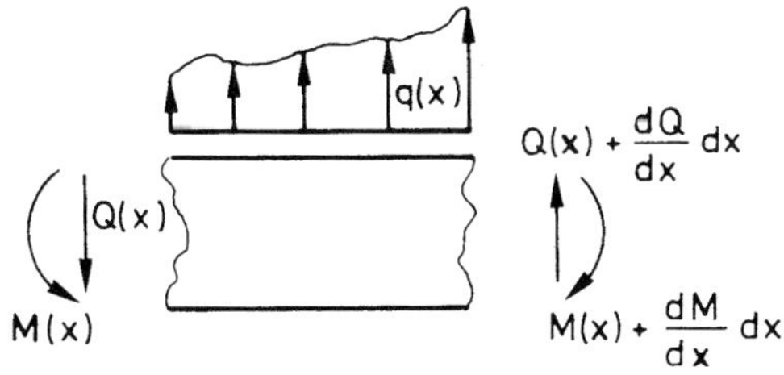

Figure 5.13 *Considerations of equilibrium for a beam element.*

$$\frac{dM}{dx}dx - \left(Q(x) + \frac{dQ}{dx}dx\right)dx - \int_{x}^{x+dx} (u - x)\, q(u)\, du = 0$$

which, for $dx \to 0$, reduces to

$$\frac{dM}{dx} = Q \quad \text{or} \quad \frac{dM_y}{dx} = Q_z \quad \text{and} \quad \frac{dM_z}{dx} = -Q_y \tag{5.8}$$

as also derived in Chapter 2, Example 2.3.2.

If a small cut-out of a thin-walled, symmetric beam is considered next, Figure 5.14, the following formula is obtained by force equilibrium, using Eq. (5.3) and Eq. (5.8):

$$\begin{aligned}\frac{\partial(h\tau)}{\partial s} &= -h(s)\frac{\partial\sigma(x,s)}{\partial x} = -h(s)\frac{\partial}{\partial x}\left\{E(s)\left[\frac{M_y(x)}{E_y}\left(z(s) - z_0\right) - \frac{M_z(x)}{E_z}y(s)\right]\right\} \\ &= -h(s)\,E(s)\cdot\left[\frac{Q_z(x)}{E_y}\left(z(s) - z_0\right) + \frac{Q_y(x)}{E_z}y(s)\right]\end{aligned} \tag{5.9}$$

In the Eq. (5.9) s is an arc length parameter and $h = h(s)$ is the thickness of the plate element. Finally, $\tau = \tau(x,s)$ is the *shear stress*. By integration

$$\begin{aligned}\tau(x,s) = &-\frac{Q_z(x)}{E_y\,h(s)}\int_s h(s)\,E(s)\left(z(s) - z_0\right) ds \\ &-\frac{Q_y(x)}{E_z h(s)}\int_s h(s)\,E(s)\,y(s)\,ds + \tau(x,0)h(0)/h(s)\end{aligned} \tag{5.10}$$

which shows that the shear stress τ at an arbitrary point in the cross-section of course is a linear combination of a contribution from each of two shear stress components $Q_z(x)$ and $Q_y(x)$.

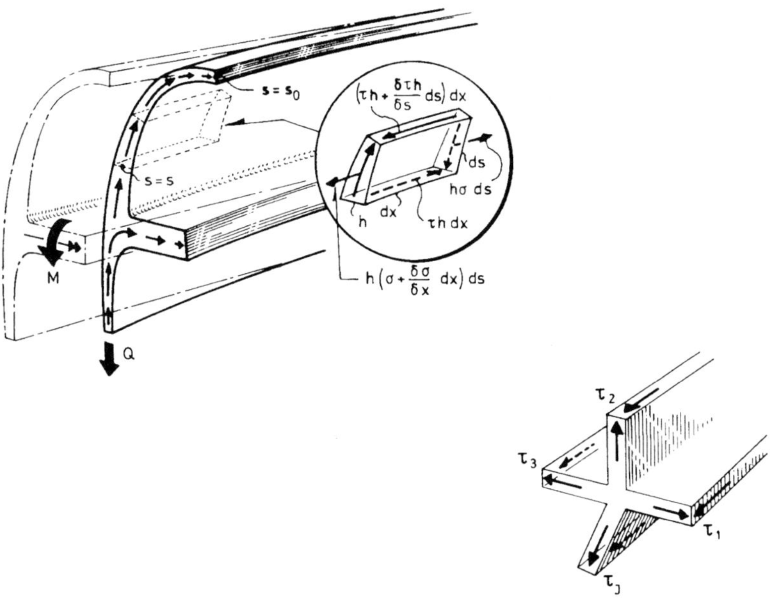

Figure 5.14 *Segment of thin-walled beam-section.*

It follows from Eq. (5.9) that $\partial(h\tau)/\partial s = 0$ if $z(s) = z_0$ and $y(s) = 0$. Hence, the maximum values of $h\tau$ occur at the neutral axes for bending, i.e. where the bending induced normal stresses are zero.

For a rectilinear, evenly thick plate element with length l (c.f. Figure 5.8) and constant modulus of elasticity for the whole section, Eq. (5.10) reduces to

$$\begin{aligned}\tau(x,s) = & -\frac{Q_z(x)}{I_y}\left[(z_1 - z_0)\,s + \frac{1}{2}s^2\frac{(z_2 - z_1)}{\ell}\right] \\ & -\frac{Q_y(x)}{I_z}\left[y_1\,s + \frac{1}{2}s^2\frac{(y_2 - y_1)}{\ell}\right] + \tau(x,0)\end{aligned} \qquad (5.11)$$

with the parameter s given by

$$s = \sqrt{(y - y_1)^2 + (z - z_1)^2}, \quad 0 \leq s \leq \ell$$

and where (y_1, z_1) and (y_2, z_2) are the endpoints of the plate element.

It is seen from Eq. (5.11) that, if the shear stress τ is known at one point of the plate element, it can be calculated at any another point. The variation is parabolic in s.

For a cross-section, composed of N evenly thick, rectilinear plate elements, Eq. (5.11) can be applied for each plate element. The constants $\tau(x, s = 0)$ for each individual plate element are determined by force equilibrium in the x-direction at the points where the plate elements are connected to each other:

$$\sum_{j=1}^{J} (h\tau)_J = 0 \qquad \text{for all points} \tag{5.12}$$

where J is the number of plate elements joining together in the point. Note that all shear stress components must be measured positive either away from or towards the point, c.f. Figure 5.14.

Especially Eq. (5.12) shows that

$$\tau = 0 \qquad \text{at a free edge} \tag{5.13}$$

provided the thickness h is not tapered to zero.

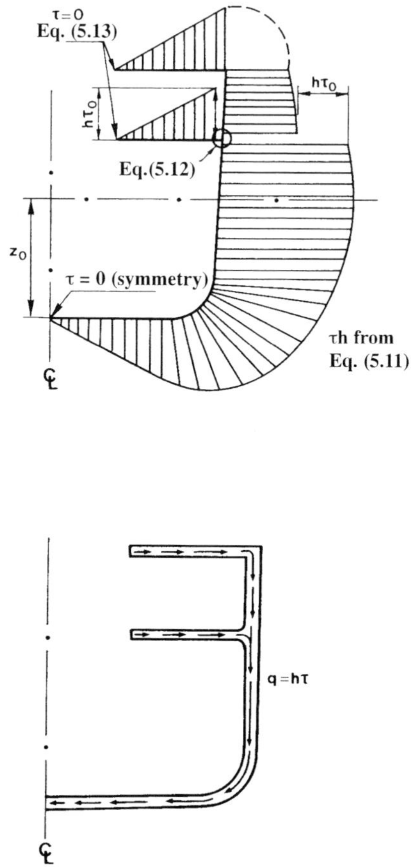

Shearing forces q = hτ in an open cross-section

Figure 5.15 *Determination of the shear stress distribution in a symmetric, open cross-section due to a vertical shearing force Q.*

The relations (5.12) and (5.13) are sufficient to determine the unknown constants ($\tau(x, s = 0)$) for each plate element in a cross-section, in which the plate elements do not form closed cells. At a symmetry line where only two plate elements join, see e.g. Figure 5.15, Eq. (5.12) implies that the shear stress is zero. Another example is given in Figure 5.15.

Example 5.2.1

Determine the shear stress distribution in a thin-walled beam with a box-shaped cross-section as shown in Figure 5.16 and subjected to a sectional shearing force Q in the symmetry line.

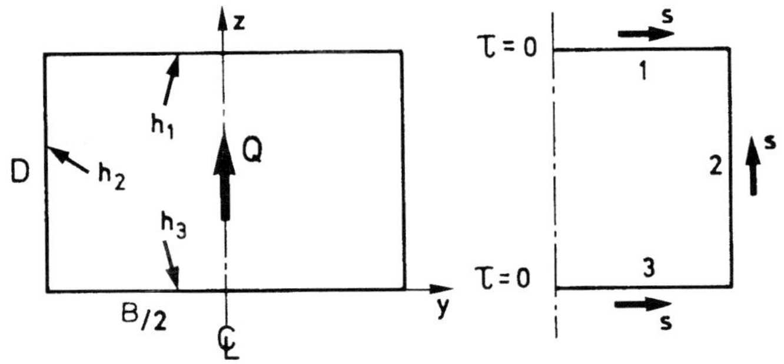

Figure 5.16 *Box-shaped cross-section.*

Because of symmetry the shear stress τ will be zero in the symmetry line. Hence, the calculation can be made solely on one half of the cross-section.

With the (y,z)-coordinate system inserted as shown in Figure 5.16, the following is obtained according to Eq. (5.5b):

$$z_0 = \left(h_1 BD + h_2 D^2\right)/\left(\left(h_1 + h_3\right) B + 2h_2 D\right)$$
$$I_y = h_1 BD^2 + \frac{2}{3} h_2 D^3 - z_0^2 \left(\left(h_1 + h_3\right) B + 2h_2 D\right)$$

If the plate elements with the plate thicknesses h_1, h_2 and h_3 are designated respectively 1, 2 and 3, the result of (5.11) is

$$\text{Plate element 1}: \ \tau_1(s) = -\frac{Q}{I_y}(D - z_0)s \quad ; \quad s = y$$

$$\text{Plate element 2}: \ \tau_2(s) = -\frac{Q}{I_y}\left[-z_0 s + \frac{1}{2}s^2\right] + \tau_2(0) \quad ; \quad s = z$$

$$\text{Plate element 3}: \ \tau_3(s) = \frac{Q}{I_y} z_0 s \quad ; \quad s = y$$

as s is measured from the symmetry line for plate elements 1 and 3 and from the lower endpoint for plate element 2. At the latter endpoint the equilibrium condition Eq. (5.12) gives

$$h_3 \tau_3(1) - h_2 \tau_2(0) = 0$$

or

$$\tau_2(0) = \frac{h_3}{h_2} \frac{QB}{2I_y} z_0$$

and thus

$$\tau_2(s) = -\frac{Q}{I_y}\left[-z_0 s + \frac{1}{2}s^2 - \frac{h_3}{h_2}\frac{B}{2}z_0\right]$$

The resulting stress distribution is shown in Figure 5.17 for $B/D = 2$ and $h_2/h_1 = 2$, $h_3/h_1 = 1.5$.

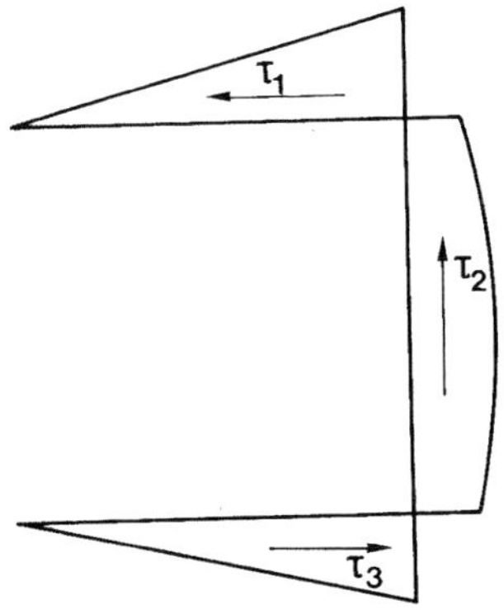

Figure 5.17 *Shear stress distribution ($B/D = 2$ and $h_2/h_1 = 2$, $h_3/h_1 = 1.5$).*

By symmetry, it is seen directly that the horizontal resultant of the shear stresses is zero, while the vertical resultant becomes

$$2h_2\int_0^D \tau_2(s)ds = Q\cdot\frac{D^2 z_0 h_2 - \frac{1}{3}h_2 D^3 + DBz_0 h_3}{I_y} = Q$$

after some rewriting.

The maximum shear stress in the side (plate element 2) is found from

$$\frac{\partial\tau_2}{ds} = 0 \Rightarrow s = z_0 \Rightarrow \tau_{2,\max} = \frac{Q}{I_y}\left(\frac{1}{2}z_0^2 + \frac{h_3}{h_2}\frac{B}{2}z_0\right)$$

and is thus found at the bending neutral axis of the cross-section.

If the cross-section contains closed cells, the Eqs. (5.12) and (5.13) are no longer sufficient to determine the unknown constants $\tau_i(0)$. This is seen by considering for instance a closed cell consisting of three rectilinear plate elements (of the same thickness). If Eq. (5.12) is written down for each plate element, then the following equations are obtained:

$$\tau_1(0) - \tau_3(0) = \Delta\tau_3 \quad ; \quad \tau_2(0) - \tau_1(0) = \Delta\tau_1 \quad ; \quad \tau_3(0) - \tau_2(0) = \Delta\tau_2$$

where the right sides are known quantities given by Eq. (5.11): $\Delta\tau_i = \tau_i(1) - \tau_i(0)$. However, these three equations are linearly dependent, and therefore $\tau_1(0), \tau_2(0)$ and $\tau_3(0)$ cannot be determined by them. So for cross-sections containing closed cells the *static considerations* Eqs. (5.9) and (5.12) are not sufficient to determine the shear stresses. In order to solve the problem the deformation of the cross-section has to be included in the analysis. Thus, opposite to the open section, the result will now depend on the *kinematic conditions*, e.g. linear elastic materials.

Consider a small rectangular plate segment of dimensions $dx \cdot ds$ and subjected to a shear stress τ. It is seen from Figure 5.18 that the decrease ϕ of the right angle due to the deformation is equal to twice the *shear strain* ε_{sx}:

$$\phi = \frac{\partial u_s}{\partial x} + \frac{\partial u_x}{\partial s} = 2\varepsilon_{xs} \tag{5.14}$$

where u_s and u_x are the deformation in the direction of the arc length s and of the beam axis x, respectively.

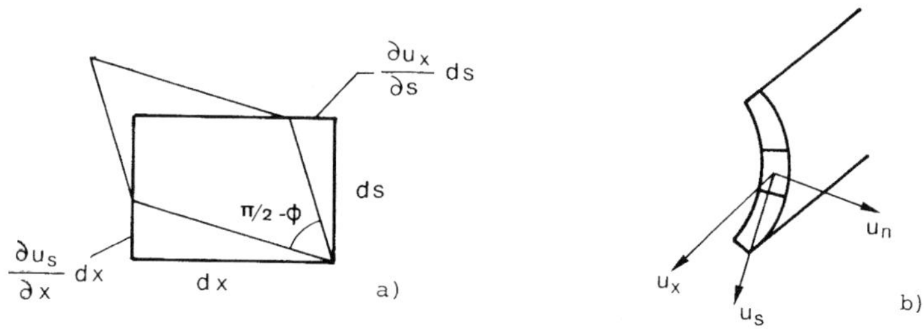

Figure 5.18 *Shear deformation of a plate element.*

If Eq. (5.14) is integrated around the closed cell,

$$\oint \phi \, ds = \frac{\partial}{\partial x}\oint u_s \, ds + \oint \frac{\partial u_x}{\partial s} ds = \frac{\partial}{\partial x}\oint u_s ds \tag{5.15}$$

as u_x is continuous in x.

Consistent with a beam theory it is assumed that the cross-section do not change shape during the deformation. Hence, the cross-section undergo a pure translation:

$$u_y(x, s) = u_y(x), \; u_z(x, s) = u_z(x)$$

and as

$$u_s(x, s) = u_y \frac{dy}{ds} + u_z \frac{dz}{ds}$$

the first integral in Eq. (5.15) becomes

$$\oint u_s ds = u_y(x) \oint dy + u_z(x) \oint dz = 0$$

so that

$$\oint \phi ds = 0$$

In order to relate the deformation ϕ to the shear stress τ, the material is assumed to be linear elastic. Hence, according to Hooke's law:

$$\phi(s) = \tau(s)/G(s) \tag{5.16}$$

where the *shear modulus* G may vary over the cross-section, $G = G(s)$. Insertion of Eq. (5.16) in Eq. (5.15) gives

$$\oint \frac{\tau(s)}{G(s)} ds = 0 \tag{5.17}$$

which is the necessary extra condition giving, together with Eq. (5.10) and Eq. (5.12) the shear stress distribution in closed cells. The following procedure, c.f. Figure 5.19, is an efficient method for determining the constants $\tau_i(0)$ in a cross-section composed of rectilinear evenly thick plate elements forming closed cells. For reasons of simplicity $G(s)$ is assumed to be constant for each cell.

(i) "Cut up" all closed cells and use Eqs. (5.10) - (5.12) for determination of all constants $\tau_i(0)$ in the thus produced open, fictitious cross-section.
This is most easily done by starting from all free edges, where $\tau_i(0) = 0$ and then apply Eq. (5.12). The resulting stress distribution is denoted $\overline{\tau}(s)$ and contains no unknown constants.

(ii) It is seen directly from Eqs. (5.9) and (5.12) that addition of a constant *shearing force* $q_j \equiv (h\hat{\tau})_j$ to all plate elements forming part of the jth closed cell gives a new stress distribution, which still fulfils Eqs. (5.11) and (5.12). For each of the total of M closed cells, $(h\hat{\tau})$ is determined by Eq. (5.17):

$$\oint_{\text{cell } j} \tau(s)\, ds = \oint_{\text{cell } j} \overline{\tau}(s)\, ds + q_j \oint_{\text{cell } j} \frac{ds}{h(s)} - \sum_{\substack{i=1 \\ i \neq j}}^{M} q_i \int_{i \cap j} \frac{ds}{h(s)} = 0 \tag{5.18}$$

where the integral $\int_{i \cap j}$ is extended over the joint plate elements of cells i and j, see Figure 5.19.

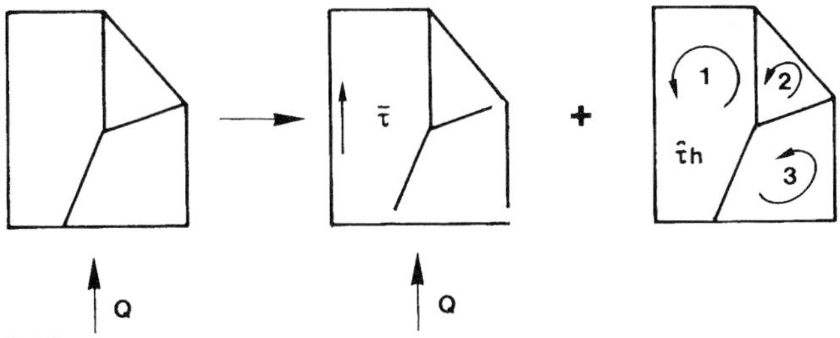

Figure 5.19 *Shear stress calculation for a beam cross-section with closed cells.*

The above M equations can be written in matrix form:

$$[\eta]\{q\} = -\{B\} \tag{5.19}$$

where $[\eta]$ is an $M \times M$ matrix with the elements

$$\eta_{jj} = \oint_j \frac{ds}{h}, \quad \eta_{ij} = \eta_{ji} = -\int_{i \cap j} \frac{ds}{h} \quad ; \quad i, j = 1, 2, \ldots, M$$

and where the vectors

$$\{q\} = \{q_1, q_2, \ldots, q_M\}^T$$

$$\{B\} = \{B_1, B_2, \ldots, B_M\}^T$$

with

$$B_j = \oint_j \bar{\tau}(s)ds$$

The solution {q} to Eq. (5.19) is

$$\{q\} = -[\eta]^{-1}\{B\} \tag{5.20}$$

When $q_1, q_2, \ldots, q_M$ have thus been determined, the shear stress in each individual plate element in the real cross-section can be written

$$\tau_i(s) = \bar{\tau}_i(s) + \frac{1}{h_i}\sum_k \delta_{ik} q_k \tag{5.21}$$

where

$$\delta_{ik} = \begin{cases} \pm 1 & \text{if plate element No. i is part of cell No. k} \\ 0 & \text{otherwise} \end{cases}$$

For δ_{ik}, +1 is used if the arc length parameter s for the plate element No."i" has the same direction as $\hat{\tau}_k$, otherwise -1 is used, c.f. Figure 5.19.

Example 5.2.2

A simplified hull cross-section with two longitudinal bulkheads is shown in Figure 5.20. It is assumed that all plate thicknesses are the same (= h) and that the distance from the centre line to the longitudinal bulkhead is $B/4$ and that the moulded depth is half the breadth B ≡ 2a. The material is the same everywhere.

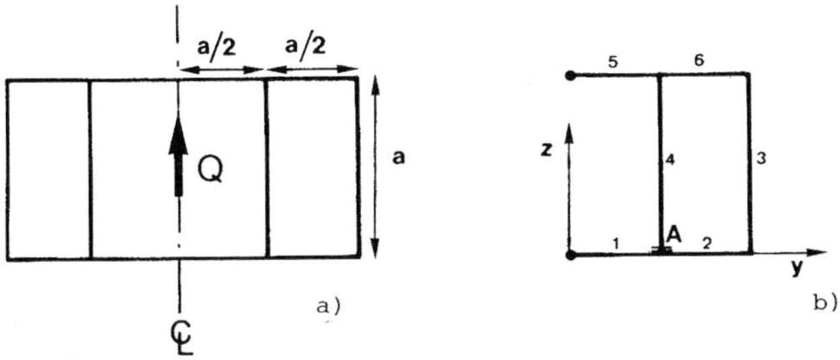

Figure 5.20 *Simplified hull cross-section with two longitudinal bulkheads.*

Due to symmetry only the cross-section to the right of the centre line plane is considered. With the (y,z)-coordinate system placed in the keel as shown in the figure, (5.5b) gives

$$z_0 = a/2$$
$$I_y = \frac{4}{3}a^3h$$

If the cut is placed in the closed cell at point A, see Figure 5.20, Eqs. (5.11)-(5.13) give

$$\bar{\tau}_1(s = y) = \beta y$$
$$\bar{\tau}_2(s = y - a/2) = \beta y$$
$$\bar{\tau}_3(s = z) = \beta(a + z - z^2/a)$$
$$\bar{\tau}_4(s = z) = \beta(z - z^2/a)$$
$$\bar{\tau}_5(s = y) = -\beta y$$
$$\bar{\tau}_6(s = y - a/2) = -\beta y$$

with

$$\beta = \frac{2\tau_0}{a}$$
$$\tau_0 = \frac{3Q}{16\,ah}$$

Eq. (5.19) then gives with $M = 1$:

$$\eta_{11} = \frac{3a}{h}$$

$$B_1 = \int_{a/2}^{a} \overline{\tau}_2(y)dy + \int_0^a \overline{\tau}_3(z)dz - \int_{a/2}^{a} \overline{\tau}_6(y)dy - \int_0^a \overline{\tau}_4(z)dz$$

$$= \frac{7a\,\tau_0}{2}$$

$$q_1 \equiv h\hat{\tau}_1 = -B_1/\eta_{11} = -\frac{7}{6}h\tau_0$$

According to Eq.(5.21) the final result then becomes

$$\tau_1 = \beta y$$

$$\tau_2 = \beta\left(y - \frac{7}{12}a\right)$$

$$\tau_3 = \beta\left(\frac{5}{12}a + z - \frac{z^2}{a}\right)$$

$$\tau_4 = \beta\left(\frac{7}{12}a + z - \frac{z^2}{a}\right)$$

$$\tau_5 = -\beta y$$

$$\tau_6 = -\beta\left(y - \frac{7}{12}a\right)$$

and it is shown in Figure 5.21.

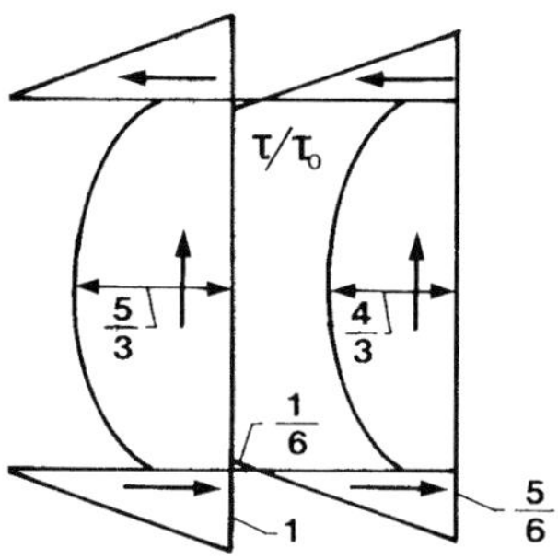

Figure 5.21 *Resulting shear stress distribution.*

The maximum stress is obtained in the longitudinal bulkhead for $z = a/2$, i. e. at the neutral axis:

$$\tau_{max} = \frac{5}{3}\tau_0 = \frac{5}{16}\frac{Q}{ah} = \frac{5}{4}\tau_{mean}$$

where

$$\tau_{mean} \equiv \frac{Q}{4ah}$$

is the mean stress defined as the shearing force divided by the total area of the vertical plate elements.

If the longitudinal bulkhead is omitted then, from Example 5.2.1:

$$I_y = \frac{7}{6}a^3h$$

and

$$\tau_{max} = \frac{15}{28}\frac{Q}{ah}$$

Hence, the addition of the longitudinal bulkheads implies a 33 per cent increase in material, a 12.5 per cent reduction of the bending normal stresses and 42 per cent reduction in maximum shear stress.

Example 5.2.3

Figure 5.22 shows the resulting shear stress distribution for the ship cross-section given in Figure 5.6. The calculations have been made by means of a computer program. The number of closed cells is $M = 13$.

The results show that the maximum shear stress for $Q = 1\text{N}$ is

$$\tau_{max} \simeq 1.6 \text{ N/m}^2 \simeq 2.4\,\tau_{mean}$$

as the vertical area is 1.535 m^2. The reason for the high maximum stress relative to the mean stress is that the vertical area here includes the area of the longitudinals. As seen from the results in Figure 5.22, the longitudinals do almost not carry any shear stresses and thus do not contribute to the shear stiffness of the cross-section. It should be mentioned that an *effective shear area* (to be defined in Chapter 6), becomes 0.7414 m^2. The mean stress based on this area becomes 1.2 N/m^2 and thus $\tau_{max}/\tau_{mean} = 1.3$.

As mentioned in Example 5.1.1 the design bending moment is approximately

$$M_y \simeq 2\ 10^9 \text{ Nm}$$

The magnitude of the maximum shear force Q can be estimated as

$$Q \simeq M_y/0.3L = 3.6\ 10^7 \text{ N}$$

so that the maximum shear stress will be

$$\tau_{\max} \simeq 1.6 \cdot 3.6\ 10^7 \text{N/m}^2 = 58 \text{MN/m}^2$$

It should be emphasised again that this maximum stress is found at the bending neutral axis where the bending stresses are zero.

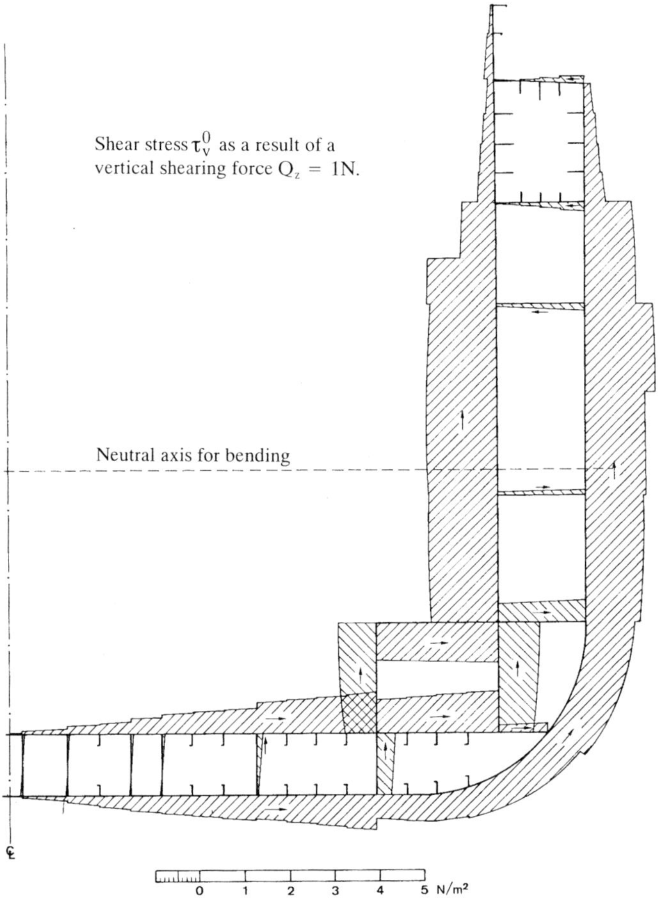

Figure 5.22 *Resulting shear stress distribution for a unit shear force.*

Example 5.2.4

In the Example 5.2.3 it was shown that the shear stresses in the longitudinal are negligible due to the free edge condition Eq. (5.13). Therefore, in hand calculations these stiffeners can be omitted by introducing an effective thickness

$$\bar{h} = h + A_s/s$$

of the associated plating, see Figure 5.23, where h is the original thickness, and s and A_s is the spacing between and area of the longitudinals, respectively. Thereby, the moment of inertia I_y and position of the neutral axis z_0 becomes approximately the same as in an exact calculation. However, Eq. (5.9) has to be replaced by

$$\frac{\partial(h\tau)}{\partial s} = -\bar{h}\frac{\partial \sigma}{\partial x}$$

as τ and σ now act on the thickness h and $\bar{h}$, respectively. Hence, the shear stress calculated by Eq. (5.11) should be multiplied by $\bar{h}/h$.

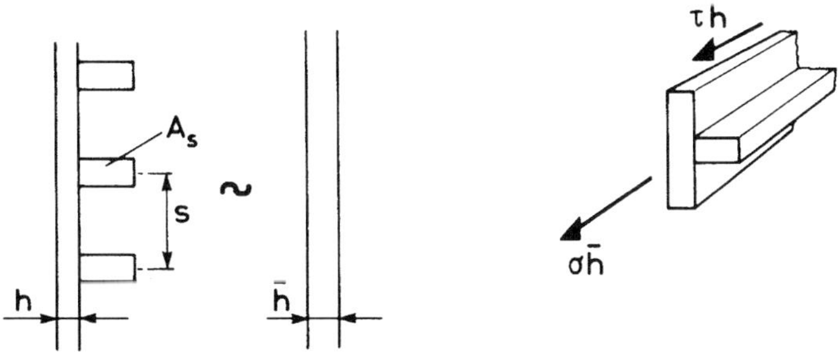

Figure 5.23 *Effective thickness approximation for shear stress calculations.*

5.3 TORSIONAL RESPONSE

If a prismatic beam is exposed to a torsional load, the deformation of the beam will consist partly of a rotation about a lengthwise torsional axis and partly of an axial deformation. These two deformation mechanisms are called *St Venant torsion* and *warping torsion*, respectively.

For prismatic beams built up of thin-walled elements which form one or more closed cells, c.f. Figure 5.19, the St Venant torsional stiffness will be dominant, so that the warping deformation can be neglected. On the contrary, if the thin-walled elements do not form large closed cells, the St Venant torsional stiffness will be so small that the torsional deformation of the beam is mainly limited by the fixation of the beam towards axial deformation.

Because of this division into two different deformation mechanisms, closed and open cross-sections are dealt with separately.

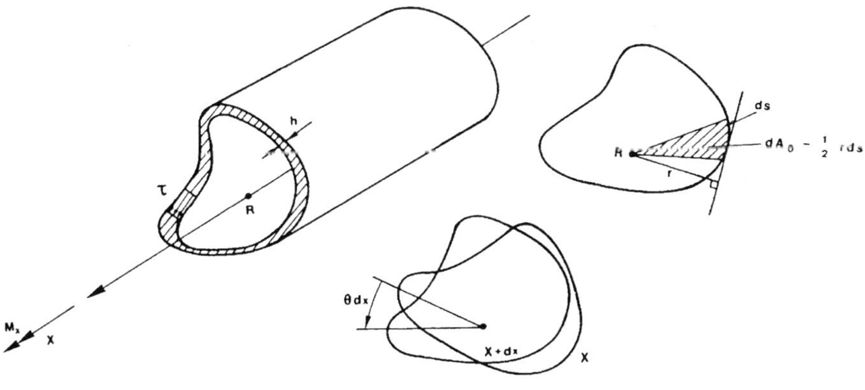

Figure 5.24 *St Venant torsion of closed pipes.*

5.3.1 Single-cell, Closed Cross-section

Figure 5.24 shows a thin-walled beam of this type. If this beam is subjected to a torque M_x, moment equilibrium yields

$$M_x = \oint \tau(s)\, h(s)\, r(s)\, ds = \oint q(s)\, r(s)\, ds \tag{5.22}$$

where the integral is extended along the complete circumference of the beam ($0 \leq s \leq \ell$). As before, the shear stress is $\tau(s)$ and the wall thickness is $h(s)$, while $q(s) \equiv \tau(s)\, h(s)$ is called the shearing force. Finally, $r(s)$ is the *moment arm* for $\tau(s)$, c.f. Figure 5.24.

As the stiffness of the cross-section towards axial deformation is neglected, axial stresses σ in the cross-section are also neglected. It follows then from the equilibrium equation (5.9) that $q \equiv h(s)\, \tau(s)$ is independent of the arc length s, and hence it is constant along the circumference of the pipe. Thus, Eq. (5.22) can be written

$$M_x = q \oint r(s)\, ds = 2A_0 q \tag{5.23}$$

where A_0 is the area circumscribed by the contour of the beam. According to Eq. (5.23), the shear stress $\tau = \tau(s)$ becomes

$$\tau(s) = \frac{M_x}{2A_0\, h(s)} \tag{5.24}$$

and it is approximately constant over the wall thickness.

As a result of the torque M_x, the pipe will be twisted the angle θ per unit of length along the beam axis (the x-axis). Hence,

$$\frac{\partial u_s}{\partial x} = \theta\, r(s)$$

where, c.f. Figure 5.18, $u_s = u_s(x, s)$ is the deformation in the direction of the arc length. The change of the angle ϕ, given by Eq. (5.14), becomes

$$\phi = \theta\, r(s) + \frac{\partial u_x}{\partial s}$$

which, by integration along the circumference of the cross-section and by use of Hooke's law: $\tau(s) = G(s)\, \phi(s)$, gives

$$\oint \frac{\tau(s)}{G(s)} ds = 2A_0\, \theta$$

or, by insertion of Eq. (5.24):

$$\theta = \frac{M_x}{4A_0^2} \oint \frac{ds}{G(s)\, h(s)} \tag{5.25a}$$

If the shear modulus G, as usually, is constant, Eq. (5.25a) is written as

$$\theta = \frac{M_x}{G\, I_d} \tag{5.25b}$$

where the *St Venant torsional constant* I_d is

$$I_d \equiv 4A_0^2 / \oint \frac{ds}{h(s)} \tag{5.25c}$$

This result is called *Bredt's formula.*

Example 5.3.1

For a (single-cell) closed box-shaped profile with the breadth B, the depth D and the constant wall thickness h the following is obtained:

$$I_d = \frac{4(BD)^2\, h}{2\,(B + D)}$$

and

$$\tau = \frac{M_x}{2BD\, h}$$

To get an impression of the order of magnitude of the deformation θ for a ship the following rather extreme expression is assumed for the torque, c.f. Example 2.3.1 in Chapter 2.

$$M_x \approx \frac{1}{8}\,\rho g\, B^2\, TL$$

where ρ is the density of the water and T is the additional draught due to the asymmetric mass distribution. If "realistic" numerical values are inserted (T= 1m, L = 200 m, B = 30 m, D = 20 m, h = 30 mm, $G = G_{steel} = 8.1 \times 10^{10}$ N/m^2), the result is

$$\theta \simeq 6\; 10^{-6}\ \mathrm{rad/m}$$

and

$$\tau \approx 6\mathrm{MN/m^2}$$

It is seen directly that the shear stress is negligible. The deformation is also very small as it corresponds to a horizontal displacement of the deck, measured from aft to fore part, of

$$\theta\, L\, D/2 \approx 12\mathrm{mm}$$

The same order of magnitude for τ and θ are found assuming a severe wave-induced torsional moment.

5.3.2 Multicell, Closed Cross-section

For a single-cell cross-section, the shear stress τ could be determined by Eq. (5.24) without considering the deformation of the cross-section. Analogously to the determination of the shear stress due to a shear force Q, this is not possible if the cross-section contains several closed cells.

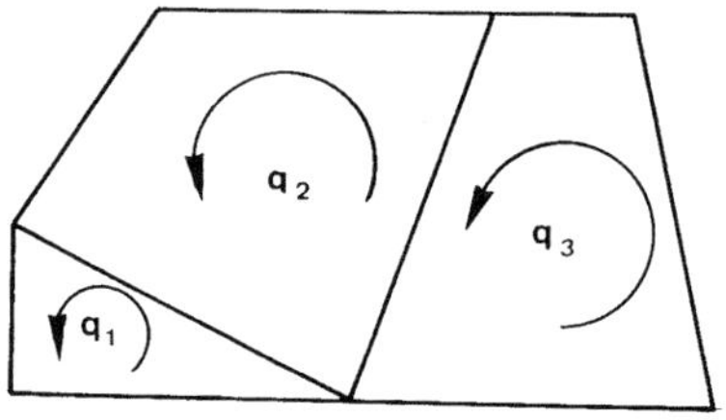

Figure 5.25 *Cross-section with several closed cells.*

If a cross-section M with closed cells is considered, c.f. Figure 5.25, it is seen that, as in the case of a single-cell cross-section, the shearing force $q = \tau(s)\,h(s)$ is constant along any plate element. In addition force equilibrium at each nodal point, Eq. (5.12), again yields

$$\sum_{J=1}^{J} q_j = 0 \qquad \text{at all points}$$

These two conditions are satisfied by assuming constant shearing forces $q_i = 1,2, .., M$ around each of the closed cells. These M unknown q_i are determined by the following equations.

Moment equilibrium gives one equation:

$$M_x = \sum_{i=1}^{M} \oint_i q_i r(s)\,ds = \sum_{i=1}^{M} q_i \oint_i r(s)\,ds = \sum_{i=1}^{M} 2q_i A_i \tag{5.26}$$

where A_i is the area which cell No. i circumscribes. This equation is sufficient if M=1 otherwise additional equations must be formulated. These are obtained by considering the deformation of the cross-section. Since the cross-section as a whole rotates about a longitudinal axis, the angular displacement θ per unit of length will be the same for all M cells. So, if the change of angle ϕ given by Eq. (5.14) is integrated around each cell, application of Hooke's law leads to

$$\oint_j \frac{\tau(s)}{G(s)}\,ds = 2A_j\,\theta; \quad j = 1,2,...,M$$

or, for constant shear modulus G:

$$q_j \oint_j \frac{ds}{h(s)} - \sum_{\substack{i=1 \\ i \neq j}}^{M} q_i \int_{i \cap j} \frac{ds}{h(s)} = 2GA_j\theta; \quad j = 1,2,...,M \tag{5.27}$$

Eqs. (5.26) and (5.27) total $M + 1$ equations for determination of the $M + 1$ unknowns $q_1, q_2, ..., q_M$ and θ.

If Eq. (5.27) is compared with Eqs. (5.18) - (5.19) it is seen that Eq. (5.27) can be written

$$[\eta]\{q\} = 2G\theta\{A\} \tag{5.28}$$

where, as earlier, $[\eta]$ is an $M \times M$ matrix with the elements

$$\eta_{jj} = \oint_j \frac{ds}{h}; \quad \eta_{ij} = \eta_{ji} = -\int_{i \cap j} \frac{ds}{h}; \quad i,j = 1,2,...M$$

and where the vectors

$$\{q\} = \{q_1, q_2, ..., q_M\}^T$$

and

$$\{A\} = \{A_1, A_2, ...A_M\}^T$$

If Eq. (5.28) is inserted in Eq. (5.26) the result is

$$M_x = 2\{A\}^T\{q\} = 4G\,\theta\{A\}^T[\eta]^{-1}\{A\}$$

or

$$\theta = \frac{M_x}{G\,I_d} \tag{5.29a}$$

with

$$I_d = 4\{A\}^T[\eta]^{-1}\{A\} \tag{5.29b}$$

The $M + 1$ unknowns are found most appropriately by first calculating $[\eta]^{-1}$, then determining θ by use of Eqs. (5.29a-b) and finally $\{q\}$ by Eq. (5.28):

$$\{q\} = 2G\,\theta[\eta]^{-1}\{A\} = \frac{2M_x}{I_d}[\eta]^{-1}\{A\} \tag{5.28a}$$

It should be emphasised that the matrix $[\eta]$ is the same as in the calculation of the shear stresses in Section 5.2. Moreover, it should be noted that the number of closed cells, even in the most complicated ship cross-sections, is usually modest ($M < 25$), so that an inversion of the matrix $[\eta]$ is an acceptable numerical method of solution by use of computers. When the shearing forces $q_1, q_2, ..., q_M$ have been determined, the stress τ_i(s) in the ith plate element is found to be

$$\tau_i(s) = \frac{1}{h_i(s)} \sum_{j=1}^{M} \delta_{ij} \, q_j \tag{5.30}$$

where

$$\delta_{ij} = \begin{cases} \pm\ 1 \text{ if plate element No.i is part Cell No.j} \\ 0 \ \text{ otherwise} \end{cases}$$

The sign of δ_{ij} is chosen to be positive, if $\tau_i(s)$ and q_j have the same direction, otherwise to be negative.

Example 5.3.2

The simplified box-shaped ship cross-section with two longitudinal bulkheads and constant plate thickness h, shown in Figure 5.20 and used in Example 5.2.2, is analysed.

In Example 5.2.2 symmetry ($\tau = 0 \ at \ C_L$) was used to reduce the problem to including only one closed cell. This cannot be done in the torsional stress calculation. However, the symmetry can be utilised to see that the shearing forces in the two outer cells must be the same but with opposite sign. If the central cell is called cell "1" and the outer cells "2", the following is obtained:

$$\eta_{11} = \frac{4a}{h}$$
$$\eta_{22} = 2 \times \frac{3a}{h} = \frac{6a}{h}$$
$$\eta_{12} = 2 \times \left(-\frac{a}{h}\right) = -\frac{2a}{h}$$

or

$$[\eta] = \frac{a}{h}\begin{bmatrix} 4 & -2 \\ -2 & 6 \end{bmatrix} \Rightarrow [\eta]^{-1} = \frac{h}{10a}\begin{bmatrix} 3 & 1 \\ 1 & 2 \end{bmatrix}$$

Besides

$$A_1 = a^2$$
$$A_2 = 2 \times \frac{1}{2}a^2 = a^2$$

Eq. (5.29b) gives

$$I_d = 4a^4 \frac{h}{10} a \,\{1, 1\} \begin{bmatrix} 3 & 1 \\ 1 & 2 \end{bmatrix} \begin{Bmatrix} 1 \\ 1 \end{Bmatrix} = \frac{14}{5} a^3 h$$

while Eq. (5.28a) leads to

$$h\begin{Bmatrix}\tau_1\\ \tau_2\end{Bmatrix} = \begin{Bmatrix}q_1\\ q_2\end{Bmatrix} = \frac{2M_x}{I_d}\frac{h}{10a}\begin{bmatrix}3 & 1\\ 1 & 2\end{bmatrix}\begin{Bmatrix}a^2\\ a^2\end{Bmatrix} = \frac{M_x}{14a^2}\begin{Bmatrix}4\\ 3\end{Bmatrix}$$

If the torsional stiffness I_d is compared with the result without longitudinal bulkhead (Example 5.3.1 with $B = 2a$, $D = a$):

$$I_d = \frac{8}{3}a^3h$$

it is seen that the longitudinal bulkheads only increase the torsional stiffness by 5 per cent, even if the increase of the material is 33 per cent. So, whereas the longitudinal bulkheads are very effective with regard to the bending and shearing rigidities, see Example 5.2.2, they have only a limited influence on the torsional rigidity.

5.3.3 Open Cross-sections

For cross-sections with closed cells the torque M_x is "carried" by shear stresses $\tau(s)$ which are constant through the plate thickness $h = h(s)$. This gives torsionally very stiff cross-sections with a torsional stiffness I_d of the order of magnitude:

$$I_d \approx b^3h$$

where b and h are respectively a characteristic cross-sectional dimension and an average plate thickness.

In the case of thin-walled cross-sections without closed cells, the torque M_x must be "carried" solely by a variation of the shear stress $\tau = \tau(s,h)$ over the plate element thickness h, as a constant value of τ cannot be found because of the presence of free edges, c.f. Eq. (5.13) and Figure 5.26 - Figure 5.27. The problem is dealt with in standard textbooks on strength of materials and only some main results will be given here as this case is not strickly relevant for ship sections.

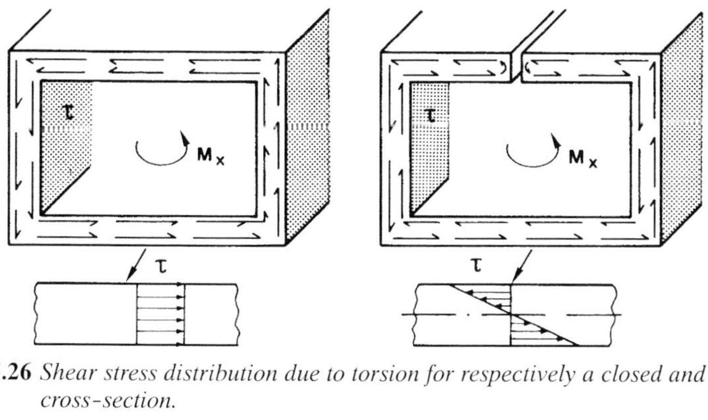

Figure 5.26 *Shear stress distribution due to torsion for respectively a closed and an open cross-section.*

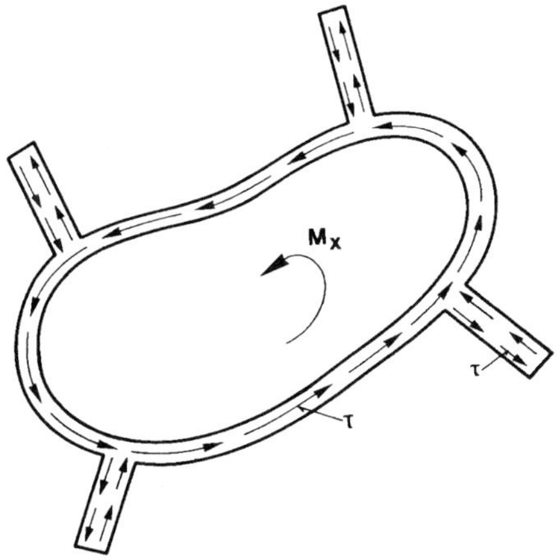

Figure 5.27 *Shear stress distribution τ in a closed cross-section with open plate elements exposed to torsion.*

As for closed cells, the angle of torsion θ per unit of length can be written

$$\theta = \frac{M_x}{GI_d}$$

but the torsional stiffness now becomes

$$I_d = \frac{1}{3} \int_{\ell} h^3 \, ds \tag{5.31}$$

where the integration is extended over all plate elements. The above value for I_d is of the order of magnitude

$$I_d \approx bh^3$$

and thus

$$\frac{(I_d)_{closed}}{(I_d)_{open}} \approx \left(\frac{b}{h}\right)^2$$

For typical ship cross-sections $b/h \approx 10^3$, so that an open cross-section will rotate in the order of magnitude of 1 million times more than the corresponding closed cross-section exposed to the same torque! This is of course physically unrealistic and is impeded by limitations in the axial (warping) deformation. Before warping deformation is considered, it should also be mentioned that the maximum transverse shear stress for an open cross-section is

$$\tau_{max} = G\,\theta h_{max} \tag{5.32}$$

and it is thus largest in the thickest plate element. In order to get a physical picture of the variation of the shear stresses over the plate elements, it may be mentioned that this variation will be the same as the variation of the inclination of a soap bubble blown over the same cross-section. If Eq. (5.32) is compared with Eq. (5.24), it is seen that the maximum shear stresses for an open cross-section are of the order of magnitude of b/h times larger than for a similar closed cross-section, again without considering limitations in the axial deformation.

Example 5.3.3

If the same box-shaped cross-section as used in Example 5.3.1 is considered but slit e.g. in the deck as shown in Figure 5.28, the result of Eq. (5.31) is

$$I_d = \frac{2}{3}\, h^3 (B + D)$$

whereas Eq. (5.32) yields

$$\tau_{max} = G\,\theta\, h = \frac{M_x\, h}{I_d} = \frac{3}{2}\frac{M_x}{(B + D)\, h^2}$$

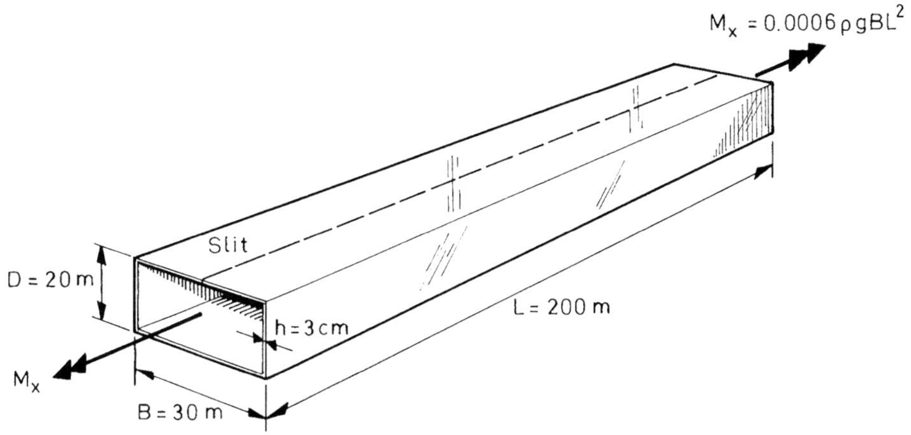

Figure 5.28 *Slit box-shaped cross-section.*

With the same numerical values as in Example 5.3.1

$$\theta = 0.3 \text{ rad/m}$$

is obtained, which corresponds to the hull girder from AP to FP rotating $\theta L = 620$ rad ≈ 100 rounds!

5.3.4 Warping Torsion

To get a realistic description of the deformation of beams with open, thin-walled cross-sections, the *warping deformation* u_x must be analysed, as well.

As shown above, the shear stresses are, as a result of a torque, equal to zero in the middle of open plate elements. According to Hooke's law: $\tau = G\phi$, the change of angle ϕ also becomes zero in the centre line of the plate elements. Thus, it is obtained from Eq. (5.14) that

$$\phi = \theta\, r(s) + \frac{\partial u_x}{\partial s} = 0 \tag{5.33}$$

The variation of the warping u_x can then be found to be

$$u_x = -\theta \int r(s)\, ds = -\theta \Omega(s) \tag{5.34}$$

where the *sector coordinate* $\Omega(s)$ is defined as

$$\Omega(s) \equiv \int r(s)\, ds \tag{5.35}$$

The warping $u_x(s,x)$ given by Eq. (5.34) and Eq. (5.35) will thus have the same variation over the cross-section for any section of a prismatic beam, as the sector coordinate Ω is the same, but the amplitude varies proportionally to θ, being the derivative of the *angular displacement* θ_x : $\theta = d\theta_x/dx = \theta_x'$. See Figure 5.29.

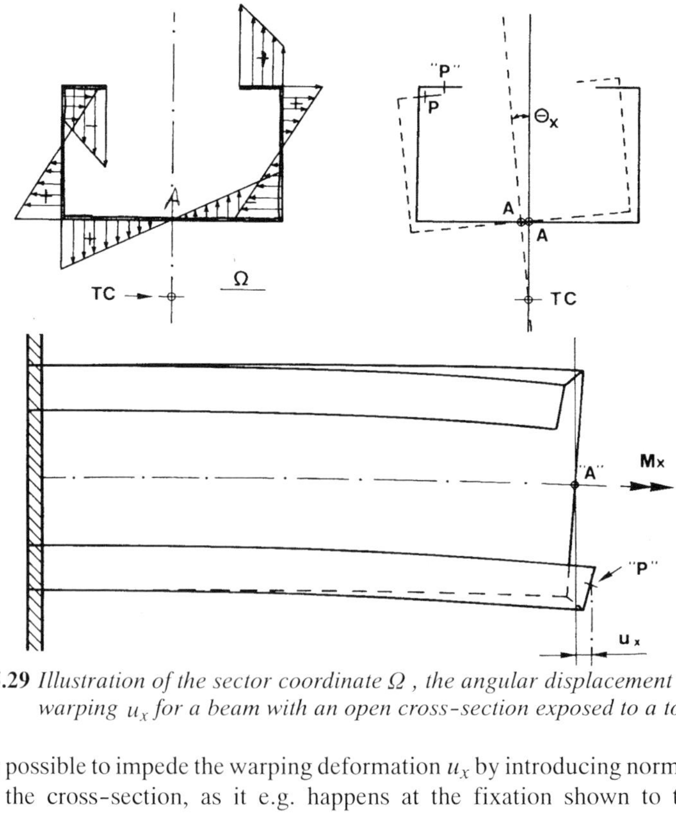

Figure 5.29 *Illustration of the sector coordinate Ω , the angular displacement θ_x and the warping u_x for a beam with an open cross-section exposed to a torque.*

It is only possible to impede the warping deformation u_x by introducing normal stresses σ_w into the cross-section, as it e.g. happens at the fixation shown to the left in

Figure 5.29. When restrictions are placed on the warping, the angular displacement of the beam will also be reduced. This may be illustrated by the idea of an empty cardboard shoe box exposed to torsion. Without a lid the angular displacement is much larger than with a lid. The lid impedes the warping of the ends of the box.

To determine the moment arm $r(s)$, c.f. Figure 5.24, and thus the sector coordinate $\Omega(s)$, the torsional centre has to be determined. According to Hooke's law, the axial stress σ_w becomes

$$\sigma_w = E\frac{\partial u_x}{\partial x} = -E\theta'\Omega(s) \tag{5.36}$$

with $\theta' \equiv d\theta/dx$. This normal stress σ_w, denoted the *warping normal stress*, must not lead to sectional bending moments nor an axial force on the cross-section, which gives

$$\begin{aligned} \int_A \sigma_w\, ydA &= -\theta' \int_\ell E\Omega(s)\, y(s)\, h(s)\, ds = 0 \\ \int_A \sigma_w\, zdA &= -\theta' \int_\ell E\Omega(s)\, z(s)\, h(s)\, ds = 0 \\ \int_A \sigma_w dA &= -\theta' \int E\Omega(s)h(s)ds = 0 \end{aligned} \tag{5.37}$$

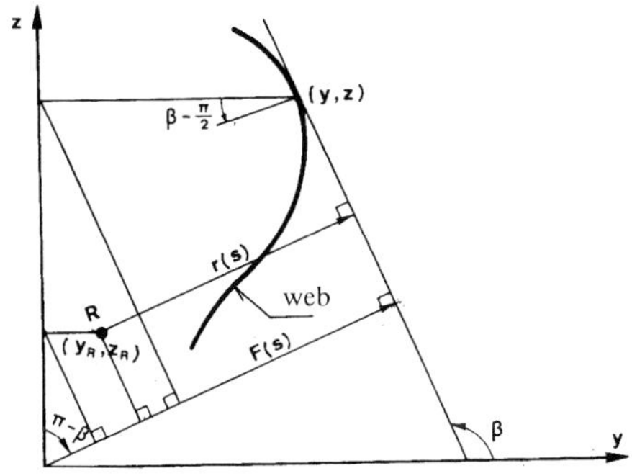

Figure 5.30 *Determination of the torsional centre R with the coordinates (y_R, z_R).*

It is seen from Figure 5.30 that the moment arm $r(s)$ measured from the *torsional centre* or the *shear centre R* can be expressed by the arm $F(s)$, measured from the (arbitrary) zero point of the yz-coordinate system, as

$$\begin{aligned} F(s) &= y\sin\beta - z\cos\beta \\ r(s) &= F(s) - y_R\sin\beta + z_R\cos\beta \end{aligned} \tag{5.38a}$$

or as $\cos\beta = \dfrac{dy}{ds}, \ \sin\beta = \dfrac{dz}{ds}$

$$r(s) = F(s) - y_R \frac{dz}{ds} + z_R \frac{dy}{ds} \tag{5.38b}$$

If Eq. (5.38b) is inserted in Eq. (5.35), the result is

$$\Omega(s) = \omega(s) - y_R z + z_R y + c \tag{5.39}$$

where c is an unknown constant to be determined and

$$\omega(s) = \int F(s)\, ds \tag{5.40}$$

can be considered to be known. If Eq. (5.39) is inserted in Eq. (5.37), the result becomes*

$$\begin{aligned}
&\int_\ell \omega\, y\, h\, ds - y_R \int_\ell y\, z\, h\, ds + z_R \int_\ell y^2\, h\, ds + c \int_\ell y\, h\, ds = 0 \\
&\int_\ell \omega\, z\, h\, ds - y_R \int_\ell z^2\, h\, ds + z_R \int_\ell y\, z\, h\, ds + c \int_\ell z\, h\, ds = 0 \\
&\int_\ell \omega h\, ds - y_R \int_\ell zh\, ds + z_R \int_\ell yh\, ds + c \int_\ell h\, ds = 0
\end{aligned} \tag{5.41}$$

which are the three linear equations to determine the torsional centre y_R, z_R and the constant c. The integral $\int_\ell$ extends over all plate elements in the cross-section. If the cross-section, as it is usual for ship hulls, is symmetric about the z-axis, Eq. (5.41) is reduced to

$$\begin{aligned}
y_R &= 0 \\
z_R &= -\int_\ell \omega\, y\, h\, ds / I_z \\
c &= 0
\end{aligned} \tag{5.42}$$

where $I_z = \int_\ell y^2\, h\, ds$ is the moment of inertia about the symmetry axis. Here use is made of $\omega(y, z) = -\omega(-y, z)$ for cross-sections which are symmetric about the z-axis.

* In the following, it is assumed that the modulus of elasticity is constant.

It is seen that

$$\int_{\ell} \Omega(s)\, h\, ds = 0 \tag{5.43}$$

When the sector coordinate $\Omega(s)$ has been determined by application of Eqs. (5.38a), (5.40), (5.41) as well as (5.39) only determination of $\theta = \theta(x)$ remains in order that the warping deformation $u_x(x, s)$ and the corresponding normal stress $\sigma_w(x, s)$ can be calculated. The determination of $\theta = \theta(x)$ is treated now.

If a segment of a plate element is looked at, considerations of equilibrium, c.f. Figure 5.14 and Eq. (5.9), imply that the warping normal stress σ_w leads to a *warping* shear stress, τ_w, given by

$$\frac{\partial h\tau_w}{\partial s} = -\,h\frac{\partial \sigma_w}{\partial x} = Eh\,\theta''(x)\,\Omega(s)$$

by use of Eq. (5.36). If the above expression is integrated, the outcome is

$$h\tau_w = E\theta''(x) \int h(s)\,\Omega(s)\,ds \tag{5.44}$$

There is a discrepancy between Eq. (5.33) and Eq. (5.44), as Eq. (5.33) assumes that there is no shear stress in the mean lines of the plate elements, while Eq. (5.44) gives precisely such stresses. This discrepancy is neglected by assuming that the contribution of the warping shear stresses $\tau_w's$ to the shear strain $(\phi/2)$ is insignificant.

By analogy with Eq. (5.22), the torque from τ_w becomes

$$\begin{aligned} M_w(x) &= \int_{\ell} h\,\tau_w r\,ds = E\,\theta'' \int_{\ell}\left[\int h\,\Omega\,ds\right] r\,ds \\ &= E\,\theta''\left[\int_{\ell} h\,\Omega\,ds \int_{\ell} r\,ds - \int_{\ell}\left[h\,\Omega \int r\,ds\right] ds\right] \end{aligned}$$

by partial integration. The expression can also be written as

$$M_w(x) = -\,\theta''(x)\,E \int_{\ell} \Omega^2\,h\,ds = -\,E\,I_{\Omega\Omega}\,\theta''(x) \tag{5.45}$$

by use of Eq. (5.35) and Eq. (5.43).

The constant

$$I_{\Omega\Omega} \equiv \int_{\ell} \Omega^2 h \, ds \tag{5.46}$$

is called the *sectorial moment of inertia* or the *warping constant*. Table 5.3 gives the values of $I_{\Omega\Omega}$ for a number of simple open cross-sections.

The sectional moment M_x thus implies two distributions of shear stresses: One due to St Venant torsion and one (τ_w) due to warping shear stresses. Each of these two distributions of shear stresses gives a resulting torque whose sum shall be the sectional moment M_x:

$$GI_d \theta - EI_{\Omega\Omega} \theta'' = M_x \tag{5.47}$$

For a prismatic beam the differential equation has the homogeneous solution for the derivative θ of the angular displacement θ_x:

$$\theta(x) = A \cosh(\alpha x) + B \sinh(\alpha x) \tag{5.48}$$

where A and B are two arbitrary constants and where

$$\alpha = \sqrt{\frac{GI_d}{EI_{\Omega\Omega}}} \tag{5.49}$$

To the solution Eq. (5.48) a particular solution dependent on M_x must be added up. As an example M_x equal to a constant is considered. Then the complete solution can be written

$$\theta(x) = \frac{M_x}{GI_d} + A \cosh(\alpha x) + B \sinh(\alpha x) \tag{5.50}$$

The constants A and B can be determined from given values of θ at the ends of the beam ($x = 0, L$). If it is assumed that the beam is fixed towards warping at the beam ends then $\theta(0) = \theta(L) = 0$, and the following is obtained:

$$A = -\frac{M_x}{GI_d}$$

$$B = \frac{M_x}{GI_d} \frac{\cosh \alpha L - 1}{\sinh \alpha L} = \frac{M_x}{GI_d} \tanh(\alpha L/2)$$

and thus, by insertion in Eq. (5.50):

$$\theta(x) = \frac{M_x}{GI_d} \left[1 - \frac{\cosh[\alpha(x - L/2)]}{\cosh(\alpha L/2)} \right] \tag{5.51}$$

Table 5.3 Sectorial moment of inertia for open cross-sections.

No.	Cross-section	Formula
1	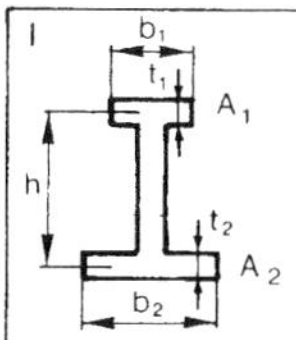	$A_1 = b_1 t_1$, $A_2 = b_2 t_2$ $I_{\Omega\Omega} = \frac{h^2}{12} \frac{A_1 b_1^2 A_2 b_2^2}{A_1 b_1^2 + A_2 b_2^2}$
2	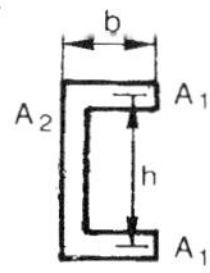	$A_1 = \frac{1}{3} \cdot$ Flange area $A_2 =$ Web area $I_{\Omega\Omega} = \frac{b^2 h^2}{24} A_1 (1 + \frac{3A_2}{6A_1 + A_2})$
3	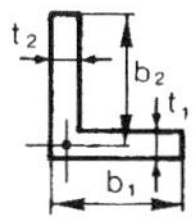	$I_{\Omega\Omega} = \frac{1}{36} (b_1^3 t_1^3 + b_2^3 t_2^3)$
4	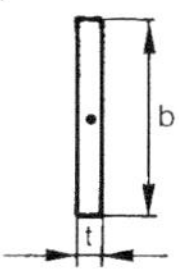	$I_{\Omega\Omega} = \frac{1}{144} b^3 t^3 \frac{b^2 - t^2}{b^2 + t^2}$
5	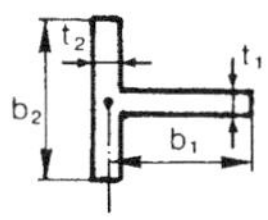	$I_{\Omega\Omega} = \frac{1}{36} (b_1^3 t_1^3 + \frac{1}{4} b_2^3 t_2^3)$
6	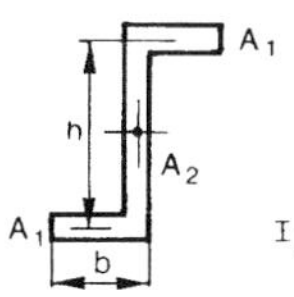	$A_1 = \frac{1}{2} \cdot$ flangeareal $A_2 =$ kropareal $I_{\Omega\Omega} = \frac{b^2 h^2}{24} A_1 (1 + 3\frac{A_2}{A_1})$

Integrating $\theta(x)$ yields total angular deflection

$$\theta_x(L) = \int_0^L \theta(x)dx = \frac{M_x}{GI_d} L \left(1 - \frac{\tanh(\alpha L/2)}{\alpha L/2}\right) \tag{5.52}$$

It follows from Eq. (5.47) that the part of the torque M_x which is taken by St Venant torsion will be

$$M_s(x) \equiv GI_d\theta = M_x\left[1 - \frac{\cosh[\alpha(x - L/2)]}{\cosh(\alpha L/2)}\right] \tag{5.53a}$$

while the warping torque becomes

$$M_w(x) \equiv -EI_{\Omega\Omega}\theta'' = M_x\frac{\cosh[\alpha(x - L/2)]}{\cosh(\alpha L/2)} \tag{5.53b}$$

These two torques are shown in Figure 5.31.

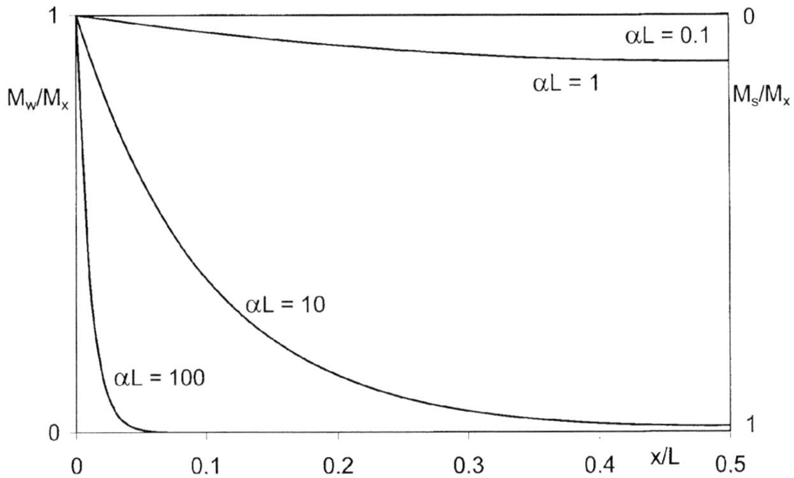

Figure 5.31 *The variation of the St Venant torque M_s and the warping torque M_w along the beam axis for a beam fixed at the ends and with different values of αL.*

It is seen that the greater dimensionless constant

$$\alpha L = \sqrt{\frac{GI_d}{EI_{\Omega\Omega}}}L \tag{5.54}$$

the less importance is the warping torsion, except at the fixations where $\theta = 0$ implies that the St Venant moment $GI_d\theta$ is zero, irrespective of the value of αL.

Finally, it should be noted that, if the beam can be freely deformed at the beam ends, it follows from Eq. (5.36), with $\sigma_w = 0$, that

$$\theta'(0) = \theta'(L) = 0$$

The solution Eq. (5.50) of the differential Eq. (5.47) then becomes

$$\theta(x) = \frac{M_x}{GI_d} = \text{constant}$$

i.e. solely St Venant torsion.

When the deformation $\theta = \theta(x)$ has been determined, the normal stresses σ_w can be determined by Eq. (5.36) and the warping shear stresses τ_w by Eq. (5.44). Both stresses σ_w and τ_w are (approximately) constant over the plate element thickness h.

It should be emphasised here that the necessity of determining the deformation of the whole beam $\theta = \theta(x)$ when the stresses in a given section is to be calculated, is a major difference of the warping stress determination compared to both the bending stress and the shear stress calculation as well as to St. Venant torsion.

Example 5.3.4

The sector coordinate $\Omega(s)$ and the sectorial moment of inertia for the cross-section shown in Figure 5.32 is determined. The plate thickness is h everywhere.

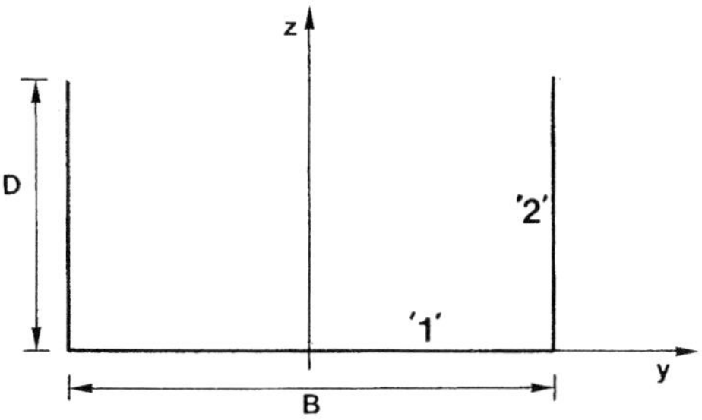

Figure 5.32 *Open box profile.*

First the torsional centre R is determined. As indices 1 and 2 designate plate elements 1 and 2, Eq. (5.38a) gives

$$F_1(s) = 0; \qquad F_2(s) = B/2$$

which, by insertion in Eq. (5.40), yields

$$\begin{aligned} \omega_1(s) &= 0 \\ \omega_2(s) &= sB/2; \qquad 0 \le s \le D \end{aligned}$$

The torsional centre is determined by Eq. (5.42):

$$\begin{aligned} y_R &= 0 \\ z_R &= -\frac{2}{I_z}\int_0^D s\frac{B}{2}\frac{B}{2}\,h\,ds = -D\frac{3D}{B+6D} \end{aligned}$$

as $I_z = B^2h(B/12 + D/2)$. Since z_R is negative, the torsional centre lies under the keel. Eq. (5.39) gives the sector coordinate, with $c = 0$

$$\begin{aligned} \Omega_1(s) &= \omega_1(s) + z_R y = z_R s \\ \Omega_2(s) &= \omega_2(s) + z_R\,B/2 = \frac{B}{2}(s + z_R) \end{aligned}$$

This variation is shown in Figure 5.33 for $B/D = 2$.

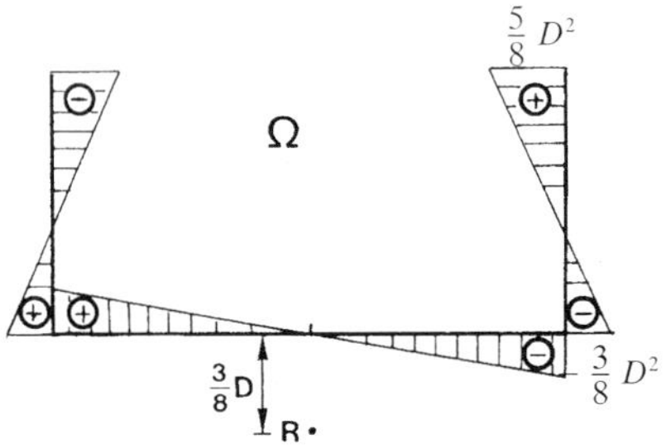

Figure 5.33 *The sector coordinate $\Omega(s)$and the torsional centre R (B/D = 2).*

According to Eq. (5.46), the sectorial moment of inertia becomes

$$I_{\Omega\Omega} = 2h\left[\int_0^{B/2} \Omega_1^2(s)\, ds + \int_0^D \Omega_2^2(s)\, ds\right]$$

$$= 2h\left[z_R^2 \frac{1}{3}\left(\frac{B}{2}\right)^3 + \left(\frac{B}{2}\right)^2 \frac{1}{3}\left\{(D + z_R)^3 - z_R^3\right\}\right]$$

$$= \frac{1}{12} h\, B^2\, D^3 \frac{2B + 3D}{B + 6D}$$

in agreement with Table 5.3 case 2.

The constant α, given by Eq. (5.49), becomes

$$\alpha = \sqrt{\frac{GI_d}{EI_{\Omega\Omega}}} = 2\frac{h}{BD} \cdot \sqrt{\frac{1}{2(1+\nu)} \frac{(B + 2D)(B + 6D)}{D(2B + 3D)}}$$

Insertion of the same numerical values as in Example 5.3.1: $L = 200$ m, $B = 30$ m, $D = 20$ m, $h = 30$ mm as well as Poisson's ratio $\nu = 0.3$ results in

$$z_R = -8\text{m} \quad ; \quad \alpha L = .026$$

The variation of the angular displacement is given by Eq. (5.51). Assuming clamped ends: $\theta(0) = \theta(L) = 0$ the largest angular variation occurs in the middle of the beam:

$$\theta(L/2) = \frac{M_x}{GI_d}\left(1 - \frac{1}{\cos h\alpha L/2}\right) = 8.4 \cdot 10^{-5} \frac{M_x}{GI_d}$$

This angular displacement per unit of length is thus much smaller than that which would occur, if the beam could be freely deformed axially at the ends of the beam.

With the same empirical torque

$$M_x \approx 2.26\ 10^8 \text{Nm}$$

as in Example 5.3.1, the following is obtained:

$$\theta(L/2) = 8.4\ 10^{-5} \cdot 4.43 \text{ rad/m} = 3.7\ 10^{-4} \text{ rad/m}$$

which is only approximately 100 times greater than found in Example 5.3.1 for the closed box-shaped cross-section. The total angular deflection, Eq. (5.52), becomes

$$\theta_x(L) = 0.05\ rad$$

corresponding to a horizontal shear of the deck of $0.05(D - z_R) = 1.4m$, which of course is larger than accepted and additional stiffening against torsional deformation is required.

5.3.5 Cross-sections with both Open and Closed Cells

A real ship cross-section contains both open plate elements and plate elements forming closed cells, see e.g. Figure 5.6.

The following calculation method is used for determining the deformation $\theta = \theta(x)$ and the corresponding stresses in such cross-sections subjected to a torque M_x:

First the St Venant torsional stresses τ_s are determined for the closed cells, neglecting all open plate elements, by application of the Eqs. (5.26)-(5.30) for a unit torque $M_x = 1$(Nm). This "stress" distribution is designated $\tau_0(s)$. Eq. (5.14) now gives

$$\phi = \theta r(s) + \frac{\partial u_x}{\partial s} = \frac{\tau_s}{G} = \frac{\tau_0(s)}{G} M_s(x) \tag{5.55}$$

which by integration gives

$$u_x(x,s) = -\theta(x)\int r(s)\,ds + \frac{M_s(x)}{G}\int \tau_0(s)\,ds$$

or, as $\theta = M_s/GI_d$:

$$u_x(x,s) = -\theta(x)\int\Big[r(s) - I_d\,\tau_0(s)\Big]ds = -\theta\,\Omega(s) \tag{5.56}$$

where the *sector coordinate* $\Omega(s)$ defined as

$$\Omega(s) \equiv \int\Big[r(s) - I_d\,\tau_0(s)\Big]\,ds \tag{5.57}$$

is a generalisation of the definition Eq. (5.35) for open cross-sections. Subsequently, the analysis and the results are the same as found for the open cross-sections:

Torsional centre:	Eqs. (5.37)-(5.42)
Warping normal stress σ_w:	Eq. (5.36)
Warping shear stress τ_w:	Eq. (5.44)
Equilibrium equation:	Eqs. (5.47)-(5.54)

Finally, it should be noted that the present analysis is the so-called classical method for torsional analysis of prismatic beams. The assumption about a prismatic beam applies only approximately to a ship's hull. Especially, the coupling of two prismatic beam segments with quite different torsional stiffness (open and closed cross-sections) must be accounted for in a rational manner. This is discussed in Section 5.3.6.

Example 5.3.5

The closed single-cell box-shaped cross-section used in Example 5.3.1 is reconsidered.

From Example 5.3.1

$$\tau_0(s) = \frac{1}{2BDh} \; (= \text{constant})$$

so that Eq. (5.57) gives

$$\Omega(s) = \int [r(s) - k] \, ds$$

with

$$k = \frac{I_d}{2BDh} = \frac{BD}{B + D}$$

For reasons of symmetry it is seen directly that the torsional centre has the coordinates $(y_R, z_R) = (0, D/2)$.

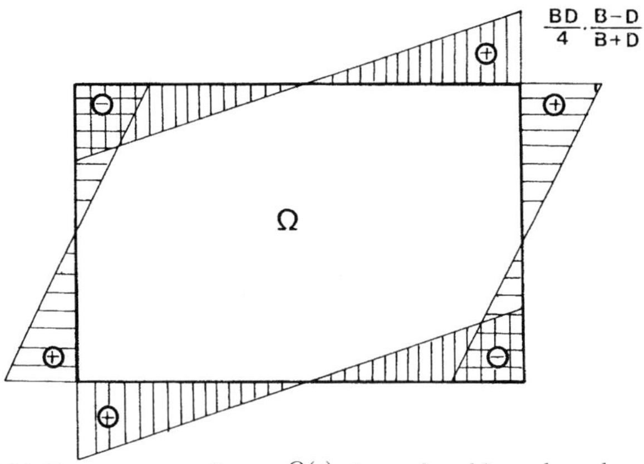

Figure 5.34 *The sector coordinate $\Omega(s)$ for a closed box-shaped cross-section.*

Thus, the sector coordinate $\Omega(s)$ becomes

$$\Omega_1(s) = \int \left[\frac{D}{2} - k\right] ds = -\frac{D}{2}\frac{B - D}{B + D}\, s \, ; \quad -\frac{B}{2} \le s \le B/2$$

$$\Omega_2(s) = \int \left[\frac{B}{2} - k\right] ds = \frac{B}{2}\frac{B - D}{B + D}\left(s - \frac{D}{2}\right); \quad 0 \le s \le D$$

where index 1 indicates the keel (and the deck with an opposite sign), and index 2 indicates the right side (and the left side with an opposite sign). The sector coordinate is shown in Figure 5.34.

The sectorial moment of inertia is obtained by Eq. (5.46):

$$I_{\Omega\Omega} = 4h\left[\int_0^{B/2} \Omega_1^2(s)\,ds + \int_0^{D/2} \Omega_2^2(s)\,ds\right]$$
$$= \frac{h}{24}\frac{[BD(B-D)]^2}{B+D}$$

It is seen that, for a quadratic cross-section ($B = D$), $I_{\Omega\Omega} = 0$. Such cross-sections are called warping-free, to which also circular, thin-walled cross-sections belong.

The constant α, given by Eq. (5.49), becomes

$$\alpha = \sqrt{\frac{GI_d}{EI_{\Omega\Omega}}} = \frac{1}{|B-D|}\cdot\sqrt{\frac{24}{1+\nu}}$$

Again the same numerical values are inserted as in Example 5.3.1. This gives

$$\alpha L = 85.9$$

It is seen from Figure 5.31 that, in this case, the warping moment - $EI_{\Omega\Omega}\,\theta''$ will be of no importance, except in a very small region around the fixed edges ($x = 0, L$). This supports the original assumption that the warping deformation can be neglected for closed, thin-walled cross-sections.

Example 5.3.6

The ship cross-section, given in Figure 5.6 and previously considered in Example 5.1.1 and Example 5.2.2, is analysed on computer.

The result is

Torsional centre R $(y_R, z_R) = (0, -7.901 \text{ m})$

St Venant torsional constant $I_d = 5.614 \text{ m}^4$

Sectorial moment of inertia $I_{\Omega\Omega} = 9752 \text{ m}^6$

Figure 5.35-Figure 5.37 show respectively the St Venant shear stress τ_s, Eq. (5.30), the warping normal stress σ_w, Eq. (5.36) and the warping shear stress τ_w, Eq. (5.44), all for a unit load:

St Venant shear stress: $\tau_s/M_s \equiv \tau_s^0$

Warping normal stress: $\sigma_w/M_\Omega \equiv \sigma_w^0$

Warping shear stress: $\tau_w/M_w \equiv \tau_w^0$

where the St Venant torque $M_s \equiv GI_d\theta$, the warping torque $M_w \equiv -EI_{\Omega\Omega}\theta''$, and the *warping bimoment* $M_\Omega \equiv -EI_{\Omega\Omega}\theta'$ have been use for the normalization of the stresses. All these factors can be calculated when $\theta(x)$ is known.

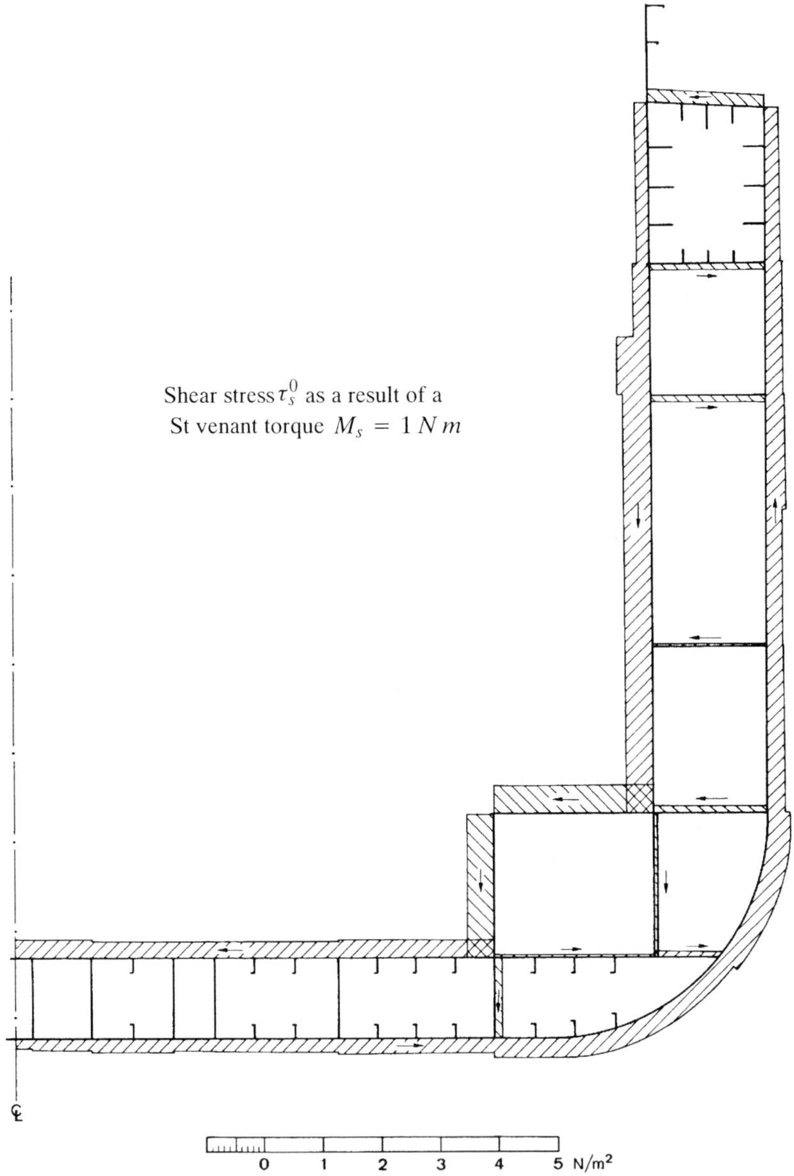

Figure 5.35 *St Venant distribution of shear stresses.*

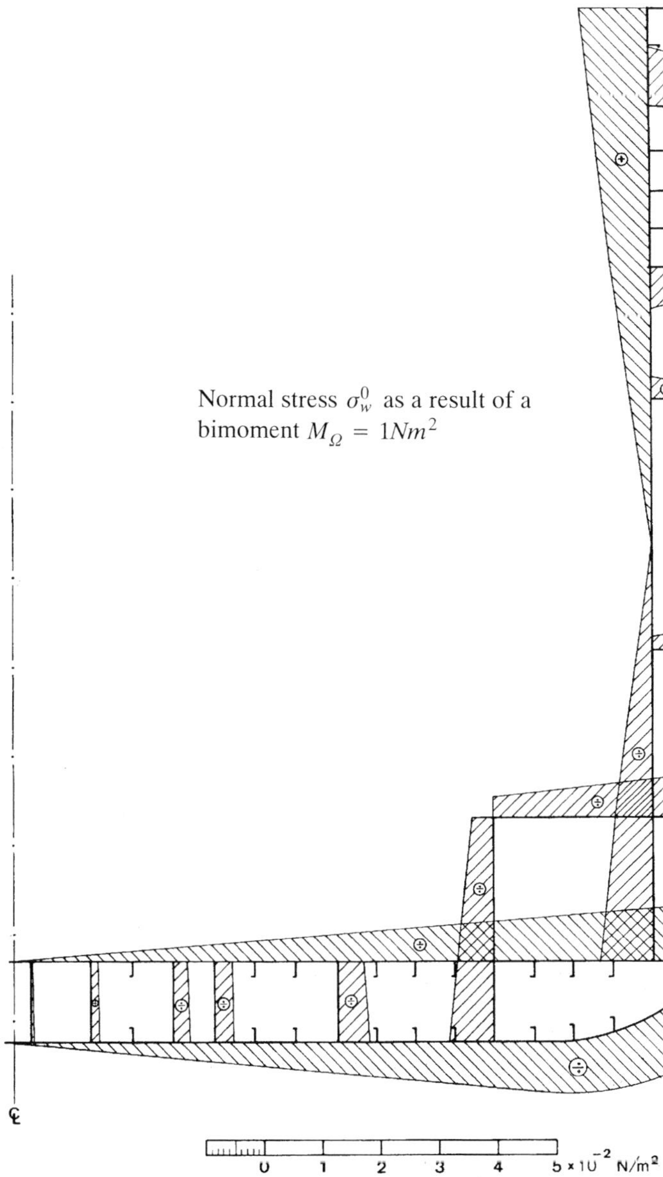

Figure 5.36 *The warping normal stress.*

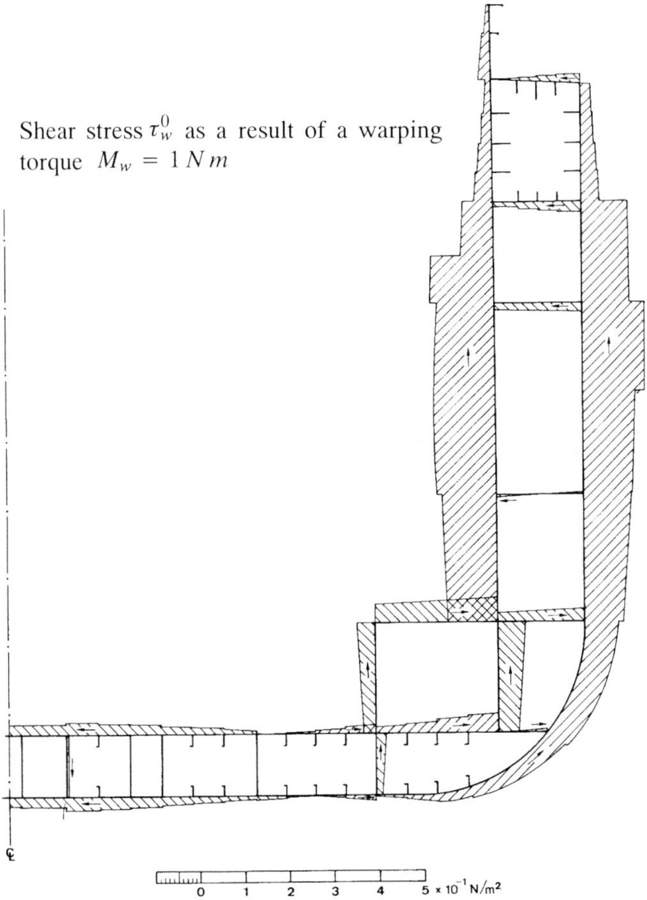

Figure 5.37 *The warping shear stress.*

To get an idea of the deformation of this container ship as a consequence of a constant representative value of the torque

$$M_x = 5.4\ 10^7\ Nm$$

it is assumed that the shown cross-section is unchanged amidships over a length of $\ell = 120$ m and that the section can be considered as fixed against axial deformations at the engine room and at the forepart. This results in

$$\alpha\ell = 1.79$$

and the angular displacement $\theta(x)$ per unit of length becomes

$$\theta(x) = \frac{M_x}{GI_d}\left[1 - \frac{\cosh[\alpha(x - \ell/2)]}{\cosh(\alpha\ell/2)}\right]$$

The total angular displacement θ_x of the section from the engine room to the forepart is obtained by integration of $\theta(x)$; Eq. (5.52)

$$\theta_x \equiv \int_0^{\ell} \theta(x)\, dx = \frac{M_x}{GI_d}\, \ell \left[1 - \frac{2}{\alpha \ell} \tanh(\alpha \ell /2) \right]$$

$$= 0.202 \frac{M_x}{GI_d} \ell = 2.9\ 10^{-3} \text{ rad}$$

As the torsional centre lies 7.901 m below the keel, the horizontal deformation U of the hatch coaming, measured from the engine room to the forepart and with a distance from keel to hatch coaming of 17.44 m, becomes

$$U \equiv \theta_x \times (17.44 + 7.901) = 73 \text{ mm}$$

When $\theta = \theta(x)$ has been determined, the corresponding stresses can be determined by Figure 5.35–Figure 5.37. As an example the warping normal stress σ_w in the hatch coaming at the passage to the engine room ($x = 0$) becomes

$$\sigma_w \approx 1.2\ 10^{-2} \times (-EI_{\Omega\Omega} \times \theta'(O))\ \text{N/m}^2$$

$$= -1.2\ 10^{-2} \times \frac{1}{\alpha} \tanh(\alpha \ell /2)\, M_x = -31\ \text{MN/m}^2$$

However, this result should be considered with some scepticism as it depends much on the modelling of the fixation.

5.3.6 Torsional Analysis of Hulls

The torque to be absorbed by a ship's hull which sails in waves, see Figure 5.38, affects to the greatest extent ships with open cross-sections, because these are "torsionally" weak and because the torque about the torsional centre is increased, as the torsional centre for open cross-sections usually lies far below the ship's bottom.

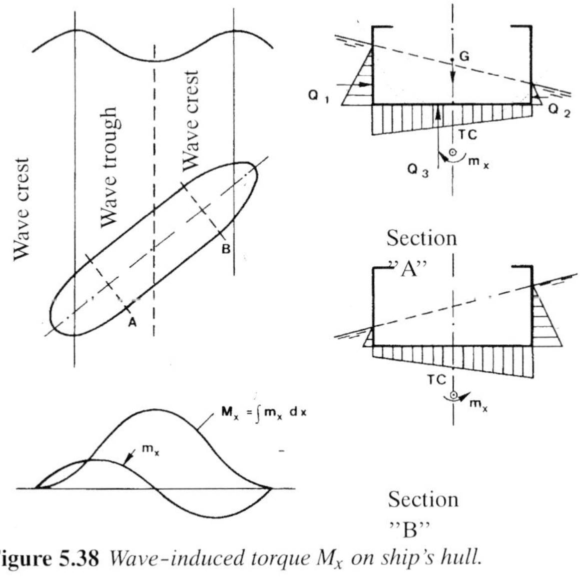

Figure 5.38 *Wave-induced torque M_x on ship's hull.*

The simplest possible torsional model of a ship's hull is obtained by neglecting the influence of possible deck beams, so that a long, almost prismatic hold is obtained, see Figure 5.39. If it can be assumed that the hold is entirely prismatic, an analytical solution of the governing differential equation, Eq. (5.47), with constant coefficients, can be determined as shown in the previous sections. As regards the influence of the ends of the ship, it is modelled either by introducing a warping stiffness or by assuming that the open area is a little longer than it actually is and subsequently assuming that the prismatic hull element is clamped towards warping at the ends. The method requires great experience, as the result with respect to both stresses and deflections depends strongly on the assumed edge conditions.

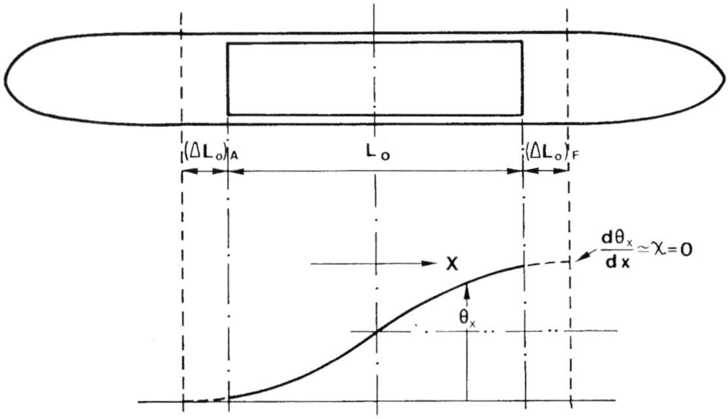

Figure 5.39 *Simple model for torsional analysis of ship's hull.*

Example 5.3.7

Spring-stiffened end conditions.

Figure 5.40 *End conditions for torsional analysis of beams.*

At the ends $x = 0$ and $x = L$ a flexibility can be introduced by the spring stiffnesses k_0 and k_L, see Figure 5.40:

$$\begin{aligned} k_0\, u_x(0,s) &= \sigma_w(0,s) \\ k_L\, u_x(L,s) &= -\sigma_w(L,s) \end{aligned}$$

From Eqs. (5.34) and (5.36) it follows that

$$k_0\theta(0) = E\theta'(0)$$

and

$$k_L\theta(L) = -E\theta'(L)$$

For a prismatic beam with $k_0 = k_L = k$ and a constant torque M_x, the solution to Eq. (5.47) with these boundary conditions becomes

$$\theta(x) = \frac{M_x}{GI_d} + A\cosh\left(\alpha(L/2 - x)\right) + B\sinh\left(\alpha(L/2 - x)\right)$$

Due to symmetry, $B = 0$ and

$$k\left(\frac{M_x}{GI_d} + A\cosh(\alpha L/2)\right) = -EA\,\alpha\,\sinh(\alpha L/2)$$

or

$$A = -\frac{M_x}{GI_d}\,\frac{1}{\cosh(\alpha L/2) + \frac{E\alpha}{k}\sinh(\alpha L/2)}$$

and hence,

$$\theta(x) = \frac{M_x}{GI_d}\left[1 - \frac{\cosh(\alpha(L/2 - x))}{\cosh(\alpha L/2) + \frac{E\alpha}{k}\sinh(\alpha L/2)}\right]$$

It is seen that for the solutions corresponding to $k \to 0$ and $k \to \infty$ free and a clamped beam are obtained, respectively. Note, that the unit for k is force/length^3. Generally, it is very difficult to assign appropriate values to k, but the classification societies have some recommended, albeit empirical values.

A model which models the whole ship and thus avoids assumptions about the edge conditions must comprise open sections as well as closed sections at the engine room and at the ends of the ship, see Figure 5.41. Torsional analysis can be performed of both the open and the closed sections. However, difficulties arise in coupling these different sections in a consistent way, because the sections warp differently. That is, the plane of the sections is deformed in such a way in the longitudinal direction that the deformation patterns do not match as the sector coordinates $\Omega(s)$ are different. A formulation minimizing this gab is given by Pedersen (1983). The method makes use of a coupling between torsion and horizontal bending. In the coupling procedure the two sections can normally not be made to cohere entirely without overlap and voids. But the additional deformations needed to make the sections compatible are of a local kind, so that the stresses and deformations which they imply quickly die out. It may be said that the St Venant assumption applies. This does not mean that the stresses corresponding to these necessary extra deformations are small. They may only be considered as local stress concentrations.

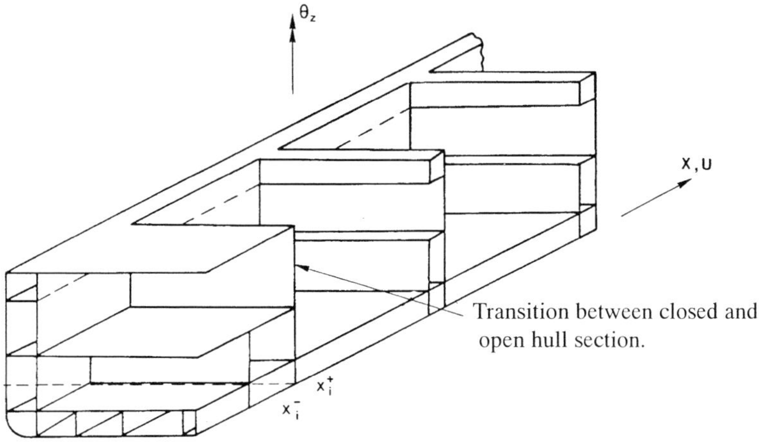

Figure 5.41 *Transition between closed and open hull section.*

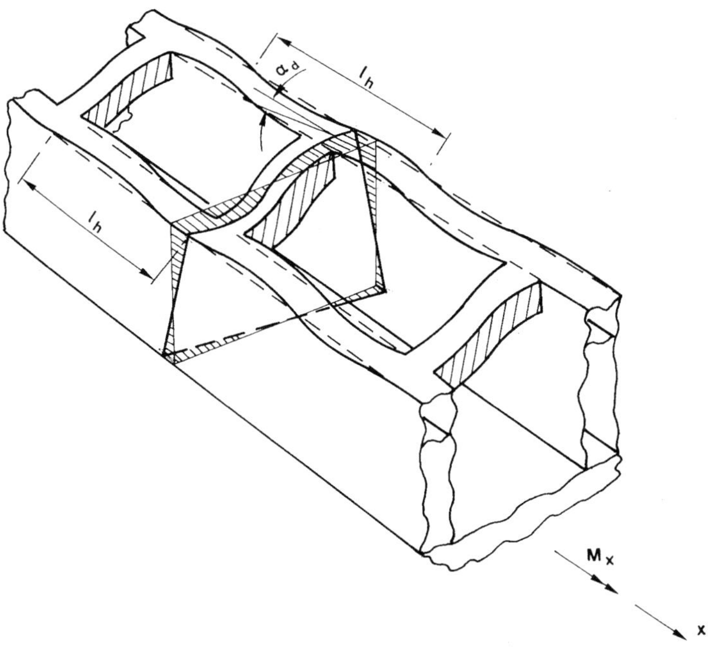

Figure 5.42 *Deformation of deck beams due to warping.*

When the hull cross-section warps, the deck beams are deformed as shown in Figure 5.42. Hence, the deck beams contribute to preventing the warping of the hull and thus to making the hull stiffer towards torsional loading. The extent of the stiffening effect depends on the moment of inertia of the deck beams for bending in the horizontal plane, the torsional stiffness of the deck beams, the value of the sector coordinate Ω at the connection of the deck beams to the hull and, finally, the degree of fixation. The flexibility at the fixation may partly be due to a common bending of the upper deck as shown in Figure 5.42 and partly to a more local deformation as shown in Figure 5.43.

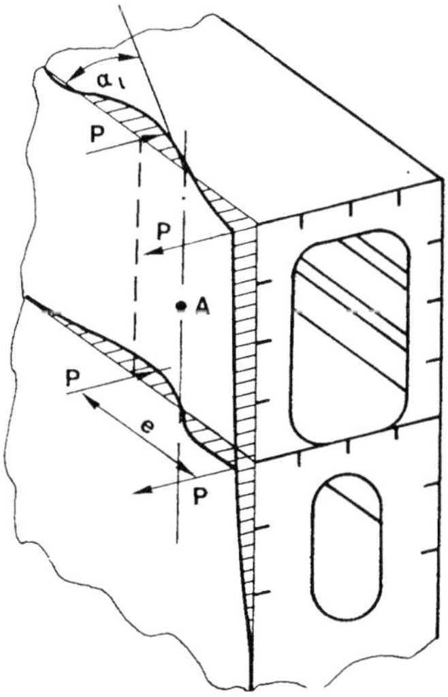

Figure 5.43 *Local deformation of importance to the degree of fixation of the deck beam.*

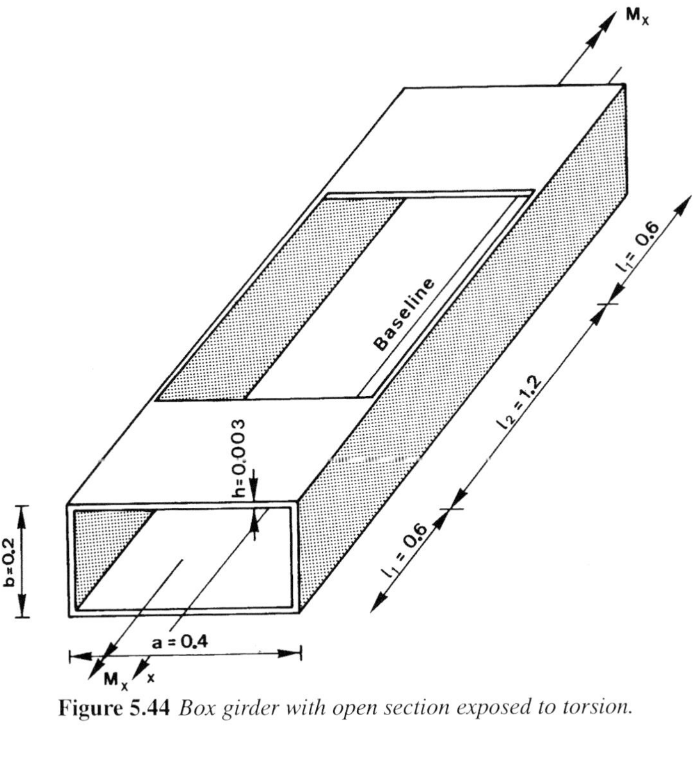

Figure 5.44 *Box girder with open section exposed to torsion.*

However, from a geometric description of the hull and the deck beam the required stiffnesses can be obtained, which can then be introduced as discontinuity conditions into the equations.

To get an impression of the accuracy which can be obtained by a beam model, an extremely simple model of a ship's hull is considered, see Figure 5.44. It is a box girder exposed to a torque. It has a torsionally weak, open central section, which to some degree is prevented from warping by two closed end sections.

The shape of the warping is given by the sector coordinates (Ω) for the two sections. These sector coordinates have been derived in Example 5.3.5 and Example 5.3.4.

It is seen from Figure 5.45 that these functions are different. But by means of the above-mentioned coupling procedure, it is in this case possible to couple bending and warping so that the two sections match. There is full compatibility in this special case.

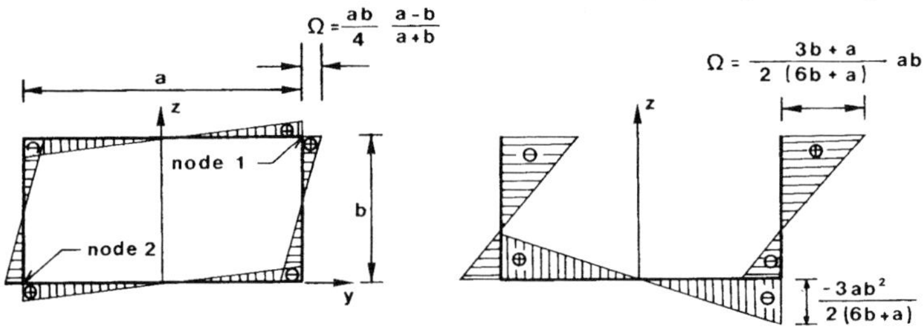

Figure 5.45 *Sector coordinates for open and closed sections.*

Figure 5.46 shows qualitatively the deformation of the box girder exposed to a torque $M_x = 10^3$ Nm, and Figure 5.47 shows how the angular displacement θ_x is distributed along the beam according to the beam theory described above with and without account for the transitional conditions between the closed and the open sections. Figure 5.47 also shows the results obtained by application of the finite element method.

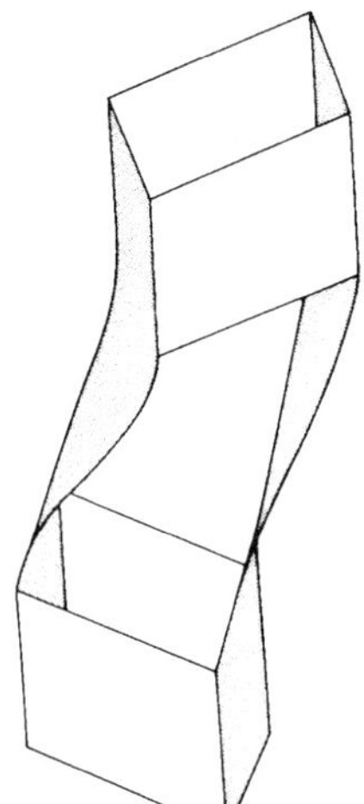

Figure 5.46 *Deformation of box girder exposed to torque.*

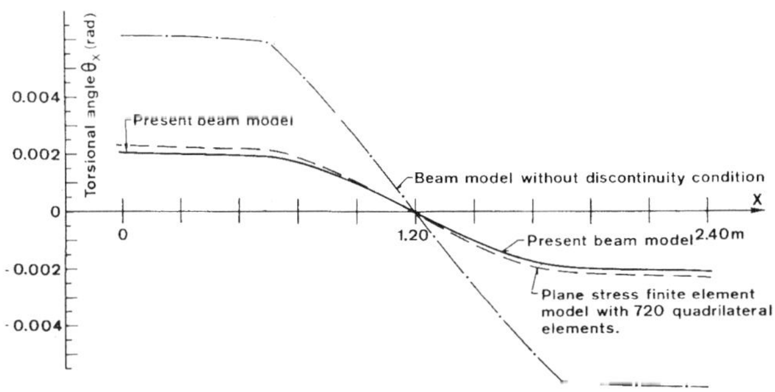

Figure 5.47 *Torsional angular displacement* θ_x *of torsionally loaded box girder with a open central section, Pedersen (1983).*

As a last example the numerical results for the container ship shown in Figure 5.48 are given. The ship is the same as used in the previous examples in this chapter.

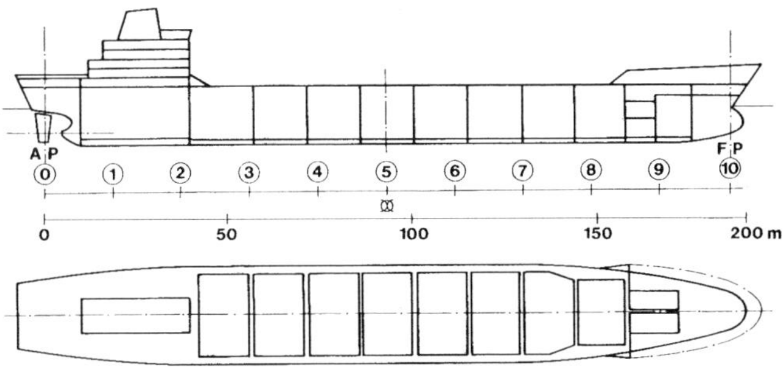

Figure 5.48 *Container ship.*

The longitudinal variation of the inner structure is depicted in Figure 5.49. The midship section is given in Figure 5.6.

The hull has three major discontinuities and seven deck beams. The distribution of sectional quantities, such as $I_{\Omega\Omega}$, along the hull girder is based on calculations for the shown sections. A torque of the form

$$M_x = 1.315\ 10^8 \sin\frac{\pi x}{L_{pp}}\ \text{Nm} \quad (0 \le x \le L_{pp})$$

is assumed, yielding the qualitative deflection shown in Figure 5.49. The characteristic S-form should be noted here. The total angular deflection is about 0.01 rad, corresponding well with the result in Example 5.3.6, taking into account the higher external torque M_x and the flexibility of the connections to the machinery space and forepeak.

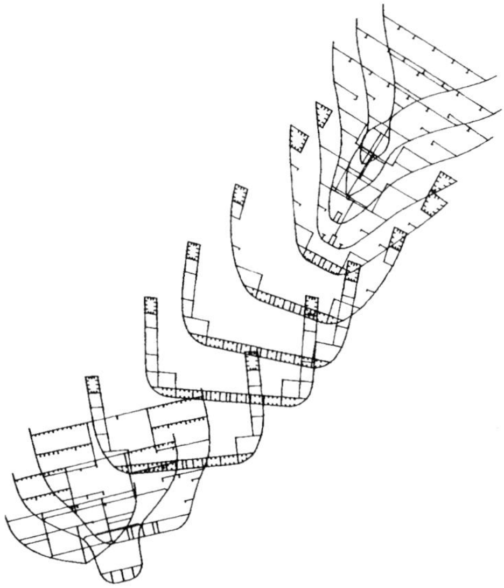

Figure 5.49 *Cross-sections of a container ship and the deformation when the hull is exposed to a torque.*

5.4 AXIAL RESPONSE

If a prismatic beam is subjected to an axial force Q_x parallel with the x-axis of the beam, the result will be that apart from axial normal stresses, which are constant over the whole cross-section, also bending stresses will arise if the axial force Q_x does not act in the bending neutral axis of the beam.

When the stresses due to an axial force Q_x are to be determined, this force is therefore replaced by the static equivalent system $\overline{Q}_x$, $\overline{M}_y$, $\overline{M}_z$ given by (see Figure 5.50):

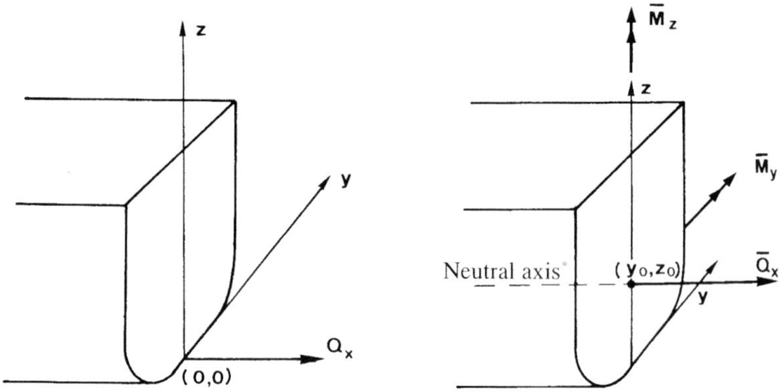

Figure 5.50 *Axial force.*

$$\overline{Q}_x = Q_x$$
$$\overline{M}_y = -z_0 Q_x \tag{5.58}$$
$$\overline{M}_z = y_0 Q_x$$

The bending stresses which follow from the sectional moments $\overline{M}_y$ and $\overline{M}_z$ are determined as in Section 5.1, where the determination of the neutral axis (y_0, z_0) also is given.

Only the contribution from the axial force $\overline{Q}_x = Q_x$ is left. This force yields a normal stress σ_a:

$$\sigma_a = \frac{Q_x}{A} \tag{5.59}$$

where A is the area of the structural members in the thin-walled cross-section.

Example 5.4.1

For the container ship shown in Figure 5.6 the cross-sectional area

$$A = 2.781 \text{ m}^2$$

has been found.

According to Archimedes' principle, the axial force in still water becomes

$$Q_x = -\rho g \int_0^T b(\zeta)\,\zeta d\zeta \approx -\rho g B \frac{1}{2} T^2$$

c.f. Chapter 2, Eq. (2.127). The design draught of the ship is $T = 8.4$ m, and if this is used together with the breadth $B = 25.92$ m

$$Q_x \approx -9.2 \text{ MN}$$

is obtained and hence the axial stress

$$\sigma_a \approx -9.2 \text{ MN}/2.781 \text{ m}^2 = -3 \text{ MN/m}^2$$

which is of no practical importance.

The axial force acts approximately $T/2 = 4.2$ m above the heel. Therefore, it give rise to a vertical bending moment, Eq. (5.58), equal to

$$M_y \simeq 9.2\text{MN} \cdot (7.238\text{m} - 4.2\text{m}) = 28\text{MNm}$$

and the corresponding longitudinal deck stress becomes, c.f. Example 5.1.1

$$\sigma = \frac{28\text{MNm}}{106.1\text{m}^4}(17.44 - 7.238)\text{m} \simeq 3 \text{ MN/m}^2$$

also of no importance.

5.5 TEMPERATURE-INDUCED RESPONSE

If there is a temperature difference $\Delta T(y, z)$ over the cross-section in a prismatic beam, a deformation of the cross-section will take place and normal stresses σ_T will arise in the direction of the beam axis (the x-axis).

Due to symmetry, plane cross-sections remain plane as for bending, and thus, analogously to the analysis of the bending stresses in Section 5.1, the *thermal strain* ε_T can be written

$$\varepsilon_T = a + by + cz \tag{5.60a}$$

while the corresponding normal stress σ_T becomes

$$\sigma_T(y, z) = E(y, z)\{a + by + cz - \alpha(y, z)\Delta T(y, z)\} \tag{5.60b}$$

where α is the thermal expansion coefficient of the material ($\alpha = 12\ 10^{-6}\ {}^{\circ}C^{-1}$ for steel). The constants a, b and c are determined by

$$\begin{aligned} \int_A \sigma_T \, dA &= Q_x = 0 \\ \int_A \sigma_T \, y dA &= -M_z = 0 \\ \int_A \sigma_T \, z dA &= M_y = 0 \end{aligned} \tag{5.61}$$

as no sectional forces and moments are applied.

If the cross-section is symmetric about the z-axis, the following is obtained:

$$\begin{aligned} a\int_A E dA + c\int_A E z dA &= \int_A E\alpha\Delta T dA \\ b\int_A E y^2 dA &= \int_A E\alpha\Delta T y dA \\ a\int_A E z dA + c\int_A E z^2 dA &= \int_A E\alpha\Delta T z dA \end{aligned} \tag{5.62}$$

Introduction of the definitions, Eqs. (5.4a) gives by solution of Eq. (5.62) and insertion in Eq. (5.60b):

$$\sigma_T(y,z) = E(y,z)\left[\frac{1}{\overline{E}}\int_A E\alpha\Delta T dA + \frac{y}{E_z}\int_A E\alpha\Delta T y dA \right. \\ \left. + \frac{z-z_0}{E_y}\int_A E\alpha\Delta T(z-z_0)\,dA - \alpha\Delta T\right] \tag{5.63}$$

where $\overline{E}$ is

$$\overline{E} = \int_A E dA$$

With constant modulus of elasticity $E(y,z) = E$ and constant thermal expansion coefficient α, the result is:

$$\sigma_T = E\alpha\left\{\frac{1}{A}\int_A \Delta T dA + \frac{y}{I_z}\int_A \Delta T y dA + \frac{z-z_0}{I_y}\int_A \Delta T(z-z_0)dA - \Delta T\right\} \tag{5.64}$$

While the stresses σ_T are usually modest (< 10 MN/m^2) for normal temperature gradients over the hull cross-section, the deformation may be relatively large.

If only the curvature of the beam about the y-axis is considered (corresponding to $\int_A \Delta T y dA = \int_A \Delta T dA = 0$), the curvature of the beam κ can be written, according to Figure 5.12

$$\kappa = \frac{d\varepsilon}{dz} = \frac{\alpha}{I_y}\int_A \Delta T(z-z_0)\,dA \tag{5.65}$$

Hence, the radius of curvature R becomes

$$R = \frac{1}{\kappa} = \frac{I_y}{\alpha\int_A \Delta T(z-z_0)dA} \tag{5.66}$$

The height δ, defined as the relative vertical deformation of the beam axis relative to a straight line through AP and FP is

$$\delta = R\left(1 - \cos(L/2R)\right) \approx \frac{L^2}{8R}$$

or

$$\delta \approx \frac{L^2}{8\,I_y}\alpha \int_A \Delta T(z - z_0)dA \tag{5.67}$$

Example 5.5.1

For the container ship shown in Figure 5.6, heating of the upper half of the cross-section by 20°C will give

$$\int_A \Delta T(z - z_0)dA \approx 200^{\circ}\mathrm{Cm}^3$$

If this is inserted in Eq. (5.67), the height δ becomes

$$\delta = \frac{185.93^2}{8 \cdot 106.1} \cdot 12\,10^{-6} \cdot 200 = 0.10\ \mathrm{m}$$

is obtained, which is of the same order of magnitude as the static deflection as a consequence of the hydrostatic still-water loads described in Chapter 2, see Example 5.1.3.

---

6 Hull Girder Vibrations

Ships are subjected by periodic or time-varying loads and therefore structural vibrations will inevitably occur. If the frequencies of the external forces are close to one of the natural frequencies of the ship, the permissible vibration levels may be exceeded. Generally, the vibration problems can be divided into three groups:

- Fatigue failure in the structure
- Destruction of electronic and mechanical equipment
- Too high noise level

For ships the most important excitation forces related to these vibration problems are

- The propeller
- The main engine
- Auxiliary machinery
- The sea

and as a result too high vibration levels may occur in

- The hull girder
- The stern and the superstructures
- Transverse frames, plate panels and plate elements
- The propeller shaft
- The main engine

The present chapter deals only with the hull girder, as local vibrations in substructures is outside the scope of the present treatment, dealing with global responses, only.

The most relevant global vibration modes are depicted in Figure 6.1. The two-noded vertical vibration mode has normally the lowest natural frequency, but very open types of ships (container ships) may have an even lower natural frequency in the torsional vibration mode. Typically, the vibration modes shown in Figure 6.1 correspond to natural frequencies in the range 0.6–6 Hz.

Relatively simple beam models can often with good accuracy be used for determination of the lowest natural frequencies for the hull girder. Such methods will be treated here.

Before discussion in more detail of these global vibrations in ships, a general method (Stodola's method) for exact determination of natural frequencies for continuous systems will be described in Section 6.1, exemplified by a Timoshenko beam.

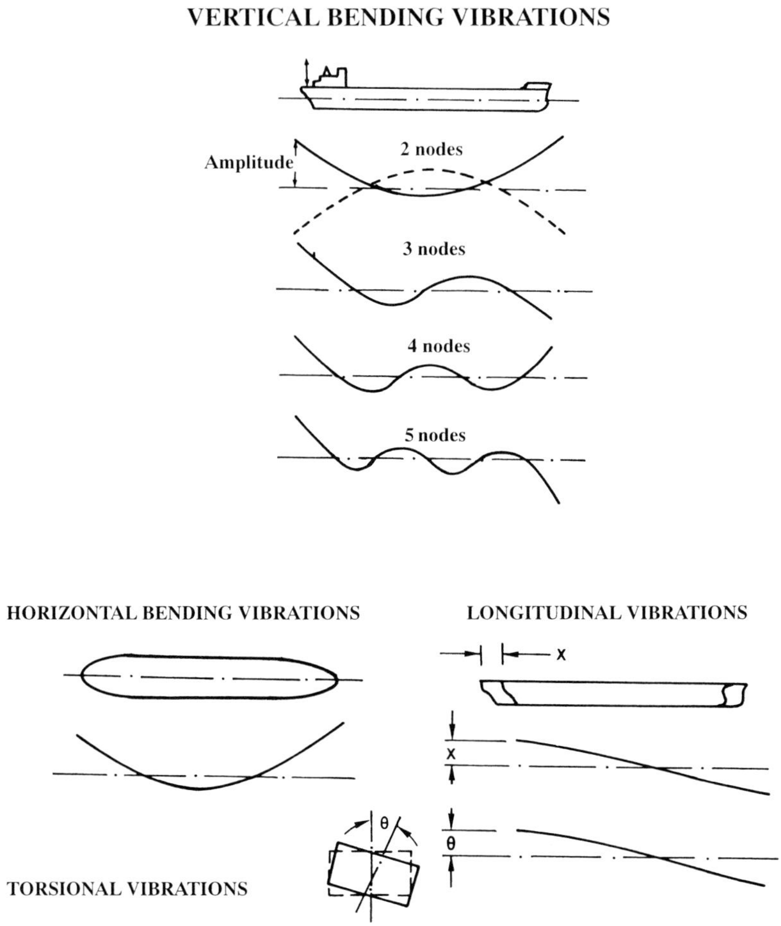

Figure 6.1 *Beam vibration modes for a ship's hull.*

To determine the global vibrations of the hull girder, the following data must be known:

- Time-varying loads on the hull girder (Section 6.2)
- The distribution of stiffness and mass of the hull girder (Section 6.3)
- Structural and hydrodynamic damping (Section 6.4)

The vibration level is determined as solution to a forced vibration problem. An efficient method for this is modal superposition, Section 6.5, where the solution sought for is expressed as a linear combination of relevant natural vibration modes. On the basis of the thus achieved results and relevant criteria, a reduction of the vibration level may finally take place as outline in Section 6.6.

6.1 NATURAL FREQUENCIES

6.1.1 Timoshenko Beam Theory

In the *Timoshenko beam theory Bernoulli's assumption* that plane cross-sections remain orthogonal to the neutral axis of the beam is replaced by the assumption that the angle between the neutral axis and the normal of the cross-section is proportional to the shear force.

This modification of the Bernoulli-Euler beam theory is needed for calculation of the higher hull girder modes, where the distance between the nodes cannot be considered to be large in relation to the cross-sectional dimensions of the hull girder.

Consider a beam with the length L, the modulus of elasticity E, the mass per unit of length m, the moment of inertia $I(x)$, the cross-sectional area $A(x)$ and the mass moment of inertia $mr^2(x)$. The linear transverse deflection $v(x)$ of the beam neutral axis can be divided into contributions from bending and shear as follows.

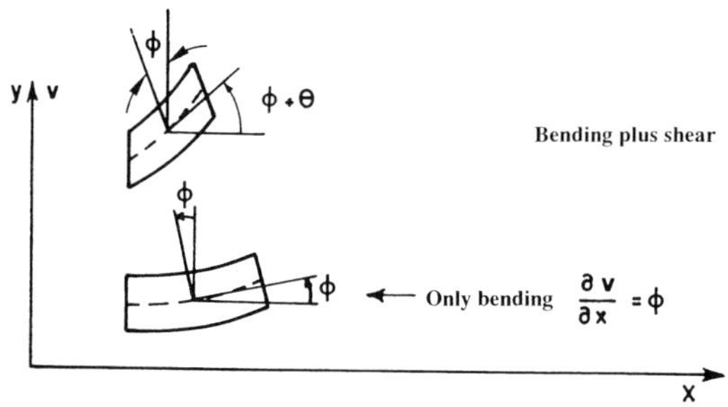

Figure 6.2 *Timoshenko beam element.*

Let ϕ be the angle which the cross-section of the beam forms with the y-axis, when only bending is considered, then, $\frac{\partial v}{\partial x} = \phi$ due to assumption in the Bernoulli beam theory. If the cross-section is subsequently exposed to shear, the cross section does not rotate further but the neutral axis change its angle with the x-axis by the angle θ, see Figure 6.2, implying:

$$\frac{\partial v}{\partial x} = \phi + \theta \tag{6.1}$$

The *constitutive equations* within linear elasticity are

$$M = EI\frac{\partial \phi}{\partial x} \quad \text{and} \quad Q = kGA\theta = kGA\left(\frac{\partial v}{\partial x} - \phi\right) \tag{6.2}$$

where M is the bending moment, Q is the shear force, G is the shear modulus $\left(G = \frac{E}{2(1+\nu)}\right)$ and k is a constant dependent on the cross-section geometry. A thorough discussion of k is given later in Section 6.3.1.

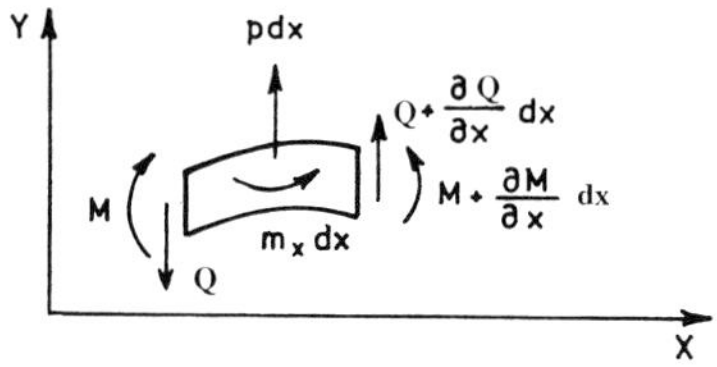

Figure 6.3 *Equilibrium for Timoshenko beam element.*

The *equilibrium conditions* yield, c.f. Example 2.3.2 and Figure 6.3.

$$\frac{\partial M}{\partial x} + m_x = -Q \quad \text{and} \quad \frac{\partial Q}{\partial x} = -q \tag{6.3}$$

Here m_x is the moment per unit of length and q is the load per unit of length. If free vibrations are assumed, i.e. no external forces, the d'Alembert principle gives

$$q = -m\ddot{v} \quad \text{and} \quad m_x = -mr^2\ddot{\phi} \tag{6.4}$$

where $(\dot{\ }) \equiv \partial(\,)/\partial t$, and r is the mass radius of gyration of the cross-section.

By insertion of the constitutive equations (6.2) in the *equilibrium equations*, a set of partial differential equations is obtained:

$$\begin{aligned} (EI\phi')' + kGA(v' - \phi) &= mr^2\ddot{\phi} \\ [kGA(v' - \phi)]' &= m\ddot{v} \end{aligned} \tag{6.5}$$

where $(\,)' \equiv \dfrac{\partial(\,)}{\partial x}$

The standard homogeneous boundary conditions for this system of equations are given in Figure 6.4:

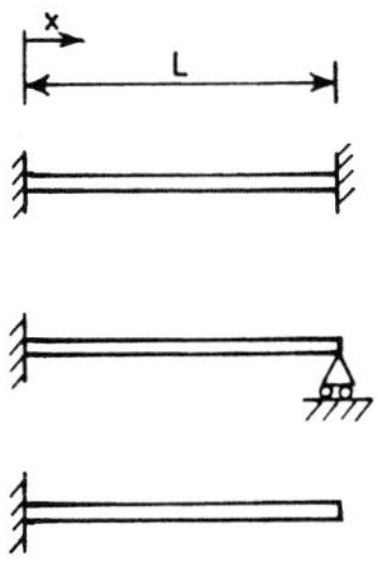

$v(0,t) = \phi(0,t) = v(L,t) = \phi(L,t) = 0$

$v(0,t) = \phi(0,t) = v(L,t) = EI\phi'(L,t) = 0$

$v(0,t) = \phi(0,t) = EI\phi'(L,t) = kGA\big(v'(L,t) - \phi(L,t)\big) = 0$

Figure 6.4 *Simple boundary conditions for Timoshenko beam element.*

The solutions to the system of equations (6.5) with a set of homogeneous boundary conditions will have the form:

$$\begin{aligned} v(x,t) &= u(x)\,\sin(\omega t + \psi) \\ \phi(x,t) &= \alpha(x)\,\sin(\omega t + \psi) \end{aligned} \tag{6.6}$$

If this solution is inserted in Eq. (6.5), the following set of ordinary differential equations is obtained:

$$\begin{aligned} -(EI\alpha')' - kGA(u' - \alpha) &= \omega^2 mr^2\alpha \\ -[kGA(u' - \alpha)]' &= \omega^2 mu \end{aligned} \tag{6.7}$$

If a set of *eigenvectors* $u_s(x)$, $\alpha_s(x)$ corresponds to a certain *natural frequency* ω_s, then

$$-(EI\alpha'_s)' - kGA(u'_s - \alpha_s) = \omega_s^2 mr^2\alpha_s \tag{6.8}$$

$$-[kGA(u'_s - \alpha_s)]' = \omega_s^2 mu_s \tag{6.9}$$

and, similarly, that $u_r(x)$, $\alpha_r(x)$ correspond to ω_r, so that

$$-(EI\alpha'_r)' - kGA(u'_r - \alpha_r) = \omega_r^2 mr^2\alpha_r \tag{6.10}$$

$$-[kGA(u'_r - \alpha_r)]' = \omega_r^2 mu_r \tag{6.11}$$

Multiplication of Eq. (6.8) by α_r, of Eq. (6.9) by u_r, Eq. (6.10) by α_s and Eq. (6.11) by u_s and subsequent integration of Eqs. (6.8) + (6.9) - (6.10) - (6.11) from 0 to L give

$$\begin{aligned} &\int_0^L \{-\alpha_r(EI\alpha'_s)' - \alpha_r kGA(u'_s - \alpha_s) - u_r[kGA(u'_s - \alpha_s)]' + \\ &+ \alpha_s(EI\alpha'_r)' + \alpha_s kGA(u'_r - \alpha_r) + u_s[kGA(u'_r - \alpha_r)]'\}dx = \\ &\left(\omega_s^2 - \omega_r^2\right)\int_0^L \left(mr^2\alpha_s\alpha_r + mu_s u_r\right)dx \end{aligned} \tag{6.12}$$

Partial integration of the left side of Eq. (6.12) yields

$$\begin{aligned} &[-\alpha_r(EI\alpha'_s) - u_r\{kGA(u'_s - \alpha_s)\} + \alpha_s(EI\alpha'_r) + u_s\{kGA(u'_r - \alpha_r)\}]_0^L = \\ &\left(\omega_s^2 - \omega_r^2\right)\int_0^L \left(mr^2\alpha_s\alpha_r + mu_s u_r\right)dx \end{aligned} \tag{6.13}$$

It is seen that, for all the boundary conditions in Figure 6.4, the left side of Eq. (6.13) will be zero. Hence, the following *orthogonality condition* is found:

$$\int_0^L \left(mr^2\alpha_r\alpha_s + mu_r u_s\right)dx = 0 \quad \text{for} \quad \omega_r \neq \omega_s \tag{6.14a}$$

Another orthogonality condition is obtained from

$$\int_0^L \left\{\alpha_r(EI\alpha_s')' + \alpha_r kGA(u_s' - \alpha_s) + u_r\left[kGA(u_s' - \alpha_s)\right]'\right\}dx$$
$$= -\omega_s^2\int_0^L \left(mr^2\alpha_r\alpha_s + mu_r u_s\right)dx = 0 \quad \text{for} \quad \omega_r \neq \omega_s \tag{6.14b}$$

These orthogonality conditions are important later, in Eqs. (6.15)-(6.17) and in Eq. (6.57).

Example 6.1.1

Determine the natural frequencies and the vibration modes for a uniform, homogeneous beam, simply supported at the ends. Note, that these boundary conditions are not relevant for the hull girder. The example only serves to illustrate the mode forms for a Timoshenko beam.

The boundary conditions are, see Figure 6.4

$$u(0) = u(L) = \alpha'(0) = \alpha'(L) = 0$$

It is from Eq. (6.7) seen immediately that $\alpha \equiv$ constant and $u \equiv 0$ is a solution giving the frequency

$$\omega_0^2 = \frac{kGA}{mr^2}$$

To find the other eigenfunctions and associated eigenfrequences α is eliminated from Eq. (6.7) yielding one differential equation in $u(x)$:

$$\frac{EI}{m}u'''' + \left(\frac{EI}{kGA} + r^2\right)u''\omega^2 - u\omega^2 + \omega^4 r^2\frac{m}{kGA}u = 0$$

with

$$u(0) = u(L) = 0$$

The remaining two boundary conditions in u follow from the last equation of Eq. (6.7)

$$\alpha' = u'' + \frac{m}{kGA}u\omega^2$$

implying that $u''(0) = u''(L) = 0$.

To this eigenvalue problem a solution of the following type

$$u = \sin\frac{n\pi x}{L}$$

satisfy the boundary conditions and insertion in the differential equation gives

$$\left(\frac{EI}{m}\right)\left(\frac{n\pi}{L}\right)^4 - \omega^2\left\{1 + \left(\frac{n\pi}{L}\right)^2\left(\frac{EI}{kGA} + r^2\right)\right\} + \omega^4 r^2\frac{m}{kGA} = 0$$

or

$$\omega_n^2 = \frac{\left\{1 + \left(\frac{n\pi}{L}\right)^2\left(\frac{EI}{kGA} + r^2\right)\right\} \pm \sqrt{\left\{1 + \left(\frac{n\pi}{L}\right)^2\left(\frac{EI}{kGA} + r^2\right)\right\}^2 - 4r^2\frac{m}{kGA}\left(\frac{EI}{m}\right)\left(\frac{n\pi}{L}\right)^4}}{2r^2\frac{m}{kGA}}$$

including the solution $n = 0$.

It is seen that, for each value of n, two different values of ω_n^2 are obtained.

The corresponding angle α_n becomes with $u = \sin\frac{n\pi x}{L}$ and $\omega = \omega_n$

$$\alpha_n = -\left\{-\left(\frac{n\pi}{L}\right)^2 + \frac{m}{kGA}\omega_n^2\right\}\frac{L}{n\pi}\cos\frac{n\pi}{L}x$$

by integration of the last equation of Eq. (6.7). It can be shown for each n that the lowest value of ω_n^2 always has the effect that the content within the braces { } becomes negative and that the largest value of ω_n^2 makes the content positive. The vibration modes will therefore be as sketched in Figure 6.5.

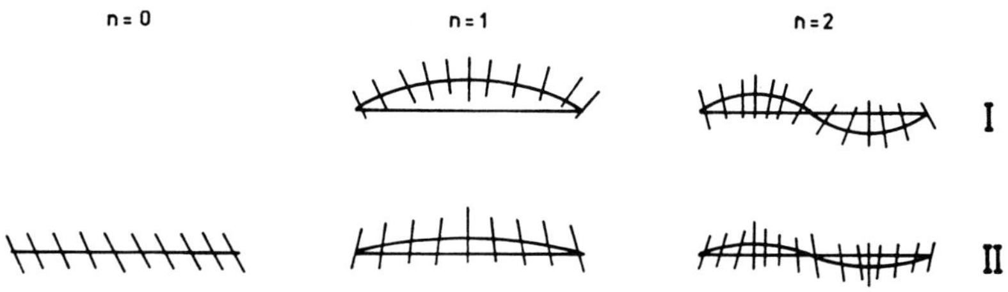

Figure 6.5 *Lowest natural vibration modes for a uniform, homogeneous Timoshenko beam, simply supported at the ends.*

From the solution for ω_n it is seen that the shear stiffness kGA and the mass moment of inertia both enter the result. For a homogeneous beam, where the mass is distributed as the structural material,

$$m = \rho A$$

and

$$mr^2 = \rho I$$

where ρ is the mass density. Hence

$$\frac{EI}{kGA} + r^2 = \frac{I}{A}\left(\frac{2\,(1 + \nu)}{k} + 1\right)$$

and, as $\nu \simeq 0.3$ and $k = 0(1)$, both terms are seen to be of equal importance.

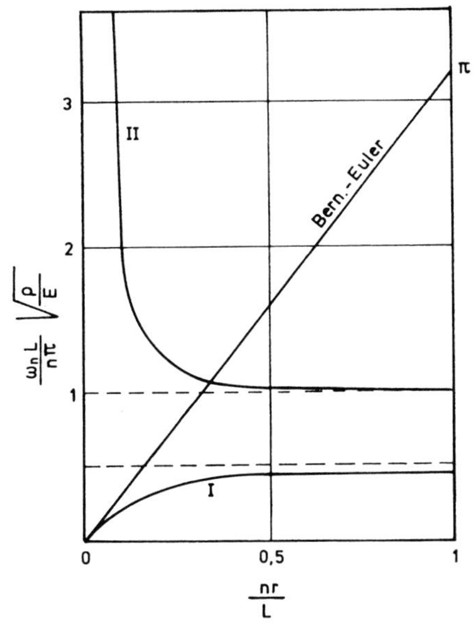

Figure 6.6 *Natural frequencies for uniform, homogeneous Bernoulli-Euler and Timoshenko beams* (k = 8/15 and G/E = 3/8, $m = \rho A$, $mr^2 = \rho I$).

The frequencies ω_n, are shown in Figure 6.6 as functions of the radius of gyration $r = \sqrt{I/A}$ and compared with the frequencies found by means of Bernoulli's beam theory $\left(\omega_n = \frac{n^2\pi^2}{L^2}\sqrt{\frac{EI}{m}}\right)$. The frequency $\omega_0 = \sqrt{\frac{kGA}{mr^2}}$ is not included as $n = 0$.

It should be noted that it is doubtful if the vibration modes corresponding to the frequencies marked II in Figure 6.6 represent real, possible vibration modes for thin-walled beams, as the assumption that the cross-section does not deform is hardly fulfilled for these vibration modes.

6.1.2 Stodola's Method

For beams with variable cross-sectional parameters, a numerical determination of the natural vibrations is needed. Although the Finite Element Method (FEM) is generally applicable for that purpose, *Stodola's method* is a very elegant alternative, Collatz (1963). The algorithm for this method is:

- Calculate $\overset{*}{v}_n$, $\overset{*}{\gamma}_n$ from the boundary value problem, given by the differential equations

$$\begin{aligned} -\left(EI(\overset{*}{\gamma}_n)'\right)' - kGA((\overset{*}{v}_n)' - \overset{*}{\gamma}_n) &= mr^2\gamma_{n-1} \\ -\left[kGA\left((\overset{*}{v}_n)' - \overset{*}{\gamma}_n\right)\right]' &= mv_{n-1} \end{aligned} \tag{6.15}$$

  and the relevant set of boundary conditions. The solution is easily obtained by successive integrations and an arbitrary initial guess (v_0, γ_0), i.e random generated numbers.

- Normalize the functions

$$\left(v_n^+, \gamma_n^+\right) = \left(v_n^*, \gamma_n^*\right)/\beta_n \quad \text{where}$$

$$\beta_n = \sqrt{\int_0^L \left\{mr^2\left(\gamma_n^*\right)^2 + m(v_n^*)^2\right\}dx} \tag{6.16}$$

- Orthogonalise the set of functions towards the previously found r eigenfunctions $\left(u_j, \alpha_j\right)$; $j = 1, 2, \ldots, r$

$$v_n = v_n^+ - \sum_{j=1}^{r} u_j \frac{\int_0^L \left\{\gamma_n^+ mr^2 \alpha_j + v_n^+ mu_j\right\}dx}{\int_0^L \left\{mr^2 \alpha_j^2 + mu_j^2\right\}dx} \tag{6.17}$$

$$\gamma_n = \gamma_n^+ - \sum_{J=1}^{r} \alpha_j \frac{\int_0^L \left\{\gamma_n^+ mr^2 \alpha_j + v_n^+ mu_j\right\}dx}{\int_0^L \left\{mr^2 \alpha_j^2 + mu_j^2\right\}dx}$$

where u_j, α_j designate the r sets of eigenfunctions which have already been determined by the iterative procedure. After repeated use of this algorithm the natural frequency ω_{r+1} can be approximated by $1/\sqrt{\beta_n}$ and the eigenfunction set (α_{r+1}, u_{r+1}) by (γ_n, v_n). The convergence is very fast.

When Stodola's method is used for determination of hull girder vibrations, the procedure must be amended to account for that the boundary conditions (moment and shear force equal to zero at both ends) imply that the ship can make rigid body motions. These rigid body motions can be determined from the condition that the resulting force and moment on the hull girder due to the d'Alembert forces Eq. (6.4) shall be zero. This gives

$$\int_0^L q dx = 0 \Rightarrow \int_0^L mu dx = 0 \tag{6.18}$$

$$\int_0^L \left[q(x - x_g) + m_x\right]dx = 0 \Rightarrow \int_0^L \left[m(x - x_g)u + mr^2 a\right]dx \tag{6.19}$$

where x_g is the centre of gravity of the ship:

$$x_g = \int_0^L mxdx \Big/ \int_0^L mdx \tag{6.20}$$

The set of functions (v_n, γ_n) given by Eq. (6.17) must be corrected by a rigid body motion $C_1 + C_2(x - x_g)$, so that Eqs. (6.18)-(6.19) are satisfied:

$$\begin{aligned} \bar{v}_n &= v_n - \left(C_1 + C_2(x - x_g)\right) \\ \bar{\gamma}_n &= \gamma_n - C_2 \end{aligned} \tag{6.21}$$

$$\begin{aligned} &\int_0^L m\bar{v}_n dx = 0 \Rightarrow C_1 = \int_0^L mv_n dx \Big/ \int_0^L mdx \\ &\int_0^L \left[m(x - x_g)\bar{v}_n + mr^2\bar{\gamma}_n\right]dx = 0 \\ &\Rightarrow C_2 = \int_0^L \left[m(x - x_g)v_n + mr^2\gamma_n\right] \Big/ \int_0^L \left[m(x - x_g)^2 + mr^2\right]dx \end{aligned} \tag{6.22}$$

If Eqs. (6.21)-(6.22) are compared with Eq. (6.17), it is seen that correction for rigid body motion corresponds to orthogonalising the set of functions (v_n, γ_n) towards the two eigenfunctions:

$$\begin{aligned} &\text{translation}: \quad (u_{-1}, a_{-1}) = (1, 0) \\ &\text{rotation}: \quad (u_0, a_0) = (x - x_g, 1) \end{aligned} \tag{6.23}$$

which both have the natural frequency zero. Thus, Eq. (6.17) only has to be modified to start at $j = -1$ rather than at $j = 1$ to take account of rigid body motions.

A similar procedure can be applied for torsional hull girder vibrations, see Pedersen (1983).

6.2 TIME-VARYING LOADS ON THE HULL GIRDER

The most common source for the generation of hull vibrations is propeller-induced forces. Formerly, the main engines (diesel) were also a considerable source of vibration problems, but better balancing of the movable parts in the large diesel engines has reduced significantly the magnitude of unbalanced vibratory forces and moments. Wave-induced forces may, for some ship types, also cause hull girder vibrations of some significance.

6.2.1 Propeller-Induced Forces

When the propeller of the ship rotates in the inhomogeneous wake field, periodic pressure forces will arise in the stern. These hydrodynamic forces will act partly on the propeller and be transferred to the hull girder via the bearings of the propeller axis and partly on the plating of the stern in the form of a pulsating water pressure, see Figure 6.7.

It is common to both types of loads that it is very difficult to calculate them by theoretical methods because of the complicated hydrodynamic flow conditions around the propeller. Therefore, it is often necessary to use model experiments and empirical formulas. Reference may be made to Breslin and Andersen (1994).

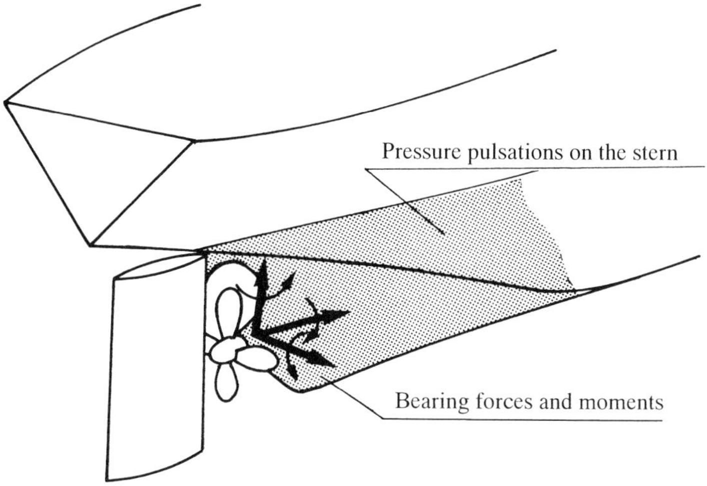

Figure 6.7 *Propeller-induced periodic forces.*

The magnitude of the periodic forces and moments transferred from the propeller to the propeller axis can in principle be determined by calculating the hydrodynamic lift L on each propeller blade. To do this, it is required that the inhomogeneous wake field around the propeller is known, which is difficult to do theoretically. The lift is determined by "lifting surface" analyses, see e.g. Breslin and Andersen (1994). To get an idea of the propeller-induced forces and moments, it is assumed in the following that the resulting lift L_j on a propeller blade No. j is known. The lift is a function of the position of the blade, given by the angle θ relative to a vertical position of the propeller blade, see Figure 6.8.

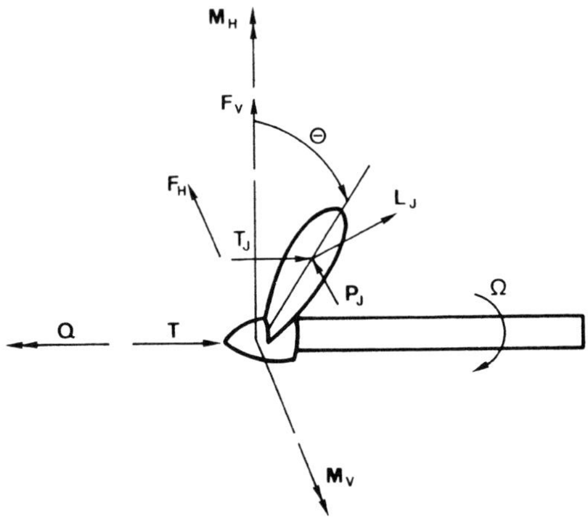

Figure 6.8 *Resulting forces and moments on the propeller.*

For each blade, the lift L_j can be divided into two force components: The blade thrust $T_j(\theta)$ and the resistance $P_j(\theta)$, having effect in respectively the direction of the propeller axis and perpendicularly to the axis of the propeller blade. The distance r from the propeller axis to the point of action of these forces must, as the lift, be determined by hydrodynamic calculations. This distance is here assumed to be independent of the blade position θ, but variation of r with θ can easily be included. The lift L_j and thus also T_j and P_j are periodic with the period 2π. Hence, T_j and P_j can be expanded in Fourier series

$$\begin{aligned} T_j(\theta) &= \frac{1}{2}a_0 + \sum_{n=1}^{\infty} a_n \cos n\theta \\ P_j(\theta) &= \frac{1}{2}b_0 + \sum_{n=1}^{\infty} b_n \cos n\theta \end{aligned} \tag{6.24}$$

It can be proved from hydrodynamics that the Fourier coefficients a_n and b_n only depend on the corresponding nth component in the wake field. In Eq. (6.24) it is assumed that the propeller axis lies in the centre line plane of the ship. If not as for ships with two propellers, there will also be sine components in the Fourier expansions due to the asymmetric wake field.

Subsequently, the resulting load components on the propeller axis at the propeller can be determined by adding up the loads T_j and P_j from the total of Z similar propeller blades:

$$
\begin{aligned}
&\text{Propeller thrust} && T(\theta) = \sum_{j=1}^{Z} T_j(\theta_j) \\
&\text{Propeller moment} && Q(\theta) = r\sum_{j=1}^{Z} P_j(\theta_j) \\
&\text{Vertical Force} && F_V(\theta) = \sum_{j=1}^{Z} P_j(\theta_j)\sin\theta_j \\
&\text{Vertical Bending Moment} && M_V(\theta) = r\sum_{j=1}^{Z} T_j(\theta_j)\cos\theta_j \\
&\text{Horizontal Force} && F_H(\theta) = \sum_{j=1}^{Z} P_j(\theta_j)\cos\theta_j \\
&\text{Horizontal Bending Moment} && M_H(\theta) = r\sum_{j=1}^{Z} T_j(\theta_j)\sin\theta_j
\end{aligned}
\tag{6.25}
$$

$$
\text{where} \quad \theta_j = \theta + \frac{2\pi}{Z}(j-1) \tag{6.26}
$$

Then Eq. (6.24) is inserted in Eq. (6.25), so that the load components T, Q, F_V, M_V, F_H, and M_H are expressed in the coefficients a_n and b_n. The expressions can be reduced considerably by application of the formulas

$$
\begin{aligned}
\sum_{j=1}^{Z} \cos\left[n\left(\theta + \frac{2\pi}{Z}(j-1)\right)\right] &= \begin{cases} Z\cos kZ\theta & \text{for} \quad n = kZ,\ k \text{ integer} \\ 0 & \text{otherwise} \end{cases} \\
\sum_{j=1}^{Z} \sin\left[n\left(\theta + \frac{2\pi}{Z}(j-1)\right)\right] &= \begin{cases} Z\sin kZ\theta & \text{for} \quad n = kZ,\ k \text{ integer} \\ 0 & \text{otherwise} \end{cases}
\end{aligned}
\tag{6.27}
$$

The validity of Eq. (6.27) follows from (with $\mathrm{i} = \sqrt{-1}$):

$$
\begin{aligned}
\sum_{j=1}^{Z} e^{in(\theta + \frac{2\pi}{Z}(j-1))} &= e^{in\theta}\frac{1 - e^{i2\pi n}}{1 - e^{i2\pi n/Z}} \\
&= \begin{cases} e^{ikZ\theta}Z & \text{for} \quad n = kZ,\ k \text{ integer} \\ 0 & \text{otherwise} \end{cases}
\end{aligned}
$$

The results become

$$T = \sum_{j=1}^{Z}\left[\frac{1}{2}a_0 + \sum_{n=1}^{\infty} a_n \cos n\theta_j\right] = Z\left[\frac{1}{2}a_0 + \sum_{k=1}^{\infty} a_{kZ} \cos kZ\theta\right]$$

$$Q = rZ\left[\frac{1}{2}b_0 + \sum_{k=1}^{\infty} b_{kZ} \cos kZ\theta\right]$$

$$\begin{aligned} F_V &= \sum_{j=1}^{Z}\left[\frac{1}{2}b_0 + \sum_{n=1}^{\infty} b_n \cos\theta_j\right]\sin\theta_j \\ &= \frac{1}{2}b_0 \sum_{j=1}^{Z} \sin\theta_j + \frac{1}{2}\sum_{n=1}^{\infty} b_n \sum_{j=1}^{Z}\left(\sin\left[(n+1)\theta_j\right] - \sin\left[(n-1)\theta_j\right]\right) \\ &= \frac{1}{2}Z\sum_{k=1}^{\infty}\left(b_{kZ-1} - b_{kZ+1}\right)\sin kZ\theta \end{aligned} \tag{6.28}$$

$$M_V = \frac{r}{2}Z\left[a_1 + \sum_{k=1}^{\infty}\left(a_{kZ-1} + a_{kZ+1}\right)\cos kZ\theta\right]$$

$$F_H = \frac{1}{2}Z\left[b_1 + \sum_{k=1}^{\infty}\left(b_{kZ-1} + b_{kZ+1}\right)\cos kZ\theta\right]$$

$$M_H = \frac{r}{2}Z\sum_{k=1}^{\infty}\left(a_{kZ-1} - a_{kZ+1}\right)\sin kZ\theta$$

It is seen from the results, Eq. (6.28), that all load components are periodic with the period $2\pi/Z$, because the same propeller configuration occurs each time a new blade gets in the same position as the preceding blade. If the propeller axis rotates with the constant frequency Ω then

$$\theta = \Omega t$$

and the load components, Eq. (6.28), will thus only contain periodic components with frequencies which are multiples of the *blade frequency* $Z\Omega$.

In addition to the propeller thrust T and the moment Q, also the vertical bending moment M_V and the horizontal force F_H have a time-independent component. These mean values may be of importance in the determination of the lay-up of the propeller axis. Moreover, it is seen that only the harmonic components of the wake field corresponding to multiples of the blade frequency $\pm\ \Omega$ enter into the expressions for F_V, M_V, F_H and M_H, while only components which are multiples of the blade frequency form part of the propeller thrust and moment.

As a rule, the most important components in the Eq. (6.28) in relation to generation of hull vibrations are the terms which vary with the blade frequency. If only these terms are kept, the result is as follows:

$$\begin{aligned} T_1 &= Za_Z \cos Z\Omega t \\ Q_1 &= rZb_Z \cos Z\Omega t \\ F_{V1} &= \frac{1}{2}Z(b_{Z-1} - b_{Z+1}) \sin Z\Omega t \\ M_{V1} &= \frac{r}{2}Z(a_{Z-1} + a_{Z+1}) \cos Z\Omega t \\ F_{H1} &= \frac{1}{2}Z(b_{Z-1} + b_{Z+1}) \cos Z\Omega t \\ M_{H1} &= \frac{r}{2}Z(a_{Z-1} - a_{Z+1}) \sin Z\Omega t \end{aligned} \tag{6.29}$$

The odd harmonic components in the wake field are usually much smaller than the even components in the wake field. It follows then from Eq. (6.29) that, for a propeller with an even number of blades, the most important periodic loads will be T_1 and Q_1 while, for a propeller with an odd number of blades, F_{V1}, M_{V1}, F_{H1} and M_{H1} will be the dominant vibratory loads.

For conventional ships, the size of the time-varying loads T_1, Q_1,... is of the order of magnitude of 5 - 20% of respectively the mean propeller thrust and moment.

The significance of the time-varying loads on the propeller is mainly that they may cause too large vibrations of the propeller axis. Their contribution to the generation of hull girder vibrations is normally considerably smaller than the contribution from the pulsating hydrodynamic forces induced on the stern as a consequence of the inhomogeneous wake field around the rotating propeller. This is especially so, if the propeller cavitates, as this effect strongly enhances the latter load but does not increase substantially the forces on the propeller.

There is no reliable theoretical method for determination of the hydrodynamic pressure induced by the rotating propeller on the stern, especially not if the propeller cavitates. However, as those loads may often lead to vibration problems, it is of great importance to be able to estimate their size and their variation with characteristic geometric quantities for the propeller and the stern. Halden (1980) gives an attempt to obtain such bases of estimation through correlation with extensive measurements.

6.2.2 Unbalanced Forces from Diesel Engine

Figure 6.9 shows a schematic cross-section of a cylinder in a diesel engine. It is seen from the figure that the vertical motion x of the piston can be written

$$x = (r + \ell) - (r\cos\theta + \ell\cos\varphi) \tag{6.30}$$

where r is the radius of the crank motion and ℓ is the length of the connecting rod. The definition of the angles θ and ϕ appear from the figure. Moreover,

$$\ell \sin\varphi = r\sin\theta \tag{6.31}$$

and

$$\theta = \omega t \tag{6.32}$$

where ω is the frequency of revolutions of the engine.

If φ is eliminated from Eq. (6.30) by use of Eq. (6.31), the result is

$$
\begin{aligned}
x(t) &= r(1 - \cos\theta) + \ell\left[1 - \sqrt{1 - \left(\frac{r}{\ell}\right)^2 \sin^2\theta}\right] \\
&\approx r(1 - \cos\theta) + \frac{1}{2}\ell\left(\frac{r}{\ell}\right)^2 \sin^2\theta \\
&= r(1 - \cos\omega t + \frac{1}{2}\frac{r}{\ell}\sin^2\omega t)
\end{aligned}
$$

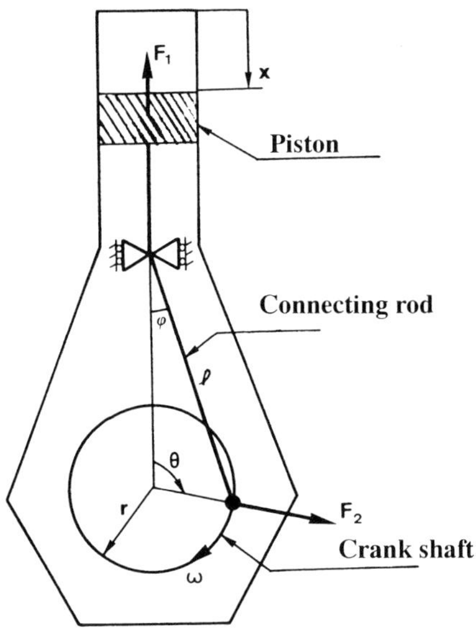

Figure 6.9 *Schematic cross-section of a cylinder.*

If the above expression is differentiated twice with respect to time, the acceleration $\ddot{x}$ is obtained:

$$\ddot{x}(t) = \omega^2 r\left(\cos\omega t + \frac{r}{\ell}\cos 2\omega t\right) \tag{6.33}$$

The resulting d'Alembert force F_2 on the parts (mass m_1) following the vertical motion of the piston becomes

$$F_1 = m_1\ddot{x}(t) \qquad \text{(positively upwards)} \tag{6.34}$$

To this must be added the centrifugal force F_2 as a result of the circular motion of the crank:

$$F_2 = m_2 r\omega^2 \tag{6.35}$$

or resolved in the vertical (F_{2V}) and the horizontal (F_{2H}) direction:

$$F_{2V} = m_2 r\omega^2 \cos \omega t$$
$$F_{2H} = m_2 r\omega^2 \sin \omega t$$

The mass m_2 is the part which follows the motion of the crank shaft.

The resulting mass forces are $F_1 + F_{2V}$ in the vertical direction and F_{2H} in the horizontal direction.

To balance these forces and associated moments for the engine as a whole, the phase shift between the ignition for the single cylinders can be chosen in an appropriate way and rotating masses can be added to the crankshaft. It is seen from Eq. (6.34) that F_1 also includes a contribution varying with twice the number of revolutions of the engine. The engine manufacturer provide today very accurate balanced engines, using various correction procedures. An excellent description may be found in MAN B&W (1992).

Resulting unbalanced forces and moments, if any, are transferred to the hull girder via the engine foundation and the bottom structure of the ship.

6.2.3 Wave-Induced Loads

Wave-induced loads on the ship are dealt with in Chapter 4. So here only a short summary will be given.

On the basis of the assumption of linearity between wave amplitudes and derived loads on the ship, it follows that the load $q = q(x,t)$ per unit of length along the hull girder (the x-axis) can be written as a sum of harmonic components, Eq. (4.97):

$$q(x,t) = \sum_{j=1}^{n} a_j \Phi_q(x, \omega_j) \cos\left(\omega_{e,j} t + \hat{\varepsilon}_j(x) + \varepsilon_{qj}\right) \tag{6.36}$$

where a_j is the wave amplitude for the wave component which has the frequency ω_j and where $\Phi_q(x, \omega_j)$ is the amplitude function, defined as the amplitude of the load in the position $x = x$, if the wave is a harmonic wave with unit amplitude and frequency ω_j. Moreover, $\omega_{e,j}$ is the frequency of encounter given as, Eq. (4.56)

$$\omega_{e,j} = \omega_j - k_j V \cos\beta \tag{6.37}$$

where k_j, V and β are respectively the wave number corresponding to the frequency ω_j ($k_j = \omega_j^2/g$ for deep-water waves), the forward speed and the angle of encounter, given as the angle between the sailing direction and the wave direction, see Figure 4.1.

In Eq. (6.36) the stochastic nature of the sea waves is included, partly through the deterministic amplitude a_j, which is determined as, Eq. (3.221)

$$a_j = \sqrt{2S(\omega_j)\,\delta\omega_j} \;\; ; \;\; \delta\omega_j = \omega_{j+1} - \omega_j \tag{6.38}$$

where $S(\omega_j)$ is the wave spectrum and $\delta\omega_j$ is the difference between the single discrete frequencies in Eq. (6.36), and partly through the statistical phase angle ε_{qj}*, which, for each j, can be chosen to be statistically independent and uniform distributed between 0 and 2π.

Hence, the linear response of the ship - which may e.g. be the vertical motion of the stern including the contribution from the elastic deformation of the hull girder - becomes, according to the central limit theorem, statistically normally distributed with a mean value of zero and a variance equal to the sum of the variances for the response calculated for each load components, Eq. (4.99). The response of the ship for each individual component can thus be considered separately without accounting for the stochastic phase angle ε_{qj}, which does not enter into the variance. Hence, the loading can be represented by a number of deterministic loads given in each individual case as a load per unit of length of the type

$$f(x, \omega_j)\,\cos(\omega_{e,j}t + \hat{\varepsilon}_j(x)) \tag{6.39}$$

Wave-induced vibrations of the hull girder only occur in relatively rare cases. The reason is that the wave amplitude a_j, which enters into $f(x, \omega_j) \equiv a_j \Phi_q(x, \omega_j)$, is normally negligibly small for frequencies of encounter $\omega_{e,j}$ of the order of the lowest natural frequency of the hull girder. In Section 6.5, the calculation of wave-induced vibrations will be treated in more detail.

6.3 STIFFNESS AND MASS DISTRIBUTION OF THE HULL GIRDER

As it is seen from Section 6.1, it is necessary to know the stiffness and mass distribution of the hull girder to be able to determine the natural frequencies and natural vibration modes of the hull girder.

6.3.1 Stiffness Distribution

The stiffness parameters do not differ from those of a static analysis of the hull girder described in Chapter 5. However, some points will be treated in more detail in the following.

The relevant stiffness parameters for vertical and horizontal vibration modes are the *bending stiffness* ($EI_y\,(x)$ for vertical vibrations and $EI_z\,(x)$ for horizontal vibrations) and the *shear stiffness* ($k_zGA(x)$ and $k_yGA(x)$). Calculation of the bending stiffness is described in detail in Section 5.1, also for the cases where the hull cross-section is built up of materials with different modulus of elasticity. However, one point will be stressed here again, namely the effectiveness of longitudinal elements which do not extend along (almost) the whole length of the ship. Such elements may be the longitudinal bulkheads in the holds, superstructures and deck areas between hatches, c.f. Figure 5.9 and Figure 6.10, adapted from Hughes (1988).

---

* $\hat{\varepsilon}_j(x)$ is the deterministic phase angle, determined by the same analysis as the amplitude Φ_q function, see Section 4.4.

In all cases the hatched areas are ineffective in relation to the bending stiffness of the hull girder. A reasonable value for the angle θ, which is a purely empirical quantity, is 15°.

The *shear stiffness* kGA, however, needs some comments. While the cross-sectional area $A = A(x)$ is easy to calculate, the calculation of the dimensionless constant k depends on some assumptions which can approximate the real three-dimensional deformation pattern with relevant beam deformation measures. These assumptions, which must also include assumptions about the applied load, cannot be postulated without objections, and therefore several calculation methods for the constant k are found in the literature.

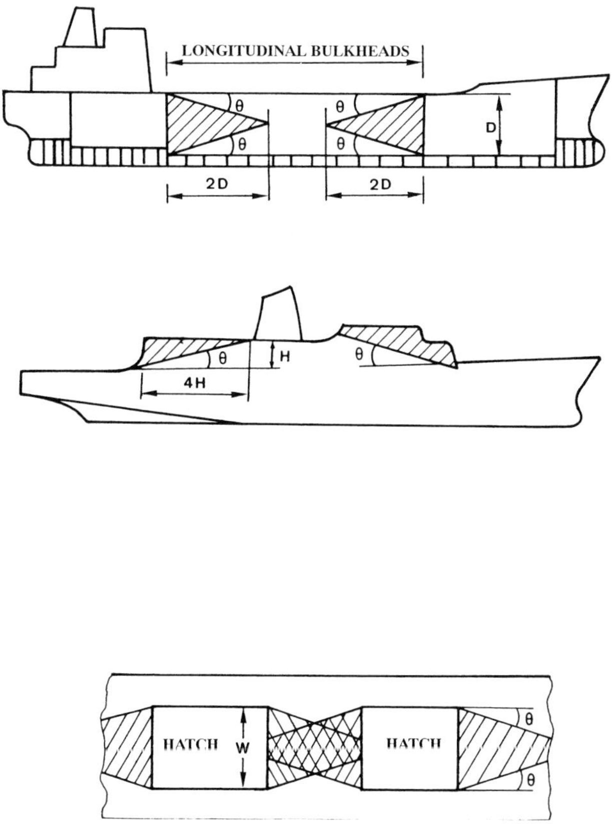

Figure 6.10 *Efficiency of longitudinal elements.*

Let us first consider a simple method which does not include the influence of the lateral contraction (corresponding to a Poisson ratio $\nu = 0$). The *shear energy* E_s per unit of length along the beam axis can be written

$$E_s = \frac{1}{2}\int_A \tau\phi dA = \frac{1}{2}\int_A \frac{\tau^2}{G} dA \tag{6.40}$$

as the shear strain ϕ and the shear stress τ are related by Hooke's law $\tau = G\phi$. Alternatively, the shear energy E_s can be expressed by the "integrated" beam quantities: the sectional force Q and the shear deformation θ:

$$E_s = \frac{1}{2} Q\,\theta = \frac{1}{2}\frac{Q^2}{kGA}$$

by use of the constitutive equations (6.2): $Q = kGA\theta$. If the two expressions for E_s are set to be equal, the following formula for the *shear coefficient* k is obtained:

$$k = \left[A \int_A \tau_0^2 dA\right]^{-1}$$

provided the shear modulus G is the same over the whole beam cross-section. The unit shear stress distribution τ_0 is defined as

$$\tau_0 = \tau/Q$$

For a thin-walled cross-section the area element dA can be written as hds, where $h = h(s)$ is the plate thickness as a function of the arc length s, and where the integration extends over all the plate elements in the cross-section. Thus,

$$k = \frac{1}{A \int_\ell \tau_0^2 hds} \tag{6.41a}$$

where the integration ℓ is over all plate element in the cross-section.

A calculation method for determination of the unit stress distribution $\tau_0 = \tau_0(s)$ for a general thin-walled cross-section is given in Section 5.2. If the individual plate element are rectilinear and of uniform thickness, then $\tau_0(s)$ will at most be a second order polynomial in the arc length s. Thus, the integral in Eq. (6.41a) can be easily calculated.

If the stress distribution $\tau_0(s)$ has not been determined, it may, as a rather rough approximation, be assumed that τ_0 is proportional to the angle $\varphi(s)$, which each plate element forms with the direction of the shear force. Thus, $\tau_0(s) = \tilde{\tau}\cos\varphi$ where the constant $\tilde{\tau}$ must be chosen so that force equilibrium in the direction of Q is fulfilled:

$$\int_\ell \tau_0 \cos\varphi \; hds = \tilde{\tau}\int_\ell \cos^2\varphi \; hds = 1$$

Eq. (6.41a) thus reduces to

$$k = \frac{1}{A\bar{\tau}^2 \int_\ell \cos^2\varphi \, hds} = \frac{1}{A}\int_\ell \cos^2\varphi \, hds \tag{6.41b}$$

The integral $\int_\ell \cos^2\varphi hds$ is often called the *projected area* A_p.

In the above calculations of k the effect of Poisson's ratio ν is neglected.

The perhaps most consistent procedure for determination of k has been given by Cowper (1966). The reduction of the three-dimensional elasticity theory to a beam theory given there is relatively complicated, and here only the result for a cross-section built up of thin-walled elements is presented, assuming the same modulus of elasticity E throughout:

$$k = \frac{2(1+\nu)I_y}{\frac{\nu}{2}(I_z - I_y) + A\int_\ell \psi\,\tau_0\, hds} \tag{6.41c}$$

where I_y and I_z are the moments of inertia about respectively the y- and the z-axis.

A corresponding expression exists for a horizontal shear force.

The function $\psi = \psi(s)$ is given by

$$\psi = 2(1+\nu)I_y\tau_0(s) + \frac{\nu}{2}\left((z^2 - y^2)\cos\varphi + 2yz\sin\varphi\right) \tag{6.42}$$

where, as above, $\tau_0(s)$ is the shear stress in the section, when the cross-section is subjected to a unit shear force in the z-direction, and where $\varphi = \varphi(s)$ is the angle between the plate element at $s = s$ and the z-axis.

For rectilinear evenly thick plate elements the integral in Eq. (6.41c) can be analytically integrated, as in this case $\tau_0 = \tau_0(s)$ is at most a second order polynomial in the arc length s and, moreover, φ is constant, and $y = y(s)$ and $z = z(s)$ vary linearly with s. Thus, $\psi = \psi(s)$ becomes a second order polynomial in s.

It is seen that for $\nu = 0$ the Eqs. (6.41a) and (6.41c) become identical.

It follows from Eqs. (6.41a) and (6.42) that if the shear stress distribution $\tau(s)$ is known, the shear coefficient k can be determined. The calculation of the shear stress distribution is described in Chapter 5, to which reference is made for comments. It should be noted that manual calculation of k according to Formula (6.41a), even for simple cross-sections, is rather comprehensive, see Example 6.3.1.

For realistic hull cross-sections the shear area kA will be of the order of magnitude of 50 - 90% of the *projected area* A_p defined previously as

$$A_p = \int \cos^2 \varphi h ds \tag{6.43}$$

provided longitudinals and similar short webs with one free end are not included in Eq. (6.43), as such elements can absorb almost no shear stresses, c.f. Example 5.2.3. and Example 5.2.4.

Table 6.1 *Shear coefficents for a container ship.*

	(6.41a)	(6.41b)	(6.41c)
k	0.257	0.552	0.267

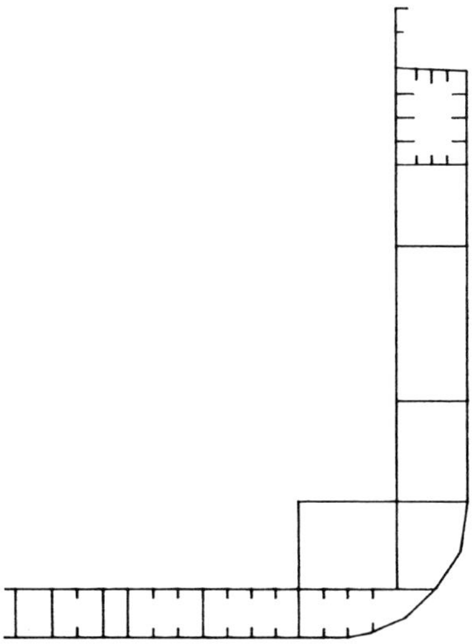

Figure 6.11 *Shear coefficient k for vertical shear for the midship section in a container ship. All longitudinals are included in the calculation of k, Jensen (1983).*

An uncertainty of 20 - 30% in the shear area is normally not acceptable. Therefore, it is recommended to use Eq. (6.41c) or Eq. (6.41a) for determination of the shear coefficient k, although this implies a calculation of the shear stress distribution τ_0. As an illustration of this, reference is made to Table 6.1, which, for a midship section in a container ship, shows that the shear area is only approximately half of the projected area which again is only about half the cross sectional area A. On the contrary, it is seen that inclusion of the effect of Poisson's ratio ν only gives a modest, approximately 4%, increase of the shear coefficient and thus also of the shear stiffness. A discussion on the various formulas for the shear coefficient k can be found in Jensen (1983).

The importance of the shear stiffness kGA compared with the bending stiffness EI grows with the number of nodes in the natural vibration mode. For the two-noded vertical natural vibration mode, the bending stiffness is normally dominant; but the shear stiffness contributes considerably to the deformation in all other vertical and horizontal natural vibration modes. While both the magnitude of the bending and the shear stiffnesses are of importance to the natural vibration modes, their variation along the hull girder will often be of less importance. Therefore, it is usually enough to calculate these stiffnesses for a few cross-sections (3–10) along the hull girder and use interpolation between these values.

Example 6.3.1

Consider the cross-section in Figure 6.12.

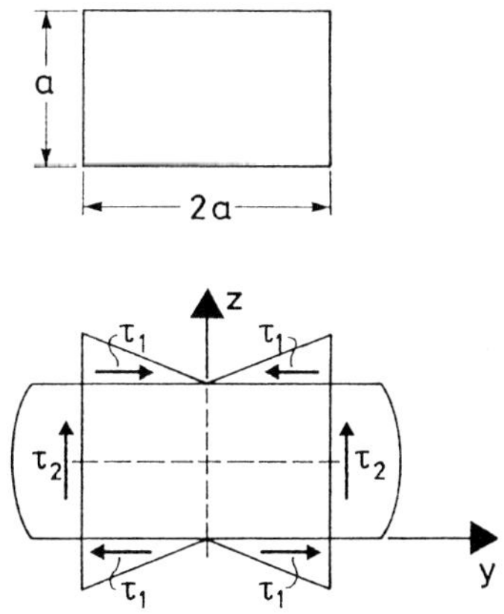

Figure 6.12 *Box-shaped cross-section with plate thickness h throughout.*

The shear stress distribution for a vertical shear force was calculated in Example 5.2.1. For the present cross-section dimensions it becomes for $Q = 1$ (N):

$$\tau_1(y) = \frac{3}{7}\frac{1}{ah}\frac{|y|}{a}$$

$$\tau_2(z) = \frac{3}{7}\frac{1}{ah}\left(1 + \frac{z}{2} - \left(\frac{z}{a}\right)^2\right)$$

with the sign convention shown in Figure 6.12.

In order to determine the shear coefficient k, the integral in Eq. (6.41a) is evaluated

$$\int_\ell \tau_0^2 h ds = 4h\left[\int_0^a \tau_1^2(y)dy + \int_0^{a/2} \tau_2^2(z)dz\right]$$

$$= \frac{36}{49}\frac{1}{ah}\left[\int_0^1 u^2 du + \int_0^{1/2} \left(1 + u - u^2\right)^2 du\right]$$

$$= \frac{183}{245}\frac{1}{ah}$$

Hence, the shear coefficient k becomes

$$k = \frac{1}{6ah\frac{183}{245}\frac{1}{ah}} = \frac{245}{1098} = 0.2231$$

Based on the projected area A_p, Eq. (6.41b), the result is

$$k = \frac{1}{6ah} 2ah = \frac{1}{3}$$

that is, 50 per cent too large!

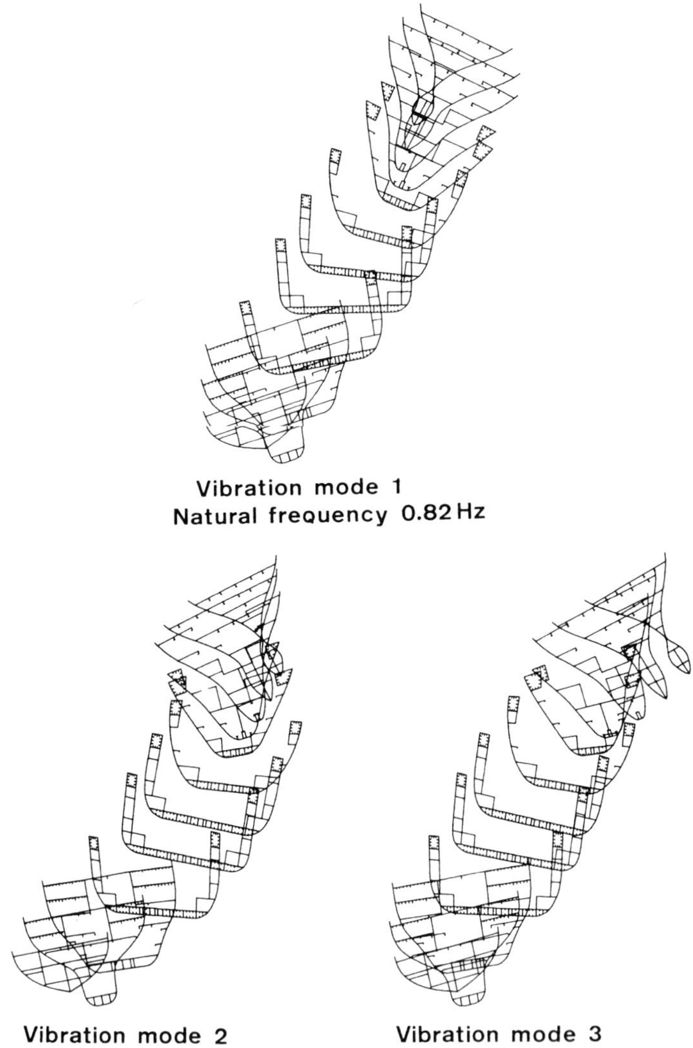

Figure 6.13 The three lowest natural vibration modes corresponding to horizontal bending – torsion models for a container ship. Pedersen (1983).

The global stiffness parameters needed for description of the torsional vibration modes are the St Venant torsional constant I_d, Eq. (5.29b) and the warping constant $I_{\Omega\Omega}$, Eq.

(5.46). The determination of both parameters is described in Chapter 5. For ships without large openings (small or no hatches) only the St Venant stiffness I_d will enter into the governing equations for torsional vibrations. These natural vibrations will, however, be unimportant, as the corresponding natural frequencies are so large that they cannot be excited by the periodic forces dealt with in Section 6.2.

If the ship has large hatch openings, the warping deformation may be of importance to the torsional vibration modes and the corresponding natural frequencies. Moreover, due to the great distance between the torsional centre - which for sections with large hatchways lies considerably under the keel - and the bending axis for horizontal bending - which lies within the section - there will be a considerable coupling between torsional vibrations and horizontal bending vibrations.

In Pedersen (1983), a beam theory is developed which can handle such coupled bending - torsion problems. Basically, the torsional deformations are modelled as in Section 5.3 whereas a Timoshenko beam model is applied for the horizontal bending deformations. As discussed in Section 5.3.6 due account must be taken to include the effect of deck strips between the hatches and to reduce the incompatibility between open and closed sections. As an example Figure 6.13 shows the three lowest natural vibration modes corresponding to horizontal bending-torsion for a container ship.

6.3.2 Mass Distribution

When the hull girder vibrates, the surrounding water will to a larger or smaller degree be forced to follow the motions of the ship. Close to the hull the motion of the water particles will be the same as the motion of the hull as regards frequency and amplitude. At a larger distance from the hull the amplitude of the water particles will quickly decrease while the frequency remains unchanged. This was also the case in the calculation of the wave-induced motions of the ship, treated in Section 4.2.

The relevant mass data entering into the calculation of the natural vibrations of the hull girder, described in Section 6.1 must therefore contain both the mass distribution of the hull girder, including the mass of the cargo, and a contribution which reflects the associated motion of water.

The determination of the mass distribution $m_s(x)$ of the hull girder can be made from knowledge of the steel weight of the ship, the equipment weight and the like and can be made precisely as described in Chapter 2 for still-water loads. In addition to the mass $m_s = m_s(x)$ per unit of length, the associated mass radii of gyration $r_y(x)$ and $r_z(x)$ for vibrations in respectively the horizontal and the vertical plane should be determined. The total mass of the ship can, as a rule, be regarded as fully effective, perhaps with the exception of large quantities of liquid cargo, which by horizontal and torsional vibrations is only partly effective. However, in the following only vertical hull vibrations will be treated.

It is shown in Example 4.2.3 that provided the excitation frequency is very high then the *added mass of water* per unit of length $m_w^\infty(x)$ by a vertical motion of a hull section can be written

$$m_w^\infty(x) = C_m(x)\,\rho A(x) \tag{6.44}$$

where ρ is the density of the water and where $A(x)$ is the submerged area of the section as measured to the still water line. The dimensionless coefficient C_m depends on the shape of the cross-section, and in Example 4.2.3 a formula for C_m is given based on the so-called Lewis transformation.

As the natural frequencies of the hull girder usually are rather high, the use of the added mass for infinite frequency is normally sufficient accurate.

Due to the short wave length of the vibration modes, it is necessary to correct for the three-dimensional water flow around the hull girder. This can be done by a factor $J = J_n$, so that

$$m_w(x,n) = J_n C_m(x)\,\rho A(x) \tag{6.45}$$

where n is the number of nodes in the vibration mode. The factor J is called the *three-dimensional reduction factor* and is defined as the relation between the kinetic energy of the real, three-dimensional flow around the hull and the kinetic energy obtained by integration of the two-dimensional flow along the hull girder. By this formulation, the factor J becomes constant for all cross-sections in the hull. In general, few analytical solutions are known for three-dimensional fluid flow, even under the present assumptions about linear potential flow and no free surface effects ($m_w = m_w^\infty$). Two solutions therefore form the basis of the determination of J for ship forms of current interest.

One method is based on the known solutions for three-dimensional flows around half-submerged elliptic bodies. This method is suitable for ships with fine lines. A good fit to these solutions is, Townsin (1969)

$$J_n = 1.02 - 3\left(1.2 - \frac{1}{n}\right)\frac{B}{L} \tag{6.46}$$

where B is the water line breadth amidships and where L is the length of the ship. The formula only applies to vertical vibrations with n = 2 - 5.

For more full-form ship types (tankers, bulkcarriers), three-dimensional solutions derived for half-submerged prismatic cylinders of finite length can be used. Such solutions are derived by Kumai (1962) and apply to cross-sections defined by the Lewis transformation. The result for vertical vibrations is, Kumai (1962):

$$\begin{aligned} J_n = {} & \frac{16}{\pi^2\left[(1+a_1)^2 + 3a_3^2\right]}\sum_m \frac{m^2}{(m^2-n^2)^2} \times \\ & \left\{\frac{(1+a_1)^2}{1 + k_m K_0(k_m)/K_1(k_m)} + \frac{9a_3^2}{3 + k_m K_2(k_m)/K_3(k_m)}\right\} \end{aligned} \tag{6.47}$$

where the summation $\sum_m$ is over even values of m, if n is odd and vice versa.

In Eq (6.46) the dimensionless wave number k_m is given by

$$k_m = m\frac{B}{4CL} \tag{6.48}$$

where the constant C is given in Example 4.2.3. Moreover, $K_i(\,)$ is the Bessel function of the second kind of order i. These functions are tabulated in standard collections of mathematical tables, as e.g. Abramowitz and Stegun (1970).

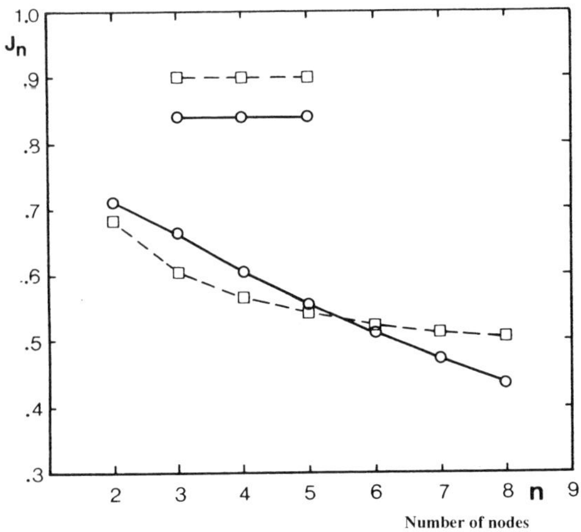

Figure 6.14 *Three-dimensional vertical reduction factor J_n for a 340,000 dwt tanker. Dash lines: Eq. (6.46), Full lines Eq. (6.47). Madsen (1978).*

Figure 6.14 shows a comparison of the two calculations, (6.46) and (6.47), of the reduction factor J, exemplified by a 340,000 dwt tanker.

It is seen from the figure that at the two-noded vibration mode the J factor represents a reduction of 30% of the two-dimensional mass of water. As the added mass of water is of the same order of magnitude as the mass of the ship, this reduction has a considerable influence on the natural frequency. It is also seen from the figure that the two formulas for J give more or less the same results. However, it should be emphasised that Eq. (6.46) is partly empirical and therefore must be used cautiously in the case of untraditional ship forms.

Theoretically, the added mass of water is zero at the ends of the ship, which cannot be represented by the definition of J used above, as it gives a mean value for the reduction, constant along the hull girder. The effect of this simplification is normally small, as the immersed sectional area A is much smaller at the ends of the ship than around amidships. But contrary to this, the largest vibration amplitudes are found at the ends of the ship. A calculation method for determination of so-called *local reduction factor* j, varying along the hull beam, is given in Madsen (1978), from where Figure 6.15 is taken.

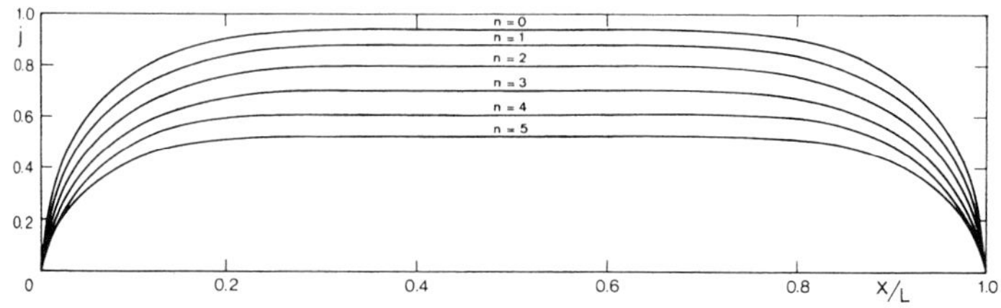

Figure 6.15 *Local reduction factors j for vertical vibrations with n nodes for half-submerged circular cylinder with L/B = 6 (n = 0 and n = 1 are rigid-body motions). Madsen (1978).*

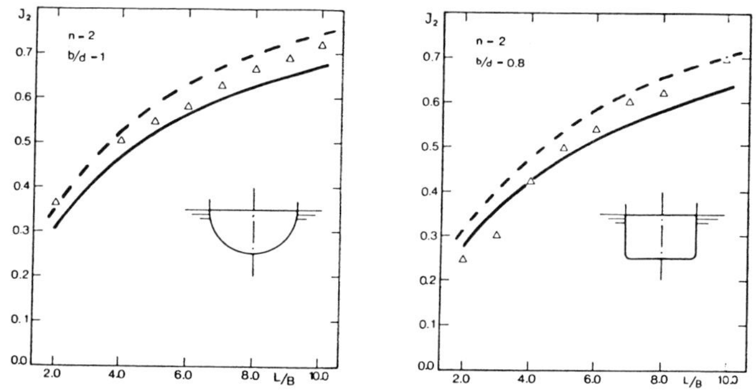

——	Fluid element method calculations
...	Calculated by Eq. (6.47)
Δ	Measurements

Figure 6.16 *Reduction factor J_2 for two-noded vibration for two different cross-sections. Madsen (1978).*

From the same reference Figure 6.16 is taken, depicting the variation of the global reduction factor J with the length to breath ratio *L*/*B*. A fluid element method for calculation of the added mass of water without free surface effects is described in Madsen (1978).

6.4 DAMPING

In the determination of the natural frequencies of the hull girder treated in Section 6.1, the influence of damping, structural as well as hydrodynamic, was neglected. This is permissible when the damping is small, and in order to illustrate this briefly, Figure 6.17 is considered, which shows the classical example of a vibrating concentrated mass m, held by a spring with the stiffness k and a viscous damper with the damping c.

It is seen from Figure 6.17 that, if the damping c is much smaller than the *critical damping* c_0, the natural frequency ω_d for the damped system will practically coincide with the the natural frequency ω_0 of the undamped system. As regards measuring technique, the *damping ratio* ξ is a somewhat inconvenient quantity. Therefore, the *logarithmic decrement* δ, defined as the natural logarithm to the relation between two successive maxima in x, is often used. The logarithmic decrement can be determined by giving the structure an impulse load ("impact") and recording the decrease of the vibration.

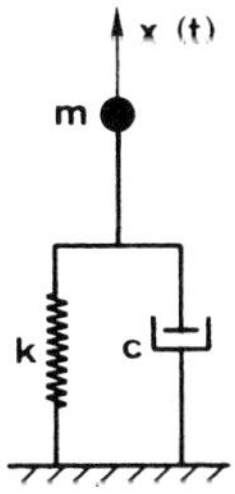

Equation of motion: $m\ddot{x} + c\dot{x} + kx = 0$

$x(0) = 0$

Solution for $c < c_0 = 2\sqrt{km}$:

$x(t) = A \exp(-\xi\omega_0 t)\sin(\omega_d t)$

$\xi = \dfrac{c}{2\sqrt{km}} = \dfrac{c}{c_0}$ = damping ratio

c_0 = critical damping

$\omega_0 = \sqrt{\dfrac{k}{m}}$ = natural frequency of undamped system

$\omega_d = \omega_0\sqrt{1-\xi^2}$ = natural frequency of damped system

$\delta = 2\pi\xi/\sqrt{1-\xi^2}$ = logarithmic decrement

Figure 6.17 *Natural vibration of damped system with one degree of freedom.*

While the damping in slightly damped vibrations may be neglected in the determination of the natural frequencies of the system, the damping will have a significant influence on the vibration amplitude around the natural frequencies. For the system shown in Figure 6.17, the *dynamic amplification factor Q* given by

$$Q(\omega) = \frac{x_o}{F_0/k} = \frac{1}{\sqrt{\left(1-\left(\frac{\omega}{\omega_0}\right)^2\right)^2 + \left(2\xi\frac{\omega}{\omega_0}\right)^2}}$$

determines the motion amplitude x_0 of the mass m when this is subjected to a periodic force $F_0 \cos \omega t$. It is seen from this expression that at resonance, $\omega = \omega_0$:

$$Q(\omega_0) = \frac{1}{2\xi}$$

so that the motion is limited at resonance in contrast to the undamped system.

The above shows that the magnitude of the damping is of decisive importance to the analysis of forced vibrations when the frequency of the external loads is close to the natural frequencies of the structure.

If ship hull vibrations are considered, damping will mainly be due to structural damping from hysteresis effects in the steel, especially as a consequence of welding. An eminent overview of the various possible effects can be found in Betts et al. (1977).

Moreover, damping takes place (by dry friction) in cargoes of grain and the like, as well as through hydrodynamic damping. However, all these effects are usually so small that they can be neglected in relation to the internal structural damping in the welded steel structure in the frequency range of interest.

Calculation of the magnitude of damping in hulls is not possible because the theoretical damping mechanisms cannot today be calculated for so complex structures as ships. Therefore, calculation of forced hull vibrations must be based on empirical formulas for the damping, derived from full-scale measurements. A number of these formulas are shown in Figure 6.18.

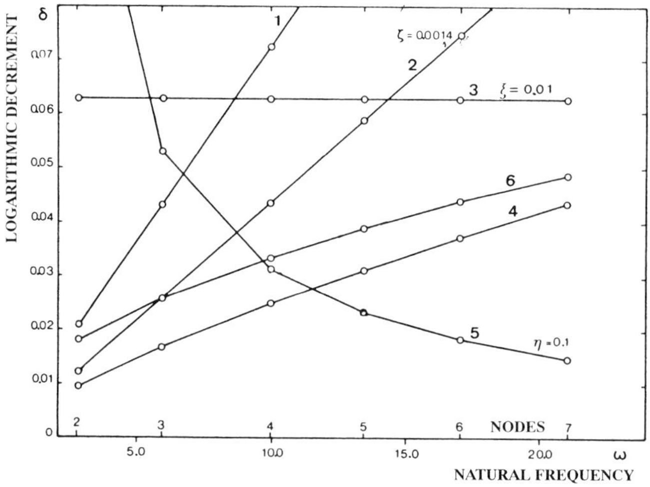

1 $\delta_n = .0073\,\omega_n$

2 $\delta_n = \zeta\pi\omega_n$

3 $\delta_n = 2\pi\xi$

4 $\delta_n = 3.5/L \cdot (\omega_n/\omega_2)^{3/4}$

5 $\delta_n = \eta\pi/\omega_n$

6 $\delta_n = .01065\omega_n^{\frac{1}{2}}$

Figure 6.18 *Examples of published values for the logarithmic decrement δ as a function of frequency ω for a 340,000 dwt tanker. Jensen and Madsen (1977).*

It is seen from Figure 6.18 that the logarithmic decrement is usually assumed to increase with the natural frequency. The variation may be reasonably well represented by an expression of the form:

$$\delta_i = a\omega_i + \frac{b}{\omega_i}$$

where a and b are constants.

For the example in Figure 6.17 such a variation of the logarithmic decrement requires that the damping

$$c \simeq 2\sqrt{km}\,\delta/2\pi = \frac{a}{\pi}k + \frac{b}{\pi}m$$

is a linear combination of the mass m and the spring stiffness k. A generalisation of the above expression is often used in the calculation of forced vibrations by means of the finite element method, not to be considered here.

In a beam formulation a more useful description of the internal damping is a linear 'viscous' damping, defined by a generalisation of the usual constitutive equations for Timoshenko beams, Eq. (6.2). Thus,

$$M = EI\left(1 + \mu\frac{\partial}{\partial t}\right)\frac{\partial\phi}{\partial x} \tag{6.49}$$

$$Q = kGA\left(1 + \varepsilon\frac{\partial}{\partial t}\right)\left(\frac{\partial v}{\partial x} - \phi\right) \tag{6.50}$$

where the constants μ and ε are normal and tangential *damping coefficients*, respectively. By insertion of these constitutive equations in the equations of motions for the Timoshenko beam, Eq. (6.5), the following eigenvalue problem for damped beam vibrations is obtained:

$$\frac{\partial}{\partial x}\left[EI\left(1 + \mu\frac{\partial}{\partial t}\right)\frac{\partial\phi}{\partial x}\right] + kGA\left(1 + \varepsilon\frac{\partial}{\partial t}\right)\left(\frac{\partial v}{\partial x} - \phi\right) = mr^2\frac{\partial^2\phi}{\partial t^2} \tag{6.51}$$

$$\frac{\partial}{\partial x}\left[kGA\left(1 + \varepsilon\frac{\partial}{\partial t}\right)\left(\frac{\partial v}{\partial x} - \phi\right)\right] = m\frac{\partial^2 v}{\partial t^2} \tag{6.52}$$

It is normally assumed that the damping parameters μ, ε are equal and independent of the cross-section of the beam:

$$\mu = \varepsilon = \text{constant} \tag{6.53}$$

This assumption is reasonable for practical calculations of hull vibrations, as it is not possible to measure the damping per section, but only to estimate the overall damping in the vibration.

6.5 MODAL SUPERPOSITION

Several methods are available for determination of forced beam vibrations:

- Numerical integration in time
- Separate solution for selected load frequencies
- Modal superposition

Numerical integration in time requires solution of initial value problems:

$$\begin{aligned}
&\frac{\partial}{\partial x}\left[EI\left(1 + \mu\frac{\partial}{\partial t}\right)\frac{\partial\phi}{\partial x}\right] + kGA\left(1 + \varepsilon\frac{\partial}{\partial t}\right)\left(\frac{\partial v}{\partial x} - \phi\right) - mr^2\frac{\partial^2\phi}{\partial t^2} = 0 \\
&\frac{\partial}{\partial x}\left[kGA\left(1 + \varepsilon\frac{\partial}{\partial t}\right)\left(\frac{\partial v}{\partial x} - \phi\right)\right] - m\frac{\partial^2 v}{\partial t^2} = -f(x,t)
\end{aligned} \tag{6.54}$$

with appropriate initial conditions for $t = t_0$.

Thus, a boundary value problem must be solved for each point of time $t > t_0$, and the method is therefore only used, if it is required only to analyse relatively short periods of time. This may be the case, if $f(x, t)$ represents a transient load ("impact"), e.g. due to water impacts on the ship. Such wave-induced, transient vibrations are called "whipping". The solution method will not be discussed here, but results are shown in Figure 4.22.

A direct solution can be applied if the load is periodic in time.

$$f(x,t) = f_0(x)\cos\Omega t$$

Then a particular solution of Eq. (6.54) may be written

$$\begin{aligned} v(x,t) &= \bar{u}_1(x)\cos\Omega t + \bar{u}_2(x)\sin\Omega t \\ \phi(x,t) &= \bar{\alpha}_1(x)\cos\Omega t + \bar{\alpha}_2(x)\sin\Omega t \end{aligned}$$

which, inserted in Eq. (6.54), gives four coupled time-independent, ordinary differential equations for determination of $\bar{u}_1(x)$, $\bar{u}_2(x)$, $\bar{\alpha}_1(x)$ and $\bar{\alpha}_2(x)$. This solution method is therefore well suited if the vibrations of the beam are only to be determined for a few load frequencies, say the blade frequency and its superharmonies.

If the first n natural frequencies ω_i of the hull girder and the corresponding natural vibration modes (deflection $u_i\,(x)$ and angular displacement $\alpha_i\,(x)$) ; $i = 1,2,3,....,n$ have been determined, *modal superposition* will be an efficient solution method.

In this method the deflection $v(x,t)$ and the angular displacement $\phi(x,t)$ are approximated by linear combinations of the known eigenfunctions $u_i(x)$, $\alpha_i(x)$:

$$\begin{aligned} v(x,t) &= \sum_{i=i}^{n} u_i(x)w_i(t) \\ \phi(x,t) &= \sum_{i=1}^{n} \alpha_i(x)w_i(t) \end{aligned} \qquad (6.55)$$

where the weight functions $w_i(t)$ are determined by insertion of Eq. (6.55) in Eq. (6.54). This gives

$$\begin{aligned} &\sum_{i=1}^{n}\Big\{\Big[\big(EI\alpha'_i\big)' + kGA(u'_i - \alpha_i)\Big]w_i + \Big[\big(\mu EI\alpha'_i\big)' + \varepsilon kGA\big(u'_i - \alpha_i\big)\Big]\dot{w}_i \\ &\quad - mr^2\alpha_i\ddot{w}_i\Big\} = 0 \\ &\sum_{i=1}^{n}\Big\{\big(kGA\big(u'_i - \alpha_i\big)\big)'\big(w_i + \varepsilon\dot{w}_i\big) - mu_i\ddot{w}_i\Big\} = -f(x,t) \end{aligned} \qquad (6.56)$$

where

$$(\,)' \equiv d/dx \quad \text{and} \quad (\dot{\;}) \equiv d/dt.$$

If the first and second equations in Eq. (6.56) are multiplied by respectively $\alpha_j(x)$ and $u_j(x)$, the following is obtained by subsequent addition of the equations and integration over the length of the beam:

$$
\begin{aligned}
\sum_{i=1} \Bigg\{ & w_i \int_0^L \Big[\alpha_j\big(EI\alpha'_i\big)' + \alpha_j kGA\big(u'_i - \alpha_i\big) + u_j\big(kGA\big(u'_i - \alpha_i\big)\big)'\Big]dx \\
& + \dot{w}_i \int_0^L \Big[\alpha_j\big(\mu EI\alpha'_i\big)' + \alpha_j \varepsilon kGA\big(u'_i - \alpha_i\big) + u_j\varepsilon\big(kGA\big(u'_i - \alpha_i\big)\big)'\Big]dx \\
& - \ddot{w}_i \int_o^L \Big[mr^2\alpha_j\alpha_i + mu_ju_i\Big]dx\Bigg\} = -\int_0^L u_j f(x,t)dx
\end{aligned}
\tag{6.57}
$$

If the orthogonality conditions (6.14a–b) are utilised and on the assumption that μ and ε are independent of x, the above expressions are reduced to

$$
\begin{aligned}
& -\omega_j^2 A_j w_j(t) - \omega_j^2 A_j \mu \dot{w}_j(t) - A_j \ddot{w}(t) \\
& + (\varepsilon - \mu) \sum_{i=1}^{n} \int_o^L \Big[\alpha_j\, kGA(u'_i - \alpha_i) + u_j\big(kGA(u'_i - \alpha_i)\big)'\Big]\, dx \\
& = -\int_0^L u_j f(x,t)dx
\end{aligned}
\tag{6.58}
$$

where

$$
A_j = \int_0^L \Big(mr^2\alpha_j^2 + mu_j^2\Big)dx \tag{6.59}
$$

It is seen that the differential equation system (6.58) for determination of $w_i(t)$; $i = 1,2,\ldots, n$ is decoupled if $\mu = \varepsilon$. As previously mentioned, the damping can only be determined empirically, and the assumption $\mu = \varepsilon$ seems to be reasonable. Thus, Eq. (6.58) is reduced to

$$
\ddot{w}(t) + \mu\omega_j^2\dot{w}_j(t) + \omega_j^2 w_j(t) = \frac{1}{A_j}\int_0^L u_j(x)f(x,t)dx \equiv F_j(t) \tag{6.60}
$$

By comparison with Figure 6.17 it is observed that the damping relation $\xi = \frac{1}{2}\mu\omega_j$ holds. The right-hand side $F_j(t)$ can be determined when the external load $f(x,t)$ is known. A particular solution $w_j^p(t)$ to Eq. (6.60) is

$$w_j^p(t) = \frac{1}{\omega_{dj}} \int_0^t F_j(\tau) \exp\left(-\frac{1}{2}\mu\omega_j^2(t-\tau)\right) \sin\left[\omega_{dj}(t-\tau)\right] d\tau \tag{6.61}$$

where

$$\omega_{dj} = \omega_j \sqrt{1 - (\tfrac{1}{2}\mu\omega_j)^2}$$

The solution sought for is then given by Eq. (6.61) and the homogeneous solution

$$w_j^h(t) = \exp\left(-\frac{1}{2}\mu\omega_j^2 t\right)\left(a \cos\omega_{dj}t + b \sin\omega_{dj}t\right) \tag{6.62}$$

where the constants a and b are determined from the initial conditions at $t = 0$.

The most difficult problem of the use of the modal superposition method for hull girder vibrations arises due to the added mass of water m_w as this mass, through the frequency and the global reduction factor J, depends on the vibration mode. Thus, the orthogonality conditions Eq. (6.14) no longer apply, and hence Eq. (6.56) cannot be reduced to the uncoupled system (6.60).

There are two ways to overcome this problem. In the first method the d'Alembert forces $- m_w\ddot{v}$ from the mass of water are regarded as an external load like $f(x,t)$. Thus, the eigenvalue problem will only comprise the natural, frequency-independent mass m_s of the ship, and the orthogonality conditions Eq. (6.14) will be satisfied. The natural vibration modes will correspond to the hull girder vibrating in air. If these natural frequencies and vibration modes are denoted by $\tilde{\omega}_i$, $\tilde{u}_i(x)$, $\tilde{\alpha}_i(x)$, the following is obtained analogously to Eq. (6.55) and Eq. (6.60):

$$\begin{aligned} v(x,t) &= \sum_{i=1}^{n} \tilde{u}_i(x) w_i(t) \\ \phi(x,t) &= \sum_{i=1}^{n} \tilde{\alpha}_i(x) w_i(t) \end{aligned} \tag{6.63}$$

where $w_i(t)$ is determined as the solution to

$$\ddot{w}(t) + \mu\tilde{\omega}_i^2 \dot{w}_i(t) + \tilde{\omega}_i^2 w_i(t) =$$

$$\frac{1}{\tilde{A}_i} \int_0^L \tilde{u}_i(x) \left[f(x,t) - m_w(x, \tilde{\omega}_i) \sum_{j=1}^{n} \tilde{u}_j(x)\, \ddot{w}_j(t) \right] dx$$

or

$$\sum_{j=1}^{n}\left(\delta_{ij} + a_{ij}\right)\ddot{w}_j(t) + \mu\tilde{\omega}_i^2\dot{w}_i(t) + \tilde{\omega}_i^2 w_i(t) = \frac{1}{\tilde{A}_i}\int_0^L \tilde{u}_i(x)f(x,t)dx \qquad (6.64)$$

where δ_{ij} is Kronecker's delta and where

$$a_{ij} = \frac{1}{\tilde{A}_j}\int_0^L \tilde{u}_i(x)m_W(x,\tilde{\omega}_i)\tilde{u}_j(x)dx$$

with

$$\tilde{A}_j = \int_0^L\left[m_s r^2\,\tilde{\alpha}_j^2 + m_s\,\tilde{u}_j^2\right]dx$$

It is seen from Eq. (6.64) that the weight functions $w_i(t)$ can no longer be found separately, but instead have to be determined by a matrix equation. This is a disadvantage of the method. The method is, however, very useful for analysis of wave-induced vibration as shown in Section 6.5.1.

An alternative method is to use the value for the added mass of water m_W, based on the two-noded vibration mode, for all higher vibration modes. In this case the calculated vibration modes and natural frequencies will normally not be quite correct with the exception of the two-noded vibration. If, however,

$$m(x,\omega_i) = m_s(x) + m_W(x,\omega_i) \approx b_i m(x,\omega_1) \qquad (6.65)$$

where ω_1 is the natural frequency corresponding to the two-noded vibration mode, it is seen from Eq. (6.7) that also the higher vibration modes will be correct, while the higher, real natural frequencies ω_i from the calculated natural frequencies $\hat{\omega}_i$ on the basis of $m(x,\omega_1)$ are found to be

$$\omega_i = \hat{\omega}_i/\sqrt{b_i} \qquad (6.66)$$

The validity of the condition Eq. (6.65) cannot be argued from a physical point of view, but it applies usually with sufficient accuracy. Thus, Eq. (6.60) still applies, if only ω_j is replaced by $\hat{\omega}_j$:

$$\ddot{w}_j(t) + \mu\hat{\omega}_j^2\dot{w}_j(t) + \hat{\omega}_j^2 w_j(t) = \frac{1}{A_j}\int_0^L u_j(x)f(x,t)dx \qquad (6.67)$$

As Eq. (6.60), this equation determines the weight functions $w_i(t)$ separately. This method is therefore easier to use than the method where the eigenfunctions of the hull girder in air are used. However, it should be noted that the assumption Eq. (6.64) introduces an error, which cannot be reduced only by including more eigenfunctions.

It should finally be mentioned that if the load $f(x,t)$ is a function of the deformation of the hull girder $f(x,t) = f(v_i, x, t)$, the right-hand side in Eq. (6.60) or (6.67) will give a coupling between the single weight functions $w_i(t)$, and in such cases it is advantageous to use Eqs. (6.63) - (6.64). This is the case with wave-induced hull vibrations. In the following, such vibrations will be treated in more detail.

6.5.1 Wave-induced Hull Vibrations

One formula for the wave-induced load $q_H(x,t)$ per unit of length along the hull girder is given by Eq. (4.58):

$$f(x) = q_H(x,t) = -\left[\frac{D}{Dt}\left(m_w(x,\omega_e)\frac{D\tilde{z}}{Dt}\right) + N(x,\omega_e)\frac{D\tilde{z}}{Dt} + \rho g B(x) z\right] \tag{6.68}$$

where $\tilde{z} = \tilde{z}(x,t)$ is a measure of the relative vertical distance between the surface of the sea and the ship at $x = x$, Eq. (4.20):

$$\tilde{z}(x,t) = v(x,t) - \kappa(x) h(x,t) \tag{6.69}$$

In Eq. (6.68) $N(x)$ is the hydrodynamic damping and $B(x)$ is the water line breadth. Furthermore, $\kappa(x)$ is the Smith correction factor and $h(x,t)$ is the wave elevation. The latter may for linear, regular, long-crested waves be written, Eq. (4.55):

$$h(x,t) = a\cos(k_e x - \omega_e t) \tag{6.70}$$

where a is the wave amplitude and where

$$\begin{aligned} k_e &= k\cos\beta \\ \omega_e &= \omega - k_e V \end{aligned} \tag{6.71}$$

Here k is the wave number and ω the frequency of the wave, while β is the heading angle. Finally, V is the forward speed.

The load given by Eqs. (6.68) - (6.69) has been derived from some assumptions which are reasonable in the wave length range relevant for determination of the rigid body motions of the ship. However, for generation of the lowest hull vibration modes much smaller wave length are required. Thus, the basic assumption in Eq. (6.68) about two-dimensional water flow around the hull girder is violated. But, for lack of anything better, Eq. (6.68) is used for determination of wave-induced hull vibration modes as well. However, the added mass of water m_w is corrected, as mentioned previously, by a global reduction factor J for three-dimensional flow.

The lowest natural vibration modes of the hull girder, calculated in air, are denoted $\tilde{u}_i(x), \tilde{\alpha}_i(x)$, while the corresponding natural frequencies are designated $\tilde{\omega}_i$. To ensure equilibrium for the ship as a whole, the rigid-body motions, Eq. (6.23), are introduced as extra eigenfunctions, both having zero natural frequency ($\tilde{\omega}_{-1} = \tilde{\omega}_0 = 0$). Moreover, all eigenfunctions are normalized, so that

$$\int_0^L \left[m_s r^2 \tilde{\alpha}_i \tilde{\alpha}_j + m_s \tilde{u}_i \tilde{u}_j \right] dx = \delta_{ij} \quad \text{for} \quad i, j = -1, 0, 1 ...., n$$

Thereby, the equilibrium equation (6.64) can be written

$$\sum_{j=-1}^{n} \left[M_{ij} \ddot{w}(t) + C_{ij} \dot{w}_j(t) + S_{ij} w_j(t) \right] = F_i(t) \quad ; \quad i = -1, 0, 1 ..., n \tag{6.72}$$

where

$$M_{ij} = \delta_{ij} + \int_0^L \tilde{u}_i(x) m_w(x) \tilde{u}_j(x) dx$$

$$C_{ij} = \mu \tilde{\omega}_j^2 \delta_{ij} + \int_0^L \tilde{u}_i(x) \left[(N(x) - V m'_w(x)) \tilde{u}_j(x) - 2 V m_w(x) \tilde{u}'_j(x) \right] dx \tag{6.73}$$

$$S_{ij} = \tilde{\omega}_j^2 \delta_{ij} + \int_0^L \tilde{u}_i(x) \left[\rho g B(x) \tilde{u}_j(x) - V(N(x) - V m'_w(x)) \tilde{u}'_j(x) + V^2 m_w(x) \tilde{u}''_j(x) \right] dx$$

and

$$\begin{aligned} F_i(t) &= a_0 \int_0^L \Big[\left(-m_w(x)\omega^2 + \rho g B(x) \right) \cos(k_e x - \omega_e t) \\ &\quad + \left(N(x) - V m'_w(x) \right) \omega \sin(k_e x - \omega_e t) \Big] \kappa(x) dx \\ &= a(F_{1i} \cos \omega_e t + F_{2i} \sin \omega_e t) \end{aligned} \tag{6.74}$$

The solution to Eqs. (6.72) - (6.74) can be written

$$w_i(t) = w_{1i} \cos \omega_e t + w_{2i} \sin \omega_e t \tag{6.75}$$

and if this assumption of solution is inserted in Eq. (6.72), two linear algebraic systems of equations are obtained, each of the order $n + 2$, for determination of the $2(n + 2)$ unknown constants w_{1i}, w_{2i} ; $i = -1,0,1...,n$. It should be noted that for $n = 0$ the usual system of equations is obtained for determination of the motions of the ship in regular waves, treated in Section 4.3.

It follows from Eq. (6.74) and Eq. (6.72) that all the weight factors w_{1i} and w_{2i} will vary linearly with the wave amplitude, such that

$$w_i(t) = a\,\Phi_i(\omega_e)\cos(\omega_e t + \varepsilon_{wi})$$

where Φ_i are the amplitude functions for the weight functions.

On the basis of the solution Eq. (6.63), all relevant sectional quantities in the hull girder can be determined, as e.g. the bending moment $M_y(x)$. The direct method is to use Eq. (6.49):

$$\begin{aligned} M_y(x,t) &= EI(x)\sum_{i=0}^{n}\tilde{\alpha}'_i(x)(w_i(t) + \mu\dot{w}_i(t)) \\ &= a\Phi_M(x,\omega_e)\cos(\omega_e t + \varepsilon_M(x,\omega_e)) \end{aligned} \qquad (6.76)$$

The bending moment can thus be written on the same form as in Section 4.3 with an amplitude given as the product of the wave amplitude a and an amplitude function $\Phi_M(x,\omega_e)$.

As regards numerical calculations, Eq. (6.76) is not suitable for determination of the sectional forces, as the truncation (finite value of n) may lead to rather large errors*. It is better to insert the solution, Eq. (6.75), in the expression for q_H, Eq. (6.68), and for the d'Alembert forces and then integrate these forces from the stern to the section $x = x$ concerned. This is the same as done in Section 4.3 and the result can of course be written on the same form as Eq. (6.76).

From the solution of the system of equations (6.72) for a number of discrete wave frequencies ω, the amplitude function $\Phi_R(\omega_e)$ for all linear responses (bending moment etc) can be determined as a function of the wave frequency ω. The natural frequencies in water is identified as the frequency of encounter where the amplitude function exhibits large peaks. The lowest natural frequency correspond to a vibration mode with two nodes, see Figure 6.1, and is often the only one of interest.

Subsequently, statistical quantities in a stationary, stochastic sea can be calculated on the basis of a given wave spectrum $S(\omega)$, as shown in Section 4.4. Here it should only be noted that the response R is normal distributed with mean value zero and a standard deviation s_R given by

* If only rigid body modes are included, i.e. n=0, then $M_y \equiv 0$, as $\tilde{\alpha}_{-1} = 0$ and $\tilde{\alpha}_0 = 1$!

$$s_R^2 = \int_0^\infty \Phi_R^2(\omega)S(\omega)d\omega \tag{6.77}$$

The wave-induced vibrations, also denoted *springing*; might increase the stress level in moderate sea state, c.f. Example 6.5.1 below. However, significant springing vibrations required either a large forward speed in head sea (container ships, fast vessels) or a very low hull stiffness (Great Lake bulk carriers).

Example 6.5.1

In Jensen & Pedersen (1981) and Jensen and Dogliani (1996) a container ship is analysed (L_{pp} = 270 m, B = 32.2 m, T = 10.85 m, C_B = 0.598). The mass distribution is approximated by a fourth degree polynomial in x, with a centre of gravity of 10.12 m abaft midship and a longitudinal radius of gyration of 0.248 L_{pp}. A transverse radius of gyration of 5 m is assumed, constant along the length of the ship. The added mass of water m_w and the hydrodynamic damping N are calculated on the basis of the body plan of the ship using Lewis transformation. Finally, the stiffness of the hull girder is approximated by $EI = 8.68\,10^{13}$ Nm^2, $kGA = 6.25\ 10^{10}$ N for the central half of the ship, with linear reduction towards the ends of the ship with the values at the ends equal to 10 per cent of the midship values.

The damping coefficient is set to $\mu = 10^{-3}$ secs and the ship sails in head sea ($\beta = 180^{\circ}$) with a forward speed V = $0.245\sqrt{gL}$.

The bending moment amidships $M \equiv M_y(x = L_{pp}/2)$ is calculated without account for the elastic deformation of the hull girder (n = 0, "rigid hull") as well as by inclusion of the two-noded vibration mode (n = 1, "flexible hull"). The two-noded natural frequency for the ship vibrating in air becomes $\tilde{\omega}_1$ = 0.8Hz.

In Figure 6.19 the amplitude function $\Phi_M(\omega_e)$ is shown as a function of the frequency of encounter ω_e. It is seen that the lowest natural frequency of the hull girder in water is approximately 4.2 rad/sec $\simeq$ 0.7 Hz, and it corresponds to wavelengths of approximately $L_{pp}/10$.

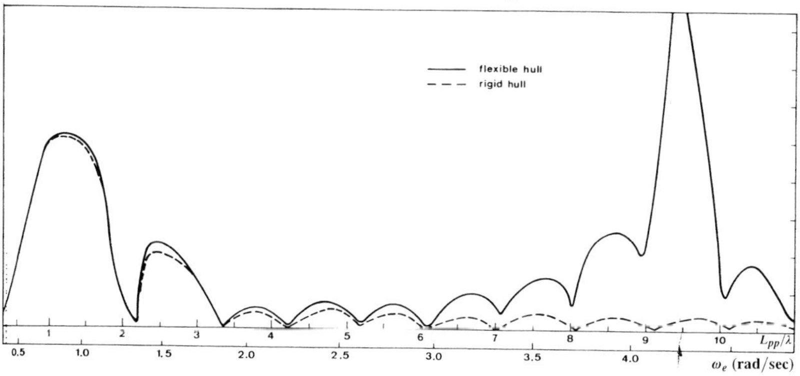

Figure 6.19 *Amplitude function for bending moment amidships $\Phi_M(\omega_e)$ as a function of the frequency of encounter ω_e and ship length to wave length ratio (L_{pp}/λ). Container ship.*

To be able to assess the importance of the flexibility of the hull girder the response spectral density c.f. Eq. (4.109)

$$S_M(\omega_e) = \phi_M^2(\omega_e)\bar{S}(\omega_e)$$

is plotted in Figure 6.20 for a sea state characterised by a Pierson–Moskowitz spectrum with $H_s = 4$ m and $T_s = 7.1$ secs.

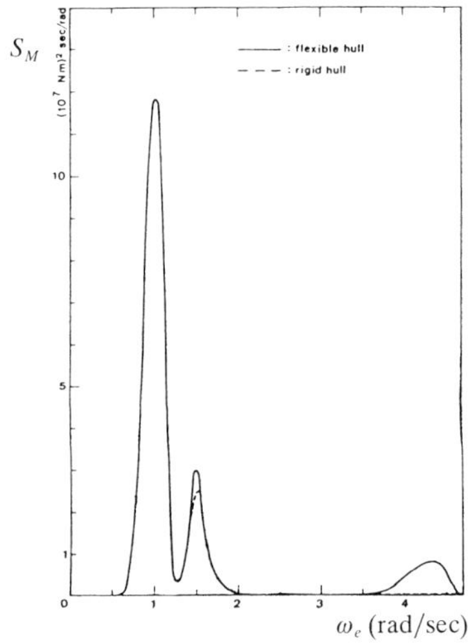

Figure 6.20 *Response spectral density $S_M(\omega_e)$ for the bending moment amidships in a container ship.*

It is seen that the most essential part of the response spectrum can be determined without taken into account to the flexibility of the hull girder. However, it should be noted that the high-frequency part around the lowest natural frequency of the hull girder may influence the fatigue strength of the hull girder.

Table 6.2

Container ship, Head Sea	2-noded vibration included ($n = 1$)
Standard values:	$EI = 8.68\ 10^{13}$ Nm^2, $\mu = 10^{-3}$ secs, $\tilde{\omega}_1 = 0.8$Hz $Fn = 0.245$, $H_s = 4$m, $T_s = 7.1$ secs
Change of standard values	s_M $10^8 Nm$
None	2.18
Rigid hull ($n = 0$)	1.99
Increased hull flexibility: EI = $7.72\ 10^{13}$ Nm^2	2.22
Reduced forward speed: $Fn = 0.20$	2.41
Reduced forward speed: $Fn = 0.15$	2.33
Increased structural damping: $\mu = 2\ 10^{-3}$ secs	2.18
Lower sea state: $H_s = 2.74$ m; $T_s = 5.88$ secs	1.01
More severe sea state: $H_s = 5.61$ m; $T_s = 8.41$ secs	4.24
Non-linear effects included	2.31

In Table 6.2 it is shown how the standard deviation s_M, given by

$$s_M^2 = \int_0^\infty S_M(\omega)d\omega$$

varies with changes in some of the parameters.

It should finally be noted that Jensen and Pedersen (1981) and Jensen and Dogliani (1996) also contain an analysis where non-linear effects are included. For the container ship these non-linear effects may be as significant as the effect of the flexibility of the hull girder, c.f. Table 6.2.

6.6 REDUCTION OF THE VIBRATION LEVEL

To judge if a ship will experience hull girder vibration problems, it is necessary to determine the vibration level, as described in the preceding sections, and to have criteria to which this level can be compared.

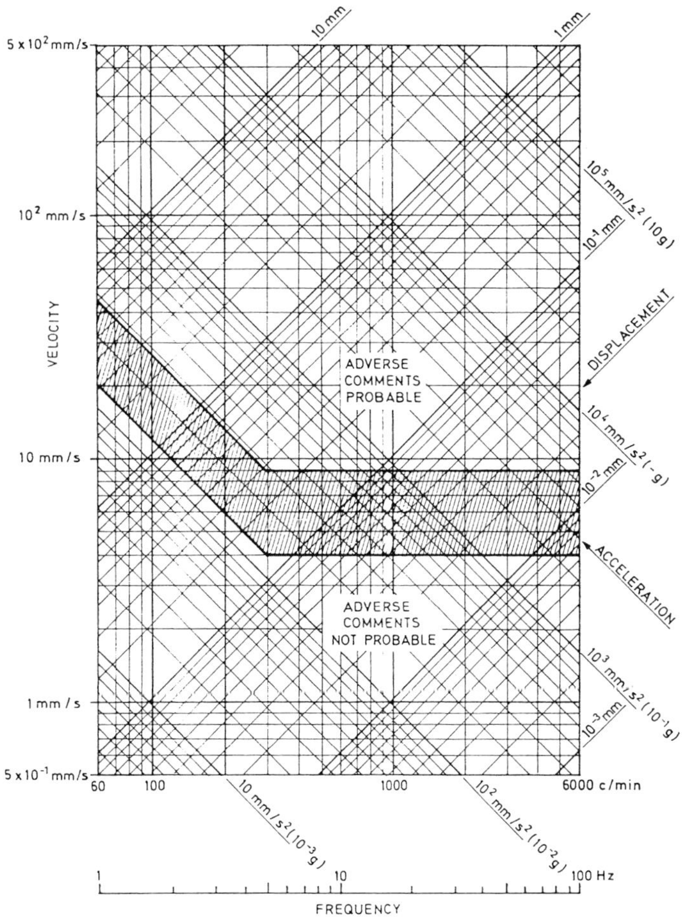

Figure 6.21 *Curves for assessment of the vibration level. (ISO 6954, "Guidelines for overall evaluation of vibration in merchant ships")*

These criteria must reflect that a too high vibration level may lead to fatigue failure (crack propagation) in the structure, to damage of the electronic and mechanical equipment, and/or adverse comments from the crew and passengers. In the high frequency range it will usually be the latter point which decides if the vibration level can be tolerated. Thus, a rational criterion for permissible vibration levels cannot be set up, as this must primarily be due to a subjective estimate. Based on a collection of data from existing merchant ships, guidelines, as shown in Figure 6.21, have been established for assessment of the vibration level. The guideline shown apply to the frequency range up to 100 Hz and reflect that, for frequencies larger than approximately 5 *Hz*, the velocity level is decisive for human subjective judgement of the vibration level, while it is rather the acceleration level for lower frequencies.

However, as regard hull girder vibrations these guidelines are of limited value as the frequency is usually below 1 Hz.

If a vibration analysis shows that the hull girder is likely to be exposed to vibration problems, there are in principle three possibilities for reduction of the vibration level:

- reduction of the amplitudes of the existing forces
- change of the frequency of the existing forces
- change of the natural frequencies of the hull girder

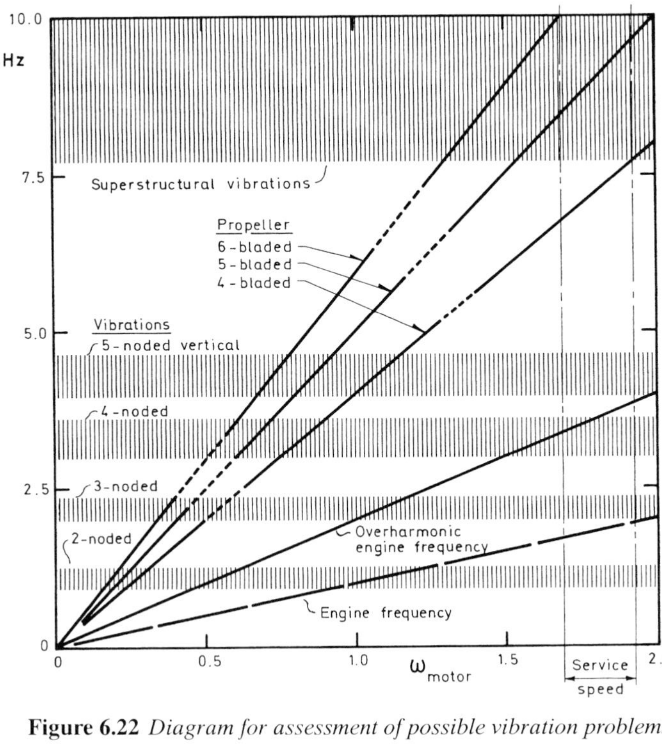

Figure 6.22 *Diagram for assessment of possible vibration problems.*

These three possibilities will be dealt with in the following, but first a practical diagram is presented, Figure 6.22, for use in the assessment of vibration problems at the design stage. In the figure are shown the lowest natural frequencies for the hull girder, the blade frequency for a number of alternative propellers with a varying number of blades, the engine frequency and its first overharmonics. As a consequence of partly the uncertainty of the calculations and partly the different possible loading conditions of the ship, the natural frequencies are indicated by frequency bands instead of lines. It is seen from the figure that the ship in question may experience problems with the first overharmonic engine frequency, as this frequency at service speed lies close to the natural frequency of the four-noded vertical hull vibration mode.

It should be noted that the classification societies issue very detailed guidelines on how to avoid vibration problems by proper design considerations. In the following only some overall guidelines are therefore given.

6.6.1 Reduction of the Amplitudes of the Forces

The most important time-varying loads are treated in Section 6.2. As regards the propeller-induced forces, vibration problems are as a rule due to the strong hydrodynamic pressure variations which may arise on the plating of the stern, if the propeller cavitates. Determination of the "force" of the cavitation is difficult and subject to large uncertainty. However, it is known that the radial velocity variation of the wake field at the propeller has a large influence on the cavitation. This velocity variation is often considerably larger at a V-shaped stern than at a U-shaped stern, and thus the latter type can be chosen to avoid vibration problems, although it may imply a less suitable hull from a propulsive and structural point of view. Other, less drastic, changes of the stern to reduce the variation of the wake field may be to increase the distance between plating and propeller, to mount fins ("vortex generators") or to mount a nozzle around the propeller. The amplitudes of the pressure variations can also be reduced by change of the propeller geometry. A number of model experiments have shown that propellers of the "skew back" type can reduce the load amplitude by 50 - 75% compared to a conventional propeller. The reason is that the propeller tip on "skew back" propellers is relatively unloaded with subsequently smaller cavitation. The disadvantages of the use of "skew back" propellers are more expensive propeller production as well as a small loss of propulsion efficiency (2 - 4%).

As regards the propeller-induced forces, transferred to the hull via the bearings of the propeller axis, these will normally only pose problems if the blade frequency is close to one of the natural frequencies of the shaft system. The force amplitudes can be reduced by choosing a propeller with another number of blades, c.f. the discussion in Section 6.1 on the dependence of the forces on the even and odd harmonic components of the wake field.

To reduce unbalanced forces from the main engines, electric or mechanical balancing mechanisms may be used which generate forces of the same magnitude and frequency as the unbalanced forces, but with in opposite phase.

6.6.2 Change of the Frequency of the Forces

Propeller-induced forces contain components, which vary in time with multiples of the blade frequency (= number of revolutions x number of propeller blades). Since the number of revolutions is usually around 1.5 - 2 Hz and the number of blades n = 3 -

5, it is mainly the blade frequency (~ 4 - 10 Hz) which is of importance to hull vibrations. If the blade frequency is close to one of the natural frequencies of the hull girder, the possibility of vibration problems may be reduced by changing the number of propeller blades.

Concerning wave-induced hull vibrations, the load frequency, which is equal to the frequency of encounter ω_e, may be changed by changing the forward speed of the ship and/or its course. This possibility is as a rule used by the captain to avoid or reduce these vibrations, as wave-induced vibrations are normally not taken into account at the design stage.

6.6.3 Change of the Natural Frequencies of the Hull Girder

The alternative to change the frequency of the load is to change the natural frequencies of the hull girder. This may be done by changing the stiffness (EI, kGA) and/or the mass distribution m_S. Change of the stiffness is normally not possible, because it is determined from strength criteria and an increase in stiffness will be expensive as regards higher steel weight of the ship and subsequent smaller cargo carrying capacity. It is also difficult to change the mass distribution. The only possibility may be to move the ballast in the ballast tanks.

7 Hull Girder Reliability

In the structural assessment procedure the calculated stresses and deformations are compared to permissible values taking into account all possible failure modes of the hull girder. Basically all relevant failure modes depend on either the yielding, ductility or fatigue behaviour of the material and hence the structural capacity depends on these material parameters. In addition the stress state depends on the modulus of elasticity E and Poisson's ratio ν as seen in the previous chapters.

The first step in the structural assessment procedure is therefore the selection of the material for the strength elements. Steel is by far the most common material for ships and numerous qualities exist with different yield strength and ductility. Usually, the higher the yield stress and ductility the more expensive the material becomes regarding handling in the ship manufacturing process. Also, the fatigue resistance is not increased proportionally implying that high tensile steel is relatively more prone to fatigue failure.

For smaller vessels, aluminium and glass-fibre reinforced panels or sandwich elements (GRP) have gained increased use. For high speed ferries, the displacement is the decisive factor to achieve the requested speed and therefore the lesser mass density of aluminium as compared to steel (approximately one-third) is very important. In the recent years the material properties (yield strength, ductility, weldability, fatigue strength) of aluminium have improved significantly with the advent of the so-called 5000, 6000 and 7000 series.

Material properties are discussed in Section 7.1 in some details from a structural point-of-view.

With the structural material chosen for the strength elements in the hull girder, the maximum loads the hull can withstand without damage or failure can be calculated. With the hull girder considered as a beam, the overall failure mode can be classified as either ultimate hull girder failure due to hull girder bending and shear or to fatigue failure in significant structural members. The former analysis involves buckling of longitudinal members in compression and therefore requires a non-linear structural analysis of the behaviour of plate panels. Such analysis is outside the scope of the present treatment and only some results from the literature are quoted.

The final step is then to compared the stresses derived in Chapter 5 and 6 due to the loads from Chapter 2 and 4, taking into account the stochastic properties of the loads and statistical uncertaincies with the failure criteria in the structural and geometrical parameters as well as the modelling uncertainties and our ignorance in the analysis procedures applied. Such hull girder reliability analysis is carried out in Section 7.3.

7.1 MATERIAL PROPERTIES

The material properties which, from a strength point-of-view, are decisive for the choise of material are yield strength, ductility and fatigue limit. These properties are to some extent independent of each other and is therefore treated in different sections below.

7.1.1 Modulus of Elasticity and Yield Strength

If a uniaxial tensile test is carried out on a test specimen of a material, *stress-strain diagrams* as shown in Figure 7.1 will be obtained qualitatively. The fully drawn curves represent experiments with low strain-rates until the test specimen fails at point C. During this load sequence, the largest stress in the material, based on the undeformed cross-sectional area of the test specimen, is found at point B and it is called the *tensile strength* σ_b. Table 7.1 shows representative* values of σ_b and the *elongation at fracture* or *fracture strain* ϵ_b. If the experiment is stopped (e.g. at point D) before fracture takes place and the load removed, the test specimen will be relieved as indicated by the broken line DE.

Table 7.1 *Strength properties for shipbuilding materials.*

	Normal strength steel	High strength steel	Extra high strength steel	Aluminium alloy 1500 series
Modulus of elasticity E (MN/m^2)	210 000	210 000	210 000	70 000
Yield stress σ_y (MN/m^2)	235-250	265-390	420-690	135-270
Tensile strength σ_b (MN/m^2)	400-490	400-650	530-940	200-340
Fracture strain ϵ_b %	15-23	20-22	20-22	5-10
Impact strength (Nm)	20-30	30-40	30-40	15-25
Fracture toughness K_{IC} MN/m$^{-3/2}$	10-100	10-100	10-100	25-50
Density (kg/m^3)	7850	7850	7850	2500

When the load is removed the stress in the test specimen becomes zero, while the strain will assume a value ϵ_p whose magnitude depends on the position of the point D. The slope $d\sigma/d\epsilon$ of the stress-strain path DE is the same as that of the path OA and equal to the *modulus of elasticity* E of the material. Thus, the residual strain (the plastic strain) will only be zero, if D is on the rectilinear path OA. Usually, the *yield stress* σ_y must not be exceeded as the deformation of the material increases strongly with the load when the yield point (point A) is exceeded.

* The uncertainty of the data is no expression of an uncertainty in this parameter, but it indicates (roughly) the limits between which this material property is found by different chemical composition. The purpose of Table 7.1 is only to indicate the magnitude of a few of the materials parameters. For further information see the rules from the classification societies, where also the required test of the materials are precisely stated.

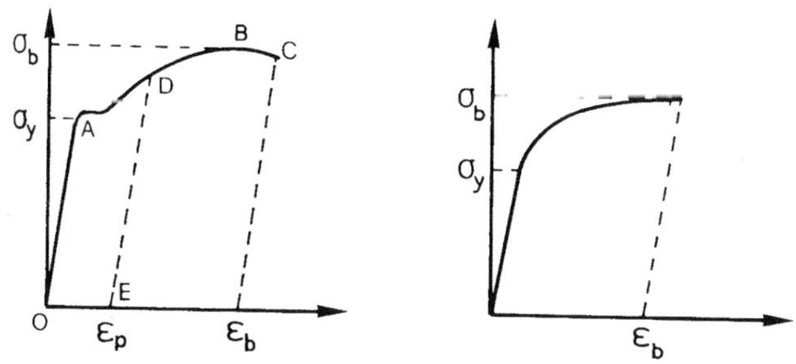

Figure 7.1 *Qualitative stress-strain diagrams for uniaxial tensile tests of steel and aluminium.*

Today, it is possible to increase the yield point and the fatigue limit of steel by means of small alloy elements and by controlling carefully the production process. Density and modulus of elasticity remain almost unchanged.

The advantage of using *high strength steel* in ships is to obtain a smaller steel weight. However, the price of the material is larger than for *normal strength steel* and the welding of high strength steel requires better quality control. Therefore high strength steel is only used if essential savings in weight can be achieved. Typically high strength steel is used for deck and keel plates in ships.

It should be noted that a weight saving by use of high tensile steel can only be achieved if the permissible stresses depend mainly on the yield stress of the material. This is not the case for fatigue failure which must be taken into account when selecting the steel quality.

An alternative method for reducing the weight of the structure is to use materials with a lower density. Aluminium with a density of approximately 1/3 of the density of steel and glass-fibre reinforced plastic with an even lower density are suitable materials.

For *aluminium* the weight advantage is partially counteracted by the fact that the modulus of elasticity E and the yield stress σ_y of aluminium are both also approximately 1/3 of the corresponding values for steel. Thus, in order to obtain the same safety against yielding and the same deformation, it is often necessary to use aluminium plating thicknesses which are about three times larger than for steel. Moreover, aluminium is a more expensive material than steel and also more complicated as regards welding. Other disadvantages of aluminium are its high thermal coefficient of expansion and its relatively low melting point. On the other hand, aluminium has better corrosion resistance, non-magnetic properties and its ductility is better at low temperatures.

These advantages and disadvantages have meant that aluminium has mainly been used as hull structure material for small craft and naval vessels (mine-sweepers). In addition, superstructures of large passenger ships have been built in aluminium, because the weight saving here leads to considerably improved hydrostatic stability due to the lower centre of gravity. Care must, however, be taken that the stresses in the superstructure imposed by vertical bending of the hull girder do not exceed what is permissible, see Example 5.1.2.

The recent improvement in the quality and yield stress of aluminium in the so-called 5000 and 6000 series has increased its use, especially for high speed ferries where the weight savings are very important.

The use of *glass-fibre reinforced plastic* (GRP) in major substructures in large ships is rare, although a larger weight saving might be obtained. One reason is the quite different form of production. While steel and aluminium sections are welded up, glass-fibre reinforced structures must be layered up in a mould.

The future development of new glass-fibre reinforced materials, such as sandwich plate elements with even better strength-weight ratios and more well defined materials properties, may increase the use of this material. Several naval vessels have today such hull structure among them the Danish Navy vessel Standardflex.

7.1.2 Ductility

The ductility of a material means its capability to absorb large, plastic deformations without initiation of fracture. Particularly two items are important, one concerning the workability of the material, the other concerning its capability to prevent propagation of cracks.

It is inevitable that there will always be cracks in a welded structure, as microcracks and slag confinements from the welding process cannot be avoided. When the structure is at sea the dynamic loads will sometimes be so great that the stress in a crack tip exceeds the tensile strength σ_b of the material. This leads to elongation of the crack but as the load is again diminished, the crack propagation usually stops. However, each material has a characteristic (critical) crack length so that, if this crack length is exceeded, the crack will continue its extension at a very great speed and result in an almost momentary brittle fracture. It is evident that the chosen material must have a *critical crack length* which is much larger than the cracks and slag confinements induced by the production process. One way to ensure this is to use a material with a large fracture strain ε_b. For ships ε_b should be larger than 13 %. In order to be able to mend these cracks as they grow, it will also be practical if they can be detected easily before they reach the critical length. This has led to the concept "leak-before-break", which e.g. for the steel in the ship's bottom means that water will penetrate the crack before the crack becomes critical.

Lamellar tearing, illustrated in Figure 7.2, is a type of fracture which have occurred previously. This fracture begins when the welded seam solidifies, because the plate has acquired a layered structure by the rolling process due to separation of non-metallic slags. Thus, the ductility of the material is small when loaded in the direction of the plate thickness.

This subsection deals with the determination of the critical crack length and equivalent quantities for a plate exposed to a static load. In the following subsection, the effect of dynamic, repeated loads will be analysed under the designation of fatigue failure.

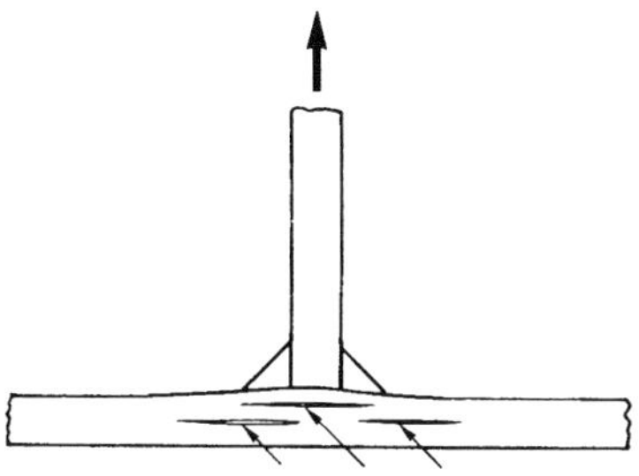

Figure 7.2 *"Lamellar tearing" of a rolled steel plate due to confinement of slags in the plate material.*

Problems with a number of ships (the Liberty ships) built about 1950 led to demands on the ductility of the material. Several of these ships broke in two as a consequence of a combination of too brittle steel, low temperature and bad design and production (large stress concentrations). At that time, no usable theoretical determination of the ductility of materials were available, so the requirements to the ductility of the materials were formulated as requirements of a so-called *Charpy - V test* on test specimens of the material concerned. In this test a drop hammer, suspended as a pendulum (see Figure 7.3), is made to hit a specimen with a given dimension and a given notch. During the experiment, it is measured how far the drop hammer swings up after it has broken the test specimen, and this result is recalculated to a lost potential energy, which is again equal to the plastic work left in the test specimen.

Contrary to aluminium alloys the ductility for carbon steel falls steeply when the temperature falls below the freezing point. This led to problems with ships for use in Arctic regions and tankers for transport of liquid natural gas (LNG) ships, where the temperature in the tanks is kept at -196 °C to large avoid boil-off of the LNG. The material for these tanks is either an aluminium alloy or a 9% nickel steel. In Figure 7.4 the importance of the content of Ni, C and $A\ell$ for the ductility is illustrated for steel alloy at different temperatures.

The *impact strength test* is an good method of controlling the ductility of given materials, but not suitable for a quantitative a priori determination of the influence of the pertinent parameters (production method, alloy content and type) on the ductility of the material.

However, a theoretical method for analysis of the ductility of materials has been developed in recent years, and it can today be used to assess the influence of crack geometry, yield stress, load distribution and temperature on the ductility of the material. The method is called *fracture mechanics* and it will be outlined in the following. For a description of the subject in more detail, see one of the many textbooks available.

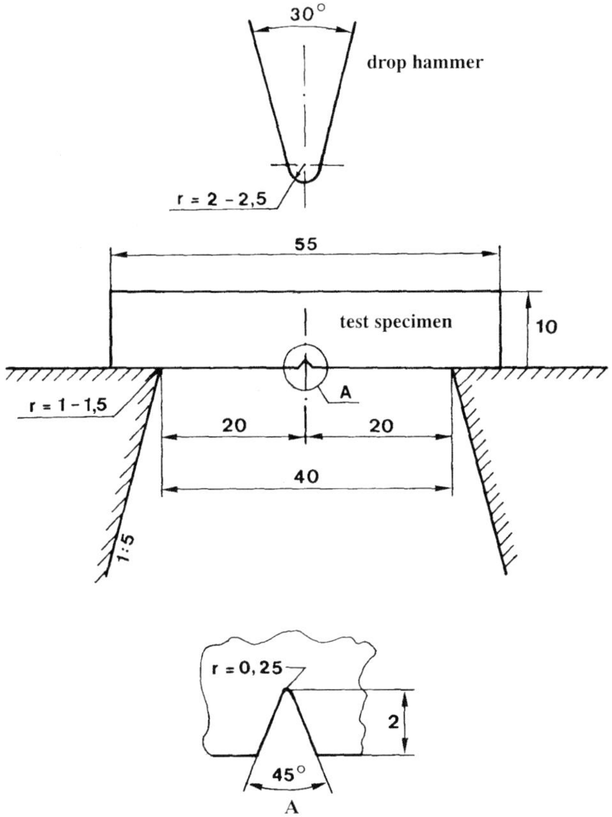

Figure 7.3 *Charpy-V test of specimens. The dimensions are in mm.*

Figure 7.5 shows a section of a plate with a single through-the-thickness crack of the length 2a.

When the plate is loaded by an inplane stress field, the stress vector $(\sigma_{11}, \sigma_{22}, \sigma_{12})$ in a small area given by the polar coordinates (r, θ) around the crack tip (at $x_1 = x_2 = 0$) can be written

$$\begin{Bmatrix} \sigma_{11} \\ \sigma_{22} \\ \sigma_{12} \end{Bmatrix} = \frac{K_I}{\sqrt{2\pi r}} \begin{Bmatrix} 1 - \sin\frac{\theta}{2}\sin\frac{3\theta}{2} \\ 1 + \sin\frac{\theta}{2}\sin\frac{3\theta}{2} \\ \sin\frac{\theta}{2}\cos\frac{3\theta}{2} \end{Bmatrix} \cos\frac{\theta}{2} + O(r) \tag{7.1}$$

if the load is symmetric as regards the x_1, x_2-plane (called Mode I). It is seen from Eq. (7.1) that the stress at the crack tip is singular because of the factor $r^{-\frac{1}{2}}$. The quantity K_I is called the *stress intensity factor* and it does not depend on (r,θ), but only on the geometry of the body, the length of the crack and the load distribution. In the simplest case where the extent of the body is much larger than the crack length 2a and where, the external load is a constant, uniaxial tensile stress (σ) in the x_2-direction, it can be shown that

$$K_I = \sigma \sqrt{\pi a} \tag{7.2}$$

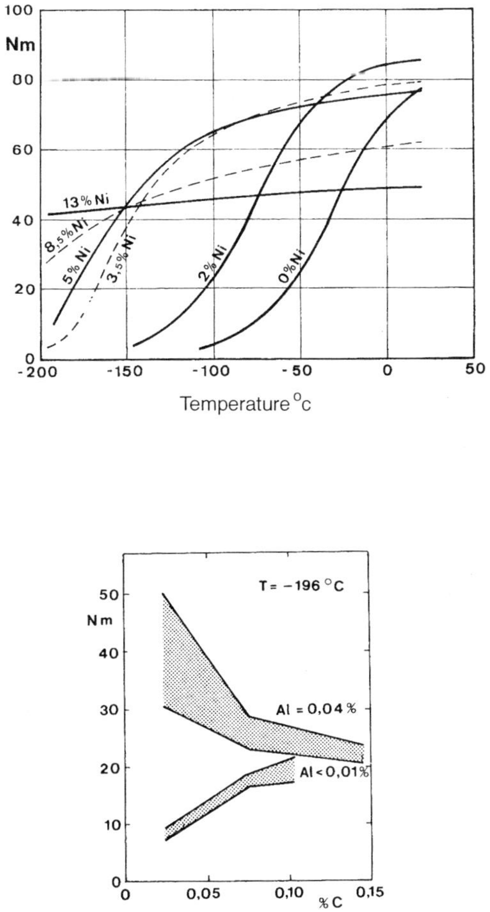

Figure 7.4 *Results of impact strength tests of steel alloys with a varying quantity of alloy element and at different temperatures T. In both figures the ordinate is the impact strength from Charpy-V tests.*

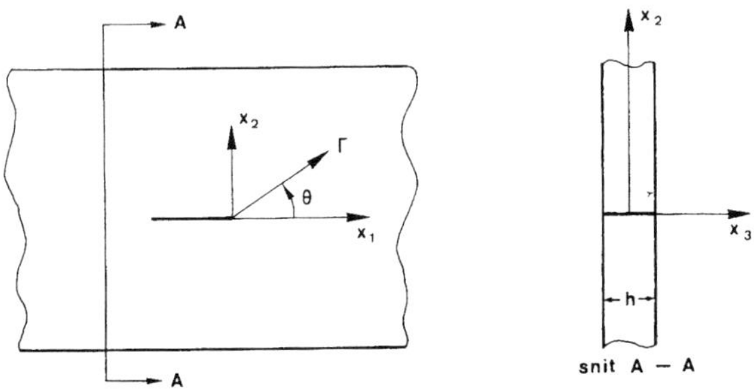

Figure 7.5 *Plate with a through-the-thickness crack of the length 2a.*

In other cases the result is written

$$K_I = \sigma\sqrt{\pi a}\, f(a) \tag{7.3}$$

where *f(a)* typically assumes a value in the range 0.5 to 3, dependent on geometry, crack length and external load. Results similar to (7.1) - (7.3) are found if the external loads lead to a shear stress state in the body. These results, which are designated Modes II and III, are not presented here, mainly because Mode I usually represents the most critical load of the crack. Results for Modes II and III can be found in textbooks on fracture mechanics.

Because of the singularity $r^{-\frac{1}{2}}$, the stress state given by Eq. (7.1) will, even by a very small external load, exceed the tensile strength of the material. This is not in complete accordance with experimental results. The main reason is that Eq. (7.1) presupposes a linear elastic material, while plastic deformations and thereby stress equalisation will actually take place in a small area around the crack tip. The order of magnitude of the plastic area can be estimated by setting $\theta = 0$ and $\sigma_{11} = \sigma_y$. This gives

$$r_p \simeq \frac{1}{2\pi}\left(\frac{K_I}{\sigma_y}\right)^2 \simeq \frac{1}{2}\left(\frac{\sigma}{\sigma_y}\right)^2 a \tag{7.4}$$

Although Eq. (7.1) cannot be used for determination of crack propagation, it leads to reasonable modelling of the stress distribution around the crack if $r_p \ll a$. This part of the fracture mechanics is called "small-scale yielding" contrary to fracture problems where the plastic area has the same extent as the crack length ("large-scale yielding").

If the load (σ) applied to a test specimen with a well defined initial crack ($2a$) is increased slowly, it is found that when the load takes a certain value (σ_c), then the crack will grow quickly with a brittle fracture as a consequence. From Eq. (7.3) a corresponding critical value K_C of K_I can be determined:

$$K_C = \sigma_c\sqrt{\pi\, a}\, f(a) \tag{7.5}$$

The reason for introducing the *stress intensity factor* K_C instead of the stress σ_c is that K_C has proved to be almost independent of crack and material geometry and of the load distribution. The only parameters affecting K_C with clear effect are the thickness h of the plate and the temperature.

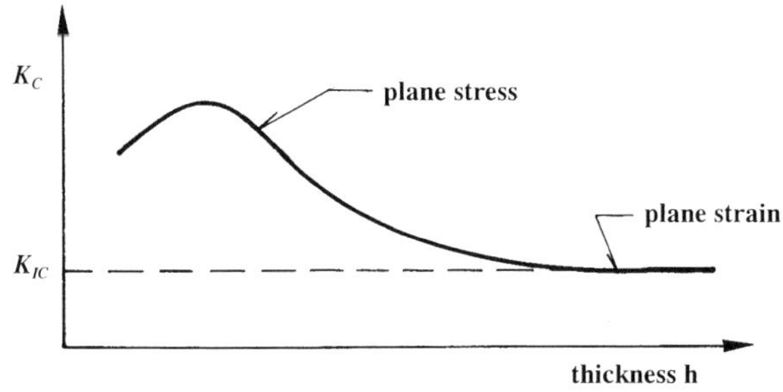

Figure 7.6 *Qualitative variation of the critical value K_c for the stress intensity factor with the material thickness.*

Figure 7.6 shows the qualitative variation of K_C with the materials thickness h. It is seen that the lowest value of K_C is found when the material thickness is large, corresponding to plane strain. This lower limit is denoted the *fracture toughness* K_{IC}, and it is tabulated in the literature on fracture mechanics for a large number of materials. The fracture toughness can be considered as a material constant like for instance the yield stress. It should be noted that K_{IC} can be very dependent on temperature. Figure 7.7 shows the qualitative variation of K_{IC} with the temperature T for typical steel alloys. The transition range which marks the transition from brittle to ductile material can, especially for steel alloys, be very small (50 °C). A comparison of Figure 7.4 and Figure 7.7 indicates that the experimental impact strength is related to K_{IC}.

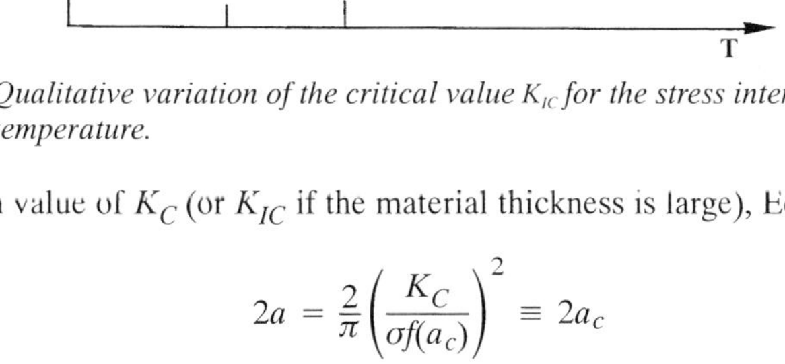

Figure 7.7 *Qualitative variation of the critical value K_{IC} for the stress intensity factor with temperature.*

With a given value of K_C (or K_{IC} if the material thickness is large), Eq. (7.3) gives

$$2a = \frac{2}{\pi}\left(\frac{K_C}{\sigma f(a_c)}\right)^2 \equiv 2a_c \tag{7.6}$$

where $2a_c$ is denoted the *critical crack length*. For a given load σ, a crack with an initial length larger than $2a_c$ will become longer. It is seen from Eq. (7.6) that if a given minimum critical crack length is used as a requirement of the ductility of the material, then the required fracture toughness K_C must be proportional to the design load. This must be taken into consideration when high strength steel is used, as such steel grades normally have the same fracture toughness K_{IC} as normal strength steel. Table 7.2 gives some examples for a_c and related parameters. The above formula offers no possibility of judging if the crack propagation will be stable or unstable, i.e. continue until fracture. This is decided by the degree of brittleness of materials. For the materials which can be treated by the fracture theory given here ("small-scale yielding"), a momentary brittle fracture will as a rule occur if $a \geq a_c(\sigma)$ and the load σ is kept during the crack propagation.

Table 7.2 *Ductile properties*

Material	T (^0C)	$\sigma_y\left(MN/m^2\right)$	$K_{IC}\left(MN/m^{3/2}\right)$	r_p (mm)	$2a_c$ (mm) $\sigma = \sigma_y/2$
$A\ell$ (6061)	20	135	33	5	120
$A\ell$ (7075)	20	310	36	0.35	8.4
Steel (AISI 4340)	0	750	33	0.05	1.2
Steel (A533-B)	93	310	200	11	130

Example 7.1.1

Typical steel qualities used in ships have a yield stress of σ_y = 270–500 MN/m^2 and, independent of this, a fracture toughness of K_{IC} = 60–120 MN/m$^{3/2}$ at 20 ^{0}C. The choice of yield stress is normally determined by the requirement that the maximum stress σ must not exceed a given percentage of σ_y. If e.g. $\sigma = 0.67\sigma_y$, Eq. (7.6) gives values for the critical crack length 2a$_c$ as shown in Table 7.3. It is assumed that $f(a_c) = 1$.

Table 7.3 *Critical crack lengths.*

σ_y / K_{IC}	270 MN/m^2	500 MN/m^2
60 MN/m$^{3/2}$	700mm (h > 120mm)	20mm (h > 36mm)
120 MN/m$^{3/2}$	280mm (h > 490mm)	80mm (h > 140mm)

The critical crack lengths presented in Table 7.3 must be multiplied by K_C/K_{IC}, if the material thickness is smaller than 2.5 $(K_{IC}/\sigma_y)^2$, being approximately the lower limit on h where a plane strain state can be assumed. This value is also given in Table 7.3 in parentheses. Only for the combination σ_y = 500 MN/m^2 and K_{IC} = 60 MN/m$^{3/2}$, the assumption $K_C = K_{IC}$ will be reasonable for plate thicknesses of current interest. As the relation K_C/K_{IC} may be considerably larger than 1, the results in Table 7.3 thus represent a rather conservative estimate of the critical crack length.

Because the critical crack length and the fracture toughness increase with decreasing material thickness it is necessary to change to a better material (larger K_{IC}), if thicker plates are to be used (at unchanged loading). As typical examples where steel with large fracture toughness will be used may be mentioned deck corners in ships.

7.1.3 Fatigue Strength

The preceding section dealt with materials with built-in defects (cracks) exposed to a static load. This led to the concept of critical crack length ($2a_c$):

$$2a_c = 2a_c(\sigma) \tag{7.7}$$

giving the maximum length which a crack may have without growing with possible fracture due to a static load σ. Thus, crack lengths smaller than $2a_c$ pose no structural problems as regards a static load.

This might not be so if the load σ varies periodically in time t about a mean stress σ_{mean}:

$$\sigma(t) = \sigma_{mean} + \frac{1}{2}\Delta\sigma \sin\omega t \tag{7.8}$$

with $\sigma(t)$ representing a tensile stress in at least some time intervals.

Even if the stress intensity factor*

$$K_{\max} = (\sigma_{mean} + \frac{1}{2}\Delta\sigma)\sqrt{\pi a}\ f(a) \tag{7.9}$$

defined by Eq. (7.3) is smaller than the critical value K_C, an initial crack of the length $2a_0$ will grow slowly with each cycle. When the crack length reaches the value $2a_c$, fracture will normally take place almost instantaneously. If the crack length is plotted as a function of time t or number of periods N in the load, a qualitative behaviour as shown in Figure 7.8 will be obtained.

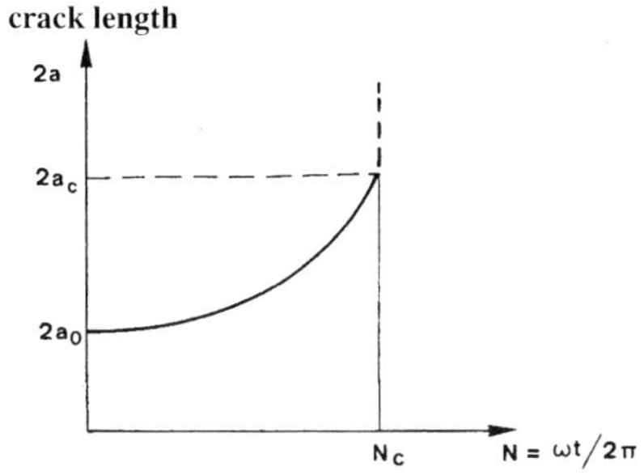

Figure 7.8 *Crack growth during cyclic loading.*

Measurements have shown that the crack propagation velocity, defined as da/dN, is essentially only a function of the *stress intensity variation* ΔK:

$$\begin{aligned}\Delta K &= K_{\max} - K_{\min} \\ &= \left(\sigma_{mean} + \frac{1}{2}\Delta\sigma\right)\sqrt{\pi a}\ f - \left(\sigma_{mean} - \frac{1}{2}\Delta\sigma\right)\sqrt{\pi a}\ f \\ &= \Delta\sigma\sqrt{\pi a}\ f(a)\end{aligned} \tag{7.10}$$

* As earlier only Mode I is considered here, for other types of cracks, as e.g. surface cracks, relevant formulas must be used.

and the variation is typically as shown in Figure 7.9.

For very low values of ΔK (range A in Figure 7.9), the crack will not grow. The limit, ΔK_l, between ranges A and B is difficult to determine experimentally and here just an example, concerning normal strength steel is given:

$$\Delta K_\ell = 7.6 - 4.3\,R \qquad \mathrm{MN/m^{3/2}} \tag{7.11}$$

where the *stress ratio* R is defined as

$$R = \frac{\sigma_{\min}}{\sigma_{\max}} = \frac{K_{\min}}{K_{\max}} = \frac{\sigma_{mean} - 0.5\Delta\sigma}{\sigma_{mean} + 0.5\Delta\sigma} \tag{7.12}$$

If Eqs. (7.11) and (7.12) are inserted in Eq. (7.10), it is seen that cracks with a length smaller than $2a_o$, where

$$a_o = \frac{1}{\pi}\left(\frac{\Delta K_\ell}{f(a)\Delta\sigma}\right)^2 \tag{7.13}$$

do not grow.

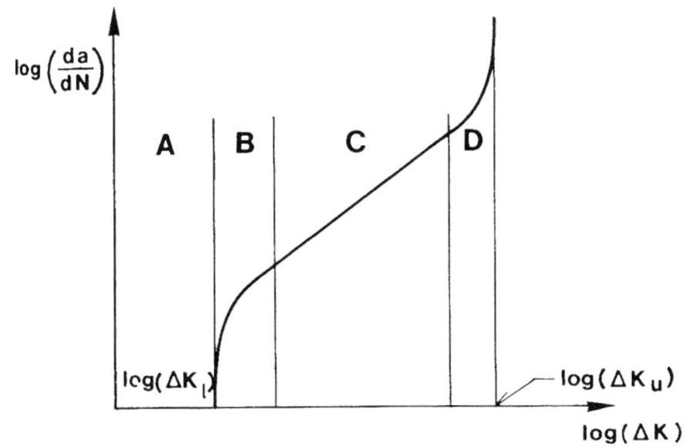

Figure 7.9 *Typical variation of the crack propagation velocity* da/dN *as a function of the stress intensity factor ΔK.*

In range B in Figure 7.9 the crack velocity is small, sometimes in the order of magnitude of one atom lattice distance per cycle. A quantitative description of the relation between ΔK and da/dN cannot be given in this area.

In a welded structure the initial crack lengths 2a and the cyclic stress variations $\Delta\sigma$ are usually so large that ΔK, given by Eq. (7.10), already from the beginning lies in range C in Figure 7.9. Extensive measurements show that in this area, a linear relation holds between $\ln(da/dN)$ and $\ln(\Delta K)$, implying

$$\frac{da}{dN} = C\,(\Delta K)^m \tag{7.14}$$

The formula is due to Paris, who also proposed the exponent $m = 4$. Later experimental results have given values of m between 1 and 15, with most of the values in the interval

3 to 4. When the crack length $2a$ has grown to a value near the critical crack length $2a_c$, ΔK will be so large that the velocity of the crack propagation rises steeply corresponding to range D in Figure 7.9. The upper bound of $\Delta K \equiv \Delta K_u$ is determined by $K_{\max} = K_C$ or, as $\Delta K = (1 - R) K_{\max}$ according to Eqs. (7.10) and (7.12):

$$\Delta K_u = (1 - R) K_C \tag{7.15}$$

As mentioned, the initial crack length $2a_i$ is normally so large that the crack propagation starts in range C. Hence, one of the most important results which the present fracture mechanics analysis can give is the remaining lifetime of the material, defined as the number of load periods N_f before fracture takes place. This value is determined by the integration of Eq. (7.14) from $a = a_i$ to $a = a_c$:

$$N_f = \frac{1}{C\left(\sqrt{\pi}\Delta\sigma\right)^m} \int_{a_i}^{a_c} \frac{da}{a^{\frac{m}{2}} (f(a))^m} \tag{7.16}$$

as the relatively few periods $\left(10^2 - 10^4\right)$ occurring in range D can be neglected in comparison with the number of periods $\left(10^7 - 10^9\right)$ in the range C.

Thus, from Eq. (7.16) the remaining lifetime of the material $T_f = 2\pi N_f/\omega$ can be determined for a given crack length ($2a_i$) in a given material (characterised by C, m) and by a given load, σ_m, $\Delta\sigma$, with the given frequency ω.

If f can be assumed to be independent of the crack length, Eq. (7.16) can be integrated to

$$N_f = \frac{a_c^{1-m/2} - a_i^{1-m/2}}{C\,(f\sqrt{\pi}\,\Delta\sigma)^m (1 - m/2)} \quad ; \quad m \neq 2 \tag{7.17}$$

The *fatigue strength of materials* (and welds) are normally described by *Wöhler curves* or *S-N curves*, which gives the number of periods N_f before fracture for a constant stress variation $\Delta\sigma$:

$$N_f = N_f(\Delta\sigma) \tag{7.18}$$

It is seen from Eq. (7.16) that

$$N_f = A\,(\Delta\sigma)^{-m} \tag{7.19}$$

or

$$\log N_f = \log A - m \log(\Delta\sigma) \tag{7.20}$$

Figure 7.10 shows qualitative S-N curves for welded details. Quantitative S-N curves can be found in e.g. Fricke et al. (1997, 1998) for welded joints in ships.

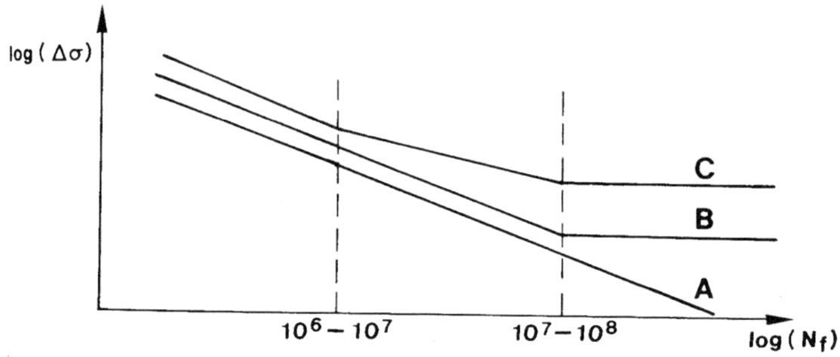

Figure 7.10 *S-N curves for welded detail.*

Curve A shows an S-N curve where Eq. (7.20) applies with a constant exponent m for all stress amplitudes. This form can be used for welds in corrosive environments. Curve A is also representative of aluminium alloys. Curve B represents an S-N curve where stress amplitudes below a level $\Delta\sigma_u$ have no influence on the fatigue strength (the crack propagation). Curve B can normally only be used for fatigue calculations for details in non-corrosive environments (e.g. in air).

Finally, curve C shows an example of an S-N curve where the exponent depends on the stress amplitude. This may be a suitable description, but usually experimental determinations of S-N curves show so large scattering of the results that such a variation in m cannot be determined. This scattering is also the reason why S-N curves are normally given with their statistical standard deviation s. The curve

$$\log N_f = \log A - m \log (\Delta\sigma) - 2 \log s \tag{7.21}$$

corresponding to the mean S-N curve corrected for twice the standard deviation is often used in design, as the probability that the real S-N curve is below this curve is 2.3 %, on the assumption that the measured results used for determination of the S-N curve are statistically normal distributed. This curve is indicated in Figure 7.11.

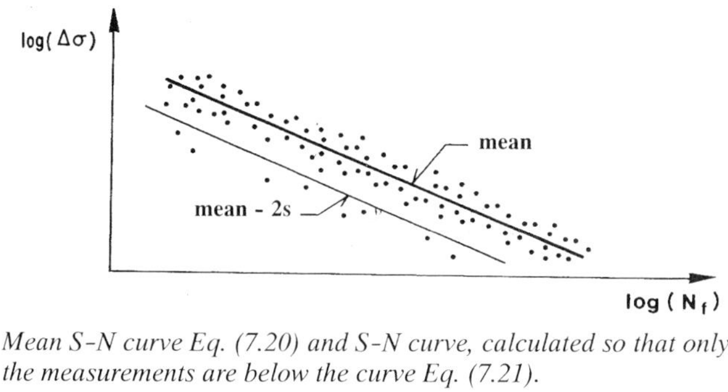

Figure 7.11 *Mean S-N curve Eq. (7.20) and S-N curve, calculated so that only 2.3% of the measurements are below the curve Eq. (7.21).*

S-N curves for test specimens made of material taken at some distance from welds are influenced by a possible mean stress σ_{mean}, as seen from the relation

$$a_c = \frac{1}{\pi}\left(\frac{K_c}{\Delta\sigma f}\right)^2 (1 - R)^2$$

A so-called Smith diagram can be used to correct the S-N curve for the influence of the mean stress.

For welds the mean stress can be assumed constant and equal to the yield stress of the material under tension. The reason is that such stress state is found around the welded seam in the form of residual stresses produced by the solidification and the subsequent shrinkage of the welded seam. Therefore S-N curves for welds are usually not corrected for the mean stress. It should be noted that the nominal stress calculated by e.g. beam theory must be corrected for stress concentrations due to weld geometry, misalignment and distortion as specified in connection to the S-N curve to be applied. Some very useful stress concentration factors are given by Ximenes et al. (1997).

So far it has been assumed that the stress variation $\Delta\sigma$ was kept constant until fracture and the S-N curves are determined on this assumption. However, real structures are usually exposed to loads of varying amplitudes. A relevant example is wave-induced loads, which will give varying stresses in the hull. Hence, it is not a constant stress amplitude which will lead to fatigue fracture, but the cumulative effect of all stress amplitudes. There is no rational theoretical calculation procedure for this effect, but an empirical relation called the *Palmgren-Miner rule*:

$$\sum_{i=1}^{n} \frac{N_i}{N_{fi}} = D, \quad (D > 0.1 - 1 \Rightarrow \text{fatigue failure}) \tag{7.22}$$

gives a description which can be used. In Eq. (7.22) N_i is the total number of load periods with stress variations close to $\Delta\sigma_i$, and N_{fi} is the number of loads before fracture at a constant stress variation $\Delta\sigma_i$, as determined from the S-N curve. To evaluate Eq. (7.22), the stress variation range $0 \leq \Delta\sigma \leq \Delta\sigma_{\max}$ is divided into an appropriate number n of discrete values $\Delta\sigma_i$, i = 1,2...,n, and the number of loads N_i within each range is determined from the load analysis. An example of this is given in Section 7.3.4 for a stochastic stress variation.

The main drawback inherent in the Palmgren-Miner rule is that it takes no account of the succession of occurrence of each individual stress variation. Measurements have proved that small stress variations occurring just after major stress variations will reduce the fatigue strength much more than if the succession of these stress variations was reversed. This sequence effect is taken into account in the rain-flow counting, peak counting and range counting methods, see e.g. Madsen et al. (1986).

Finally, it should be mentioned again that high strength steel usually has no better fatigue properties than normal strength steel, as it is rather the ductility of the material than its yield stress which determines the fatigue life of the material, especially if the number of periods before fracture is larger than $10^4 - 10^5$.

A detailed and comprehensive description of the assessment of fatigue strength related to ship structures is given by Fricke et al. (1997, 1998) and strongly recommended for further readings.

7.1.4 Corrosion Resistance

By corrosion is meant a chemical decomposition of materials under the influence of the environment. For ships it is especially the destructive effect of sea water on steel alloys which poses problems, but also possible corrosion in tanks and pipes with oil, water and chemical liquids should be considered.

The corrosion process is in principle the same for all metals. It is here exemplified by steel in contact with water. The chemical process is as follows:

$$Fe + \frac{1}{2}O_2 + H_2O \rightarrow Fe\,(OH)_2\,(\rightarrow Fe\,(OH)_3, nH_2O) \qquad (7.23)$$

i.e. iron + oxygen + water $\rightarrow$ rust where rust is a mixed ferri - oxide - hydroxide.

Thus, rusting requires admission of both oxygen and water to the iron in steel alloys.

If an unprotected ordinary steel plate is placed in sea water, the plate will becoming 0.5-1.5 mm thinner annually due to corrosion. As a ship is designed for 20 years of operation, such a corrosion rate is unacceptable.

To limit the corrosion the following methods can be used, either together or separately:

- Painting
- Cathodic protection
- Inhibitors
- Alloy materials

Cathodic protection is obtained by attaching a material with a lower electrochemical potential than iron into electric contact with the steel. Thus, the corrosion of the steel will cease at the expense of the other material (the sacrificial anode), usually zinc or aluminium. The process is illustrated in Figure 7.12.

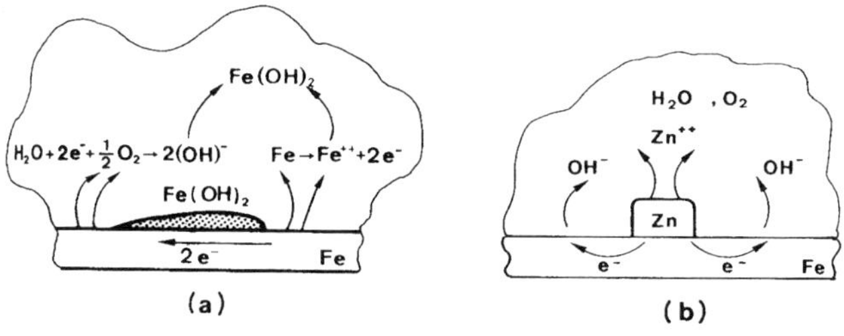

Figure 7.12 *Cathodic protection.*

It is seen from the reaction (7.23) that oxygen must be present in order that corrosion can take place. Corrosion in tanks with a limited quantity of liquid can therefore be prevented by addition of oxygen-consuming substances (inhibitors, e.g. nitrite and borate) in so far as this is allowable in consideration of the later use of the liquid. Alternatively, the access of the oxygen to the iron in the steel can be impeded either by physical separation (with paint or epoxy) or by addition of alloy materials to the steel (stainless steel).

Finally, it should be mentioned that corrosion problems for aluminium are minor, but also that welds and bolted joints between iron and aluminium will lead to quick corrosion of aluminium if the joints are exposed to sea water.

From a structural point-of-view corrosion poses two problems. One concerns the reduction of plate thickness as a consequence of corrosion. This is usually taken care of in the design by providing the structural elements exposed to corrosion with an increase of thickness, not included in the structural analysis.

The other problem concerns the fatigue strength of the material, as this changes due to corrosion. These changes may be described as a change of the parameters C and m in Eq (7.14), which gives the crack propagation velocity da/dN as a function of the stress intensity variation ΔK. For instance cathodic protection may lead to increased crack propagation velocity and hence lower fatigue strength. The reason for this seems mainly to be hydrogen separated by the cathodic protection.

7.2 HULL GIRDER STRENGTH

The two main failure modes of the hull girder is fatigue failure and collapse of the hull girder due to an excessive vertical bending moment. In the fatigue failure analysis linear elastic stresses are compared to the capacity of the structural element given by their S-N curves, Eq. (7.22). This analysis is considered later in Section 7.3.4.

The collapse strength of the hull girder can, however, only be determined by a non-linear elastic-plastic structural analysis, taking into account buckling failure of the individual plate panels in compression. Different methods exist to calculate the hull girder collapse strength ranging from a full non-linear finite element analysis to ingenious engineering models. Among the latter is a method proposed by Caldwell (1965). Basically the collapse is assumed to take place between two adjacent transverse frames, which are assumed undeformed such that plane sections remain plane. With this assumption the deformation of all longitudinal strength elements are coupled together and from the force-elongation curve for the longitudinal and girders with attached plating the moment-curvature relation for the hull girder until collapse can be established. Since the method was proposed in 1965 significant improvements have been suggested, notably Smith (1977), Rutherford and Caldwell (1990), Yao and Nikolov (1991, 1992) and Hansen (1996). A detailed discussion is outside the scope of the present treatment and only some results from the literature are given below. In the analysis due account is taken of imperfections in the structure as specified by the out-of-straightness of stiffners and platings and by welding-induced residual stresses. The results are compared to the *first-yield moment* M_y, defined as

$$M_y = \sigma_y W_y \tag{7.24}$$

where W_y is the elastic section modulus, Eq. (5.7), is the deck or bottom, whichever smallest. An applied moment larger than M_y will thus lead to longitudinal stresses larger than the yield stress σ_y. For comparison also the *plastic moment* M_p is used. This moment corresponds to a longitudinal stress state with $\sigma = \sigma_y$ throughout the section. Hence

$$M_p = \sigma_y W_p \tag{7.25}$$

where W_p is the *plastic section modulus,* given as

$$W_p = \frac{A}{2}(e_1 + e_2) \tag{7.26}$$

Here A is the area of the longitudinal structural members in the cross-section and e_1, e_2 are the vertical distances to the geometrical centre of the areas $\frac{A}{2}$ above and below the horizontal axis, that divide the area A into two equal parts. The two moments M_y and M_p are illustrated in Figure 7.13. For typical ship sections W_p is 15-25 per cent larger than W_y.

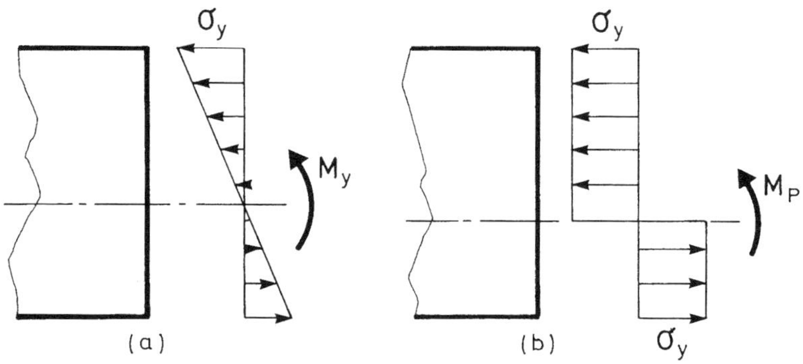

Figure 7.13 *Stress distribution σ for the first-yield moment M_y and the plastic moment M_p.*

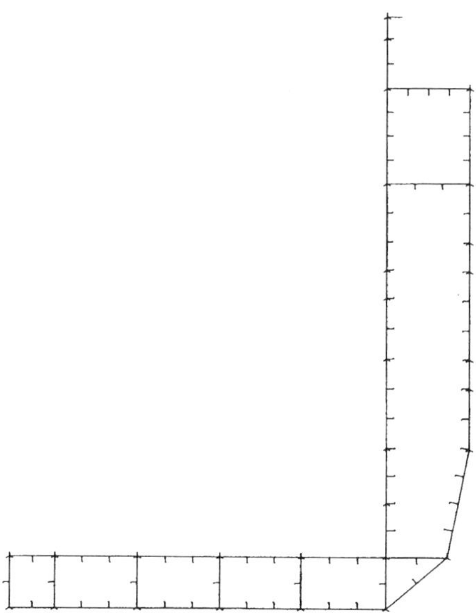

Figure 7.14 *Midship section of the container ship.*

A collapse analysis of a container ship was performed in Hansen (1996). The ship is 190 m long with a deadweight of 34,150 tons. The midship section is shown in Figure 7.14. The frame spacing for the bottom structure is 1.634 m and 3.268 m for the side structure. The material parameters and imperfections are chosen to be those in Table 7.4 and the sectional parameters are given in Table 7.5.

Table 7.4 *Material parameters and imperfections for the container ship.*

σ_y	356 MN/m^2	Yield stress for upper side structure
	236 MN/m^2	Yield stress for lower side and for bottom structure
σ_R	$\sigma_y/10$	Residual stress
δ_p	$t/10$	Plate imperfection (t = plate thickness)
δ_s	$l/500$	Stiffener imperfection (l = frame spacing)

Table 7.5 *Cross-sectional parameters for the container ship.*

Sectional area A	1.43 m^2
Moment of inertia I_y	54.6 m^4
Position of elastic neutral axis (from BL)	6.64 m
Plastic moment M_p	2.32 GNm
Position of plastic neutral axis (from BL)	7.39 m
Elastic section modulus (deck) W_y	4.98 m^3
Elastic section modulus (bottom) W_y	8.23 m^3
Elastic moment (first yield) M_y	1.77 GNm

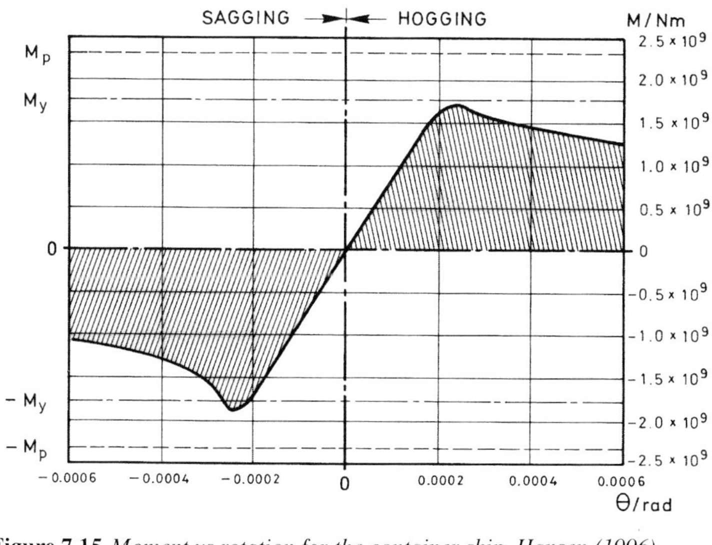

Figure 7.15 *Moment vs rotation for the container ship. Hansen (1996).*

The applied moment as function rotation is shown in Figure 7.15. The rotation θ is related to the curvature κ of the hull girder by $\theta = \frac{l}{2}\kappa$ where l is the frame spacing of the side structure (3.268 m). Collapse in sagging begins by compression failure of the extreme deck stiffeners but later in the loading history, collapse is dominated by tension failure of the bottom structure. Collapse in hogging is purely due to compression failure of the bottom structure.

Normally the hogging collapse moment is expected to be numerically larger than the sagging collapse moment. However, this was not found in this analysis.

A collapse analysis of a VLCC was also performed in Hansen (1996). The VLCC tanker is 320 m long with a deadweight of 280,000 tons and it is built as a single plating construction with a frame spacing of l = 5.316 m. The midship section is shown in Figure 7.16 and the material parameters and imperfections are chosen to be those given in Table 7.6. The sectional parameters and imperfections are given in Table 7.7.

The moment vs rotation is shown in Figure 7.17. Collapse in sagging is purely controlled by compression failure of the deck stiffeners, while collapse in hogging is controlled by both compression failure of the bottom stiffeners and yielding of the deck stiffeners.

Table 7.6 *Material parameters and imperfections for the VLCC.*

σ_y	310 MN/m^2	Yield stress
σ_R	$\sigma_y/10$	Residual stress
δ_p	$t/20$	Plate imperfection (t = plate thickness)
δ_s	$l/200$	Stiffener imperfection (l = stiffener spacing)

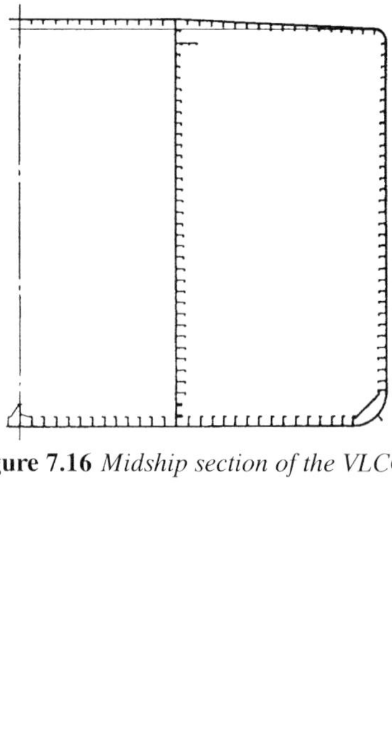

Figure 7.16 *Midship section of the VLCC.*

Table 7.7 *Cross-sectional parameters for the VLCC.*

Section area A	3.54 m^2
Moment of inertia I_y	531.7 m^4
Position of elastic neutral axis (from BL)	14.0 m
Plastic moment M_p	12.3 GNm
Position of plastic neutral axis (from BL)	12.6 m
Elastic section modulus (deck) W_y	32.6 m^3
Elastic section modulus (bottom) W_y	38.6 m^3
Elastic moment (first yield) M_y	10.1 GNm

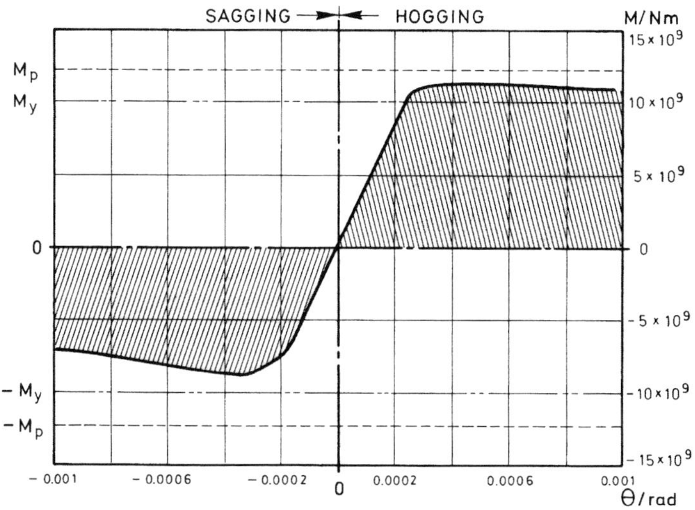

Figure 7.17 *Moment vs rotation for the VLCC. Hansen (1996).*

From Figure 7.15 and Figure 7.17 it is seen that the ultimate moment capacity, i.e. the collapse strength, of the ships considered is close to the first-yield moment M_y both in sagging and hogging. This is generally so for most types of merchant ships, see Figure 7.18, taken from ISSC (1994). However, this is not a design criteria, but a result of a proper selection of dimensions of the individual plates and stiffeners. The principle behind this selection is mainly govern by the local buckling strength, a topic not covered here.

No.	$L \times B \times D$ (*in Meter*)	M_Y (*in Tonf* $\cdot$ *m*)	M_P/M_Y	M_{FS}/M_Y	M_{US}/M_Y	M_{FH}/M_Y	M_{UH}/M_Y
1	215 × 32.2 × 17.8	0.5405×10^6	1.1534	0.9204	0.9813	0.9251	1.1031
2	217 × 32.26 × 18.3	0.5592×10^6	1.2316	0.8498	0.9312	0.9408	1.1296
3	276 × 45 × 24.2	0.1344×10^7	1.1584	0.8702	0.9400	0.9470	1.0803
4	247.4 × 36.2 × 21.8	0.7917×10^6	1.2260	0.9502	1.0213	0.9553	1.1698
5	315 × 57 × 30.8	0.2193×10^7	1.2681	0.9134	0.9234	1.0119	1.1172
6	315 × 58 × 30.4	0.2098×10^7	1.4147	0.9214	0.9473	1.0119	1.2312
7	162 × 30 × 16.2	0.2912×10^6	1.6326	0.8908	1.0251	1.0261	1.4214
8	315 × 52 × 23.45	0.1951×10^6	1.1886	0.9846	1.0359	0.9800	1.1220
9	180 × 32.26 × 30.55	0.4892×10^6	1.6492	0.4563	0.7482	0.9663	1.2122
10	230 × 32.2 × 21.5	0.6624×10^6	1.3439	0.9674	1.0482	0.8284	1.0652

M_Y: Initial Yielding Strength M_P: Fully Plastic Bending Moment
M_{FS}: Initial Member Collapse Strength under Sagging Condition
M_{US}: Ultimate Strength under Sagging Condition
M_{FH}: Initial Member Collapse Strength under Hogging Condition
M_{UH}: Ultimate Strength under Hogging Condition

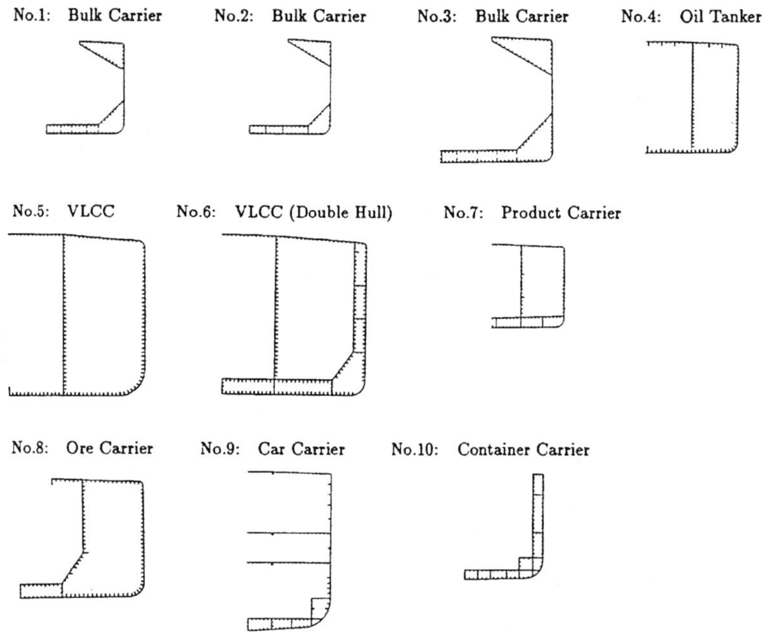

Figure 7.18 *Collapse strength of various types of ships. ISSC (1994).*

7.3 STRUCTURAL RELIABILITY

In the structural assessment procedure the load and the strength are compared taking into account the inevitable uncertainty in both results. The aim is to assure that the probability of failure is sufficiently low, given the economical and environmentally consequences of such failure. Previously, most assessment methods were deterministic in the sense that an assumed upper limit of the load (σ_ℓ) was compared with an assumed lower limit of the strength $(\sigma_s)^*$. The requirement, the *so-called limit state format*, was then simply

$$(\sigma_s)_{lower} \geq (\sigma_\ell)_{upper} \tag{7.27}$$

However, rational values of these upper and lower limits could only seldom be defined for real structures. With the construction of larger, more complicated and environmentally sensitive structures (nuclear power plants, large bridges, dams, offshore oil production plants etc.) the need for better, more rational methods was obvious. Out of this has developed within the last 30 years the very powerful structural reliability theory. Several excellent textbooks are available, Madsen, Krenk an Lind (1986), Melchers (1999) to mention just two very eminent books. The present treatment will only outline some of the pertinent features of structural reliability theory with the textbooks mentioned above as obvious choices for further readings.

Basically, structural reliability assessment can be divided into four levels of different complexities.

Level 1 is the lowest level. It corresponds roughly to Eq. (7.27) but with the rather arbitrary lower and upper limits replaced by characteristic values derived from the assumed probability distributions of σ_s and σ_ℓ. Safety factors are multiplied on these characteristic values to account for known and unknown (!) uncertainty in our mathematical modelling of the load and strength processes. The safety factors should also reflect the consequences of a failure. Various codes have been formulated in this so-called partial safety factor format and the ship classification society rules for ships and offshore structures can be considered among them. It is the most common procedure today.

Level 2 methods apply the basis statistical knowledge (mean, variance and correlation) of the pertinent parameters in the load and strength models to arrive at a nominal *probability of failure*, usually expressed in terms of a reliability or safety index to avoid too much focus on the number itself.

Level 3 methods aim at a direct calculation of the probability of failure using the full joint probability distribution of all relevant parameters. The result becomes strongly dependent on the modelling of the tails of the probability distributions involved. This tail sensitivity makes the use of Level 3 methods very difficult in practice as for instance the tail in the response is rather uncertain.

Level 4 methods are extensions of the Level 3 methods and include the consequences of a failure in economical and environmental terms. Thereby these methods may be considered as a decision support tool as it can be used to compare the overall (life cycle) cost of various alternatives.

In the following Level 1, 2 and 3 methods will be discussed and exemplified by problems related to hull girder collapse.

Although the procedures can be applied for both fatigue failure and for ultimate limit state failure; i.e. hull girder failure due to an excessive bending moment the focus is on the latter failure mode. Fatigue failure is only considered in Section 7.3.4.

---

* The use of σ to denote the load and strength variable is taken because often stresses are compared. However, other measures might of course be used, e.g. bending moment as long as corresponding values of load and strength are compared.

7.3.1 Level 1 - Partial Safety Factors

It is assumed that the load σ_ℓ and the strength σ_s are statistically independent stochastic variables. Their probability density functions are denoted $p_s(\sigma_s)$ and $p_\ell(\sigma_\ell)$, respectively, and they are assumed known, derived from say, a combination of theoretical analyses and measurements. An example can be a Rayleigh distribution for the load and a normal distribution for the strength.

In the partial safety factor method the model uncertainty is divided into two main groups: The *objective uncertainty* and the *subjective uncertainty*. The first group comprises the stochastic description of the load and the strength as specified by the probability density functions $p_\ell(\sigma_\ell)$ and $p_s(\sigma_s)$. The subjective uncertainty covers the remaining uncertainty, notably the expected modelling accuracy of the load and strength analysis and the consequences of a failure. The subjective uncertainty cannot be calculated but has to be assessed from past experience and the economical, environmental and social consequences a failure may infer.

The objective uncertainty is described by characteristic values, $\sigma_{\ell c}$ and σ_{sc}, representing values of the load and strength which seldom will be exceeded from below and above, respectively:

$$\begin{aligned} P(\sigma_\ell > \sigma_{\ell c}) &= \alpha_\ell \ll 1 \\ P(\sigma_s < \sigma_{sc}) &= \alpha_s \ll 1 \end{aligned} \tag{7.28}$$

where α_ℓ, α_s typically can be of the order 0.01-0.05 when the period considered covers the expected lifetime of the structure. The ultimate limit state format, Eq. (7.27) can thus be written

$$\sigma_{sc} \geq \sigma_{\ell c} \tag{7.29}$$

provided the analysis is completely in accordance with the physical realities. This is never so for real complicated structures as e.g. the hull girder. Hence, safety factors have be to included to account for the subjective uncertainty. In the simplest case one *central safety factor* $\gamma > 1$ can be applied:

$$\sigma_{sc} \geq \gamma \sigma_{\ell c} \tag{7.30}$$

where γ normally is factorized as, say,

$$\gamma = \gamma_c \gamma_d \gamma_\ell \gamma_s$$

to account individually for different uncertaincies:

γ_s: model uncertainty in the strength analysis ($\gamma_s = 1.0 - 1.2$)

γ_ℓ: model uncertainty in the load analysis ($\gamma_\ell = 1.0 - 1.6$)

γ_d: consequences due to loss of serviceability ($\gamma_m = 1.0 - 1.4$)

γ_c: consequences due to loss of safety ($\gamma_c = 1.0 - 1.4$)

Subsequently, the responsible authorities must specify values for these safety factors. Typical values are indicated above in the parentheses and the actual values depend on the analysis method, the material quality, the inspection and repair procedures during operation and the consequences of failure. The partial safety factor method is illustrated in Figure 7.19.

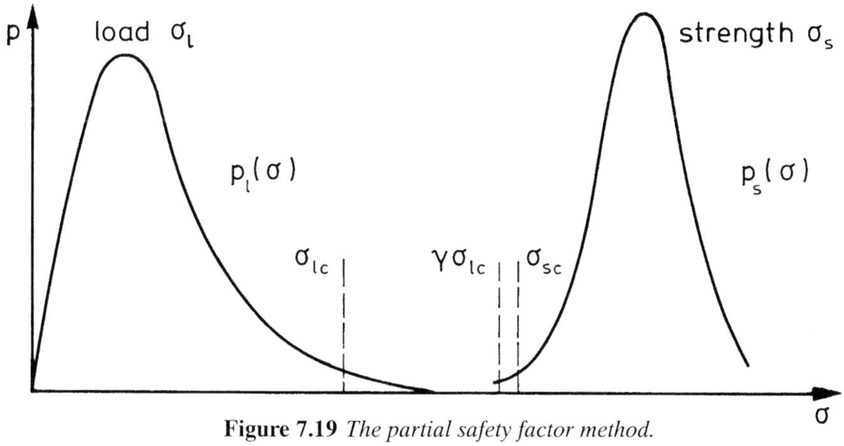

Figure 7.19 *The partial safety factor method.*

In real applications, a large number of different load cases are to be combined, e.g. still water loads and wave loads. The model uncertainty may be different for each load case, necessitating different partial safety factors. The same holds for the strength as e.g. material yielding in tension is much better understood than compression failure of plate panels. Thus different collapse mechanisms may have different partial safety factors. A general ultimate limit state formula is therefore for each failure mode

$$\sigma_{sc} \geq \gamma_c \gamma_d \gamma_s \left\{ \sum_{i=1}^{n_j} \gamma_{\ell i} \sigma_{\ell ci} \right\}_j \quad ; \quad j = 1, 2, ... \tag{7.31}$$

where j is the number of load combinations (e.g. various operational conditions (ballast laden), collision, grounding, docking), i the number of load types (e.g. still water loads, wave loads, collision load). Many design codes and standards are today based on this method. It is easy to apply but the main problem is the selection by the authorities of appropriate values of the partial safety factors. However, by breaking down the central safety factor in Eq. (7.30) in various subelements, Eq. (7.31), the experience from one type of structure can more easily be transferred to another structure because on an element level they might look alike.

<u>Example 7.3.1</u>

The Danish Code of Practice DS 449 (1984) is based on the partial safety factor method. The various partial safety factors named partial coefficients for the action (i.e. loads) and the material parameters (i.e. the strength) are given in Figure 7.20.

Table 5.2.1 Action combinations and partial coefficients for actions

	action combination								
	use	failure				failure	failure	accident	
type of action		2.1				2.2	2.3	3.1	3.2
		a	b	c	d	a,b,c,d			
permanent action									
dead load	1.0	1.0	1.0	1.0	1.0	0.9	1.15	1.0	1.0
dead load of soil and hydro-static water pressure	1.0	1.0	1.0	1.0	1.0	1.0	1.0	1.0	1.0
variable action									
imposed action	1.0	1.3[2]	1.0	1.0	1.0	as 2.1 a,b,c,d	0	1.0	1.0
natural action									
waves and current	ψ[1]	0.75	1.3	1.0	0.75		0	1.0[3]	0
wind	ψ[1]	0.5	1.0	1.3	0.5		0	1.0[3]	0
snow and ice	ψ[1]	0.5	0.5	0.5	1.3		0	0	0
action from deformation	1.0	1.0	1.0	1.0	1.0	1.0	0	0	0
accidental action	0	0	0	0	0	0	0	0	1.0

Table 5.2.2 a. Partial coefficients for steel parameters

action combinations 2 and 3.1

material parameters		safety class normal	high
yield stress	f_y	1.09	1.21
tensile strength	f_u	1.34	1.48
punching strength	τ_g	1.21	1.34
modulus of elasticity	E	1.34	1.48
coefficient of friction μ:			
normal friction joints		1.09	1.21
unlimited slip possible		1.34	1.48

Figure 7.20 *Partial safety factors in the Danish Code of Practice (1984).*

For a complete description of the use of the values in the figure, the code should be consulted. Here only a few comments are given.

The code does not consider explicitly serviceability but only failure. The consequences of a failure are divided into two safety classes: normal and high as applied to the partial coefficient for strength γ_S. It should be emphasised that γ_S represents the uncertainty in the analysis model, for instance the approximate description of buckling failure of panels. This is the reason for the large safety factor on the modulus of the elasticity E, which of course in itself is a very accurately known parameter.

7.3.2 Level 2 - Second Moment Reliability Method

The partial safety factor format makes use of the probability distributions of the load and strength to define some characteristic values, Eq. (7.28), depending mainly on the lowest statistical moments of the load and strength processes. Thereby, the method does not give any information of the probability that a failure will occur, but rather define a safe set of load and strength parameters, Eq. (7.31).

If the probability distributions of the load and strength are known exactly, then the probability of failure P_f can be calculated as, see Eq. (3.56)

$$P_f = P(\sigma_s < \sigma_\ell) = \int_{-\infty}^{\infty} \int_{-\infty}^{\sigma_\ell} p_s(\sigma_s)\, p_\ell(\sigma_\ell)\, d\sigma_s\, d\sigma_\ell$$
$$= \int_{-\infty}^{\infty} p_\ell(\sigma_\ell)\, F_s(\sigma_\ell)\, d\sigma_\ell = 1 - \int_{-\infty}^{\infty} F_\ell(\sigma_s)\, p_s(\sigma_s)\, d\sigma_s \tag{7.32}$$

provided the load and strength processes are statistically independent. The result is illustrated in Figure 7.21.

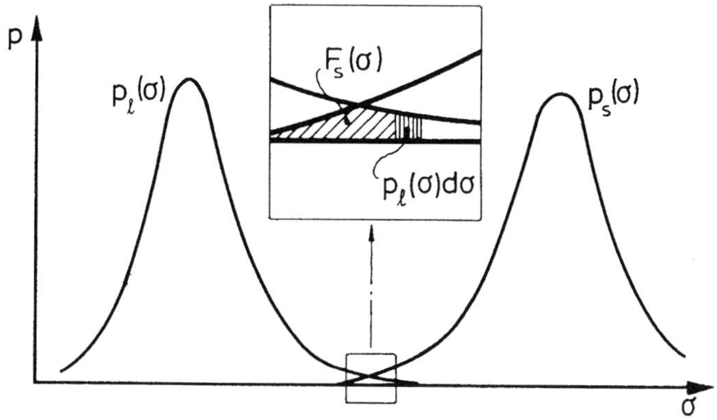

Figure 7.21 *Illustration of the calculation of the probability of failure, Eq. (7.32). (Note that the common area below both curves is not the probability of failure).*

Rather than considering the load and strength individually in the calculation of the probability of failure, a safety margin σ_m can be defined

$$\sigma_m = \sigma_s - \sigma_\ell \tag{7.33}$$

The probability of failure P_f can then be written

$$P_f = P(\sigma_m < 0) = \int_{-\infty}^{0} p_m(\sigma_m) d\sigma_m = F_m(0) \tag{7.34}$$

An analytical solution is readily obtained if both the load and strength are normal distributed. Then also the safety margin becomes normal distributed, see e.g. Section 3.1.7, with mean value

$$\mu_m = \mu_s - \mu_\ell$$

and variance

$$s_m^2 = s_s^2 + s_\ell^2$$

Hence, from Eq. (3.18)

$$P_f = F_m(0) = \Phi\left(\frac{0 - \mu_m}{s_m}\right) = \Phi\left[-\frac{\mu_s - \mu_\ell}{\sqrt{s_s^2 + s_\ell^2}}\right] \tag{7.35}$$

or

$$P_f = \Phi(-\beta_c) \tag{7.36}$$

with the *safety index* β_c defined as, Cornell (1969)

$$\beta_c = \frac{\mu_m}{s_m} = \frac{\mu_s - \mu_\ell}{\sqrt{s_s^2 + s_\ell^2}} \tag{7.37}$$

In the Level 2 methods all variables are assumed normal distributed and hence alone characterized by their mean and standard deviation. Therefore the name: *Second moment methods*.

The border line between the safe and failure region is given by the limit state or failure function

$$G(\sigma_s,\sigma_\ell) = \sigma_s - \sigma_\ell = \sigma_m = 0 \tag{7.38}$$

with the property

$$\begin{aligned} G(\sigma_s,\sigma_\ell) &< 0 \quad \text{for } \sigma_s, \sigma_\ell \in F \\ G(\sigma_s,\sigma_\ell) &> 0 \quad \sigma_s, \sigma_\ell \in S \end{aligned} \tag{7.39}$$

where F and S denote the failure and safe sets of combinations of σ_s, σ_ℓ.

Generally, both the load σ_ℓ and strength σ_s are function of other stochastic variables, e.g. geometrical and material parameters. These basic variables are denoted X_i and hence Eq. (7.38) can be written

$$G(\underline{X}) = 0 \tag{7.40}$$

where $\underline{X}$ is a vector containing the basic variables.

If the basic variables are statistically independent and normal distributed with mean μ_i and standard deviation s_i and, furthermore the limit state function is a linear combination of X_i

$$G(\underline{X}) = a_0 + \sum_{i=1}^{n} a_i X_i = 0 \tag{7.41}$$

then the safety margin σ_m

$$\sigma_m = a_0 + \sum_{i=1}^{n} a_i X_i$$

is normal distributed with mean value

$$\mu_m = a_0 + \sum_{i=1}^{n} a_i \mu_i$$

and variance

$$s_m^2 = \sum_{i=1}^{n} (a_i s_i)^2$$

The safety index β_c, Eqs. (7.36), (7.37), becomes

$$\beta_c = \Phi^{-1}(P_f) = \frac{\mu_m}{s_m} = \frac{a_0 + \sum_{i=1}^{n} a_i \mu_i}{\sqrt{\sum_{i=1}^{n} (a_i s_i)^2}} \tag{7.42}$$

as a natural generalization of Eq. (7.37). The result can be given a geometrical interpretation if the basic variables X_i are transformed to standard normal distributed variables Z_i (with zero mean and unit variance):

$$Z_i = \frac{X_i - \mu_i}{s_i} \tag{7.43}$$

Expressing X_i in terms of Z_i and substition in the limit state function, Eq. (7.41) yields

$$\begin{aligned} G(\underline{Z}) &= a_0 + \sum_{i=1}^{n} a_i \mu_i + \sum_{i=1}^{n} a_i s_i Z_i = 0 \\ &\equiv b_0 + \sum_{i=1}^{n} b_i Z_i \end{aligned} \tag{7.44}$$

and thus β_c is simply the distance from the origin $\underline{Z} = \underline{0}$ to the hyperplane $G(\underline{Z}) = 0$, as measured in the n-dimensional coordinate system $Z_1, Z_2, ..., Z_n$. This is illustrated in Figure 7.22 for $n = 2$.

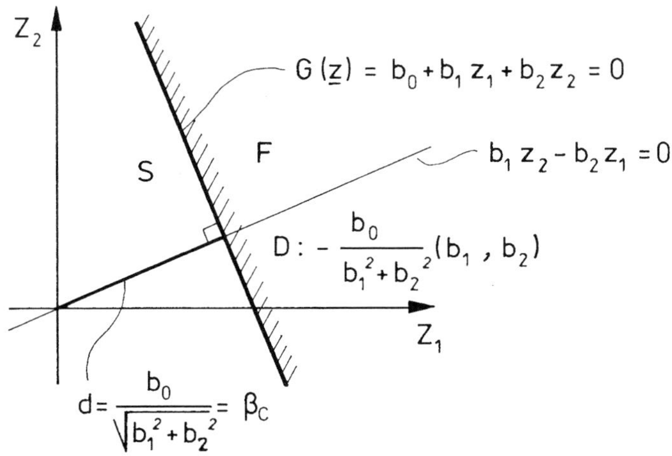

Figure 7.22 *Illustration of limit state function $G(\underline{Z})$, design point D and safety index β_c in a normalized two-dimensional space Z_1, Z_2.*

Extension to correlated normal distributed basic variables X_i is straightforward, see e.g. Madsen et. al (1986), as $G(\underline{Z}) = 0$ still is a line in n-space. However, if $G(\underline{Z}) = 0$ is a non-linear function of $\underline{Z}$, then $G(\underline{Z})$ is a curved surface in n-space. It is not obvious then to define a proper safety index. One way is to linearize the limit state function around a point $\underline{z}_0$:

$$G(\underline{Z}) \simeq G(\underline{z}_0) + \sum_{i=1}^{n} \left.\frac{\partial G}{\partial z_i}\right|_{\underline{z}_0} (Z - z_{0i}) \qquad (7.45)$$

and then use the result, Eq. (7.42), derived for linear limit state functions. Unfortunately, the result depends on both $\partial G/\partial z_i$ and the choice of linearization point. Thus for instance the limit state function

$$G(\sigma_s, \sigma_\ell) = \sigma_s^n - \sigma_\ell^n = 0 \quad ; \quad n > 1$$

will yield different results for different n, see also Example 7.3.2. The most obvious linearization point is the mean value: $\underline{z}_0 = \underline{\mu}$, but this point might be far away from the *design point* i.e the *most probable failure point* and certainly not on the limit state function. It is, however, often used and the corresponding safety index is called the *mean-value first-order second-moment safety index* and becomes from Eq. (7.45)

$$\beta_{FO} = \frac{G(\underline{\mu})}{\sqrt{\left[\sum_{i=1}^{n} \left.\frac{\partial G}{\partial z_i}\right|_{\underline{\mu}} s_i\right]^2}} \tag{7.46}$$

For a linear limit state function, $\beta_{FO} = \beta_C$ as readily seen from Eqs. (7.42) and (7.46).

The lack of invariance with respect to the limit state function and linearization point is solved by defining a new safety index β_{HL}, Hasofer and Lind (1974), as the shortest distance in the normalized $\underline{Z}$ space from the origin to the limit state surface, Figure 7.23.

$$\beta_{HL} = \min_{G(\underline{Z})=0} \sqrt{\left(\underline{Z}^T\underline{Z}\right)} \tag{7.47}$$

For a linear limit state surface $\beta_{HL} = \beta_c$ and the design point D, is easily determined as shown in Figure 7.22.

It should be stressed that only for a linear limit state surface P_f is exactly equal to $\Phi(-\beta_{HL})$. However, for other limit state functions, the error introduced by taking $P_f = \Phi(-\beta_{HL})$ is usually small if $\beta_{HL} > 3$, because $\Phi(-u)$ decreases rapidly for $u > 3$:

$$\Phi(-u) \simeq \frac{1}{\sqrt{2\pi}\, u} e^{-\frac{1}{2}u^2} \quad ; \quad u \gtrsim 3 \tag{7.48}$$

This implies that the contribution to P_f from the failure set F decreases rapidly with the distance $u > \beta_{HL}$.

In the general case where $G(\underline{Z}) = 0$ is a curved surface, the design point D must be determined numerically as the point on $G(\underline{Z}) = 0$ closest to the origo. This problem can be formulated as an optimization problem similar to Eq. (7.47), see e.g. Madsen et. al (1986).

The derivatives of the safety factor β_{HL} with respect to the variables Z_i are denoted the *sensitivity factors*. A small sensitivity factor imply that β_{HL} does not vary much with changes in this variable, which then can be taken as deterministic in order to reduce the computational effort. Before doing so, a transformation back to the physical variables X_i is necessary.

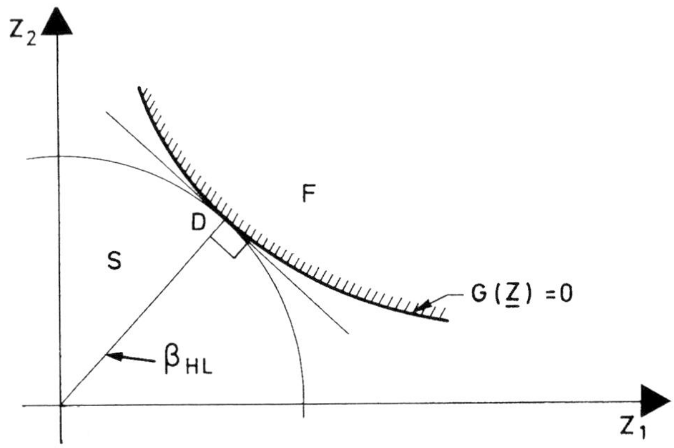

Figure 7.23 *Definition of the Hasofer-Lind safety index* β_{HL}.

Although the *Hasofer-Lind safety index* may have some deficits for highly curved limit state functions, it generally provides a reasonable measure of the probability of failure.

The second moment (Level 2) methods are very convenient provided the basic variables are normal distributed. For marine structures, where the wave loads are far from normal distributed, the probability of failure can only be estimated by a Level 3 analysis.

Example 7.3.2

In order to illustrate the difference between the safety indices β_{FO} and β_{HL} consider a hull girder loaded by a deterministic (still water) bending moment $M = \mu_m$. The limit state is assumed to be yielding in a deck panel with a stochastic yield stress Y and section modulus W. Both parameters are taken as normal distributed with mean values μ_y, μ_w and standard deviations, s_y, s_w, respectively.

The limit state function can be written in an infinite number of ways, satisfying Eqs. (7.38) and (7.39). Here two are considered:

$$G_1(Y, W) = YW - \mu_m = 0$$

and

$$G_2(Y, W) = Y - \frac{\mu_m}{W} = 0$$

The first corresponds to comparison of bending moments and the latter of stresses. Both limit state functions are non-linear and the mean value first-order second-moment safety index β_{FO} is obtained by linearization around the mean values, that is by Eq. (7.45) with $z_{01} = \mu_y$, $z_{02} = \mu_w$:

$$G_1(Y, W) \simeq \mu_y \mu_w - \mu_m + \mu_w(Y - \mu_y) + \mu_y(W - \mu_w)$$

and

$$G_2(Y, W) \simeq \mu_y - \frac{\mu_m}{\mu_w} + (Y - \mu_y) + \frac{\mu_m}{\mu_w^2}(W - \mu_w)$$

Hence from Eq. (7.46)

$$\beta_{FO1} = \frac{\mu_y\mu_w - \mu_m}{\sqrt{(\mu_w s_y)^2 + (\mu_y s_w)^2}} = \frac{\gamma - 1}{\gamma\sqrt{\nu_y^2 + \nu_w^2}}$$

and

$$\beta_{FO2} = \frac{\mu_y - \frac{\mu_m}{\mu_w}}{\sqrt{s_y^2 + \left(\frac{\mu_m}{\mu_w^2}s_w\right)^2}} = \frac{\gamma - 1}{\gamma\sqrt{\nu_y^2 + \left(\frac{\nu_w}{\gamma}\right)^2}}$$

where the coefficients of variation, Eq. (3.14)

$$\nu_y = \frac{s_y}{\mu_y} \quad ; \quad \nu_w = \frac{s_w}{\mu_w}$$

and the safety factor of the means

$$\gamma = \frac{\mu_y\mu_w}{\mu_m}$$

have been introduced. As $\gamma > 1$ it is clear that β_{FO1} and β_{FO2} are different, although the physical problem is the same.

The invariant safety index β_{HF}, Eq. (7.47), will be the same for both formulations as it does not make use of the derivative of the limit state function. The value can be obtained by replacing Y, W by the standard normal distributed variables:

$$Z_1 = \frac{Y - \mu_y}{s_y} \quad , \quad Z_2 = \frac{W - \mu_w}{s_w}$$

yielding

$$G_1(Z_1, Z_2) = (s_y Z_1 + \mu_y)(s_w Z_2 + \mu_w) - \mu_m = 0$$

or

$$G_2(Z_1, Z_2) = G_1(Z_1, Z_2)/(s_w Z_2 + \mu_w) = 0$$

and hence for both cases, the distance from (0,0) to $G(Z_1, Z_2) = 0$ becomes:

$$\sqrt{\underline{Z}^T\underline{Z}} = \sqrt{Z_1^2 + Z_2^2}$$

$$= \left\{Z_1^2 + \left[\frac{\mu_m}{s_w(s_y Z_1 + \mu_y)} - \frac{\mu_w}{s_w}\right]^2\right\}^{\frac{1}{2}}$$

$$= \left\{Z_1^2 + \left[\frac{1}{\gamma\nu_y\nu_w(Z_1 + 1/\nu_y)} - \frac{1}{\nu_w}\right]^2\right\}^{\frac{1}{2}}$$

The Hasofer–Lind safety index β_{HL} is defined as the minimum value of $\sqrt{\underline{Z}^T\underline{Z}}$. For given values of ν_y, ν_w and γ the solution is easily found by e.g. plotting $\sqrt{\underline{Z}^T\underline{Z}}$ as a function of Z_1.

As an example

$$\nu_y = 0.10\,,\ \nu_w = 0.05 \text{ and } \gamma = 1.5$$

yields

$$\begin{aligned}\beta_{FO1} &= 2.981\\ \beta_{FO2} &= 4.000\\ \beta_{HL} &= 3.144 \quad (\text{for } Z_1 = Z_{1D} = -2.94,\ Z_2 = Z_{2D} = -1.10)\end{aligned}$$

A significant difference between the safety indices is clearly seen and only β_{HL} can be used to estimate P_f through $P_f \simeq \Phi(\beta_{HL})$. The design point D, i.e. the most probable values of Y and W if failure occur become

$$\begin{aligned}Y_D &= s_y Z_{1D} + \mu_y = \mu_y(\nu_y Z_{1D} + 1) = 0.706\mu_y\\ W_D &= s_w Z_{2D} + \mu_w = \mu_w(\nu_w Z_{2D} + 1) = 0.945\mu_w\end{aligned}$$

corresponding to a yield stress 29.4 per cent below its mean value and a section modules 5.5 per cent below its mean value.

7.3.3 Level 3 - Probability of Failure

In the Level 3 methods the "real" probability distributions of the basis variables X_i are used. Otherwise, the analysis is generally the same as in the Level 2 methods. If only a few variables are needed to represent the load and strength, a direct calculation of the probability of failure P_f may be possible. The simplest case is the one where the load σ_ℓ and the strength σ_S are defined with their probability distributions. Then Eq. (7.32) gives the probability of failure as a convolution integral, easily carried out either analytically or numerically. Example 7.3.3 illustrates this case for a Rayleigh distributed load σ_ℓ and a normal distributed strength σ_S.

In general, the probability of failure can be written

$$P_f = P(G(\underline{X}) \le 0) = \int \cdots \int_{G(\underline{x}) \le 0} p_{\underline{X}}(\underline{x})\, dx_2 \ldots dx_n \tag{7.49}$$

for any limit state function $G(\underline{x}) = 0$ depending on any number of basic variables X_i. Here $p_{\underline{X}}(\underline{x})$ is the joint probability density function of the basic variables. The number of integrals to be carried out is equal to the number n of basic variables. For a large number of variables it might be computationally very difficult to carry out all these integrals. A much more effective method is then often to make simulations of the behaviour of the system. This is called *Monto Carlo simulations*. Basically, all basic variables X_i are sampled a large number N of times and each time the limit state function $G(\underline{X})$ is evaluated. From the definition (7.49) it then follows that

$$P_f \simeq \frac{N^-}{N} \tag{7.50}$$

where N^{-} is the number of times where $G(\underline{X})$ turns out to be less than or equal to zero. In order to get a reasonable accuracy in P_f, N has to be large. This is clear for instance that if P_f is of the order 10^{-4} then at least 10^{+5} simulations must be carried out. Various improvements of this direct sampling procedure can be made, see e.g. Melchers (1999) for a very detailed description.

The Monto-Carlo simulation requires simulations of random variables with given distributions. Usually, a computer can generate a random number U uniformly distributed between 0 and 1. A basic variable X with the known probability distribution $F(x)$ can then be generated from

$$F(x) = P(X < x) = P(U < u) = u \tag{7.51}$$

implying that

$$x = F^{-1}(u) \tag{7.52}$$

Thus for instance a realization of a Rayleigh distributed variable X, Eq. (3.43), can be obtained as

$$X_{Rayleigh} = \alpha\sqrt{-\ln(1-U)} \tag{7.53}$$

For a normal distributed variable X, the inverse probability distribution, Eq. (7.52), is not available in analytical form. Then a more convenient generation of realizations of X is

$$X_{normal} = \sqrt{-2\ln U_1}\,\cos(2\pi U_2) \tag{7.54}$$

where U_1 and U_2 are two independent realizations of U, see Example 7.3.4. Another normal distributed variable independent of X_{normal} above is obtained by replacing cosine with sine.

Finally, the use of the Hasofer-Lind safety index β_{HL}, Eq. (7.47), can be generalized to cover non-normal distributed basic variables. In that case a method for transformation of a general set of non-normal dependent variables to a set of standard normal distributed independent variables is needed. The so-called Rosenblatt transformation, see e.g. Madsen et. al (1986) is here a very powerful procedure.

From the value of β_{HL} the probability of failure P_f can then be estimated as

$$P_f \simeq \Phi(-\beta_{HL}) \tag{7.55}$$

corresponding to a linearization of the limit state function $G(\underline{Z}) = 0$ around the design point D. This is called a *first-order reliability method* (FORM) A *second order reliability method* (SORM) is obtained by retaining also the second order terms in the Taylor series expansion of $G(\underline{Z}) = 0$ around the design point. Breitung (1984) has derived an expression for the probability of failure in the SORM analysis. Basically, Eq. (7.55) is modified by a term involving the curvature of the limit state surface around the design point as expected from the applied series expansion.

A final remark is that the present treatment is just a short outline of a very important field within structural design. The reader is strongly recommended to consult the textbooks mentioned previously for further readings.

Example 7.3.3

In a stationary sea the individual peaks in the wave load response is Rayleigh distributed, Eq. (3.242), provided the response is linear and narrow-banded:

$$F_\ell(\sigma_\ell) = 1 - \exp\left(-\frac{1}{2}\left(\frac{\sigma_\ell - \bar{\sigma}_\ell}{s_\ell}\right)^2\right) \quad ; \quad \sigma_\ell > \bar{\sigma}_\ell$$

Here s_ℓ is the standard deviation of the wave response and $\bar{\sigma}_\ell$ is the corresponding deterministic still water response.

The strength is approximated by a normal distribution

$$p_s(\sigma_s) = \frac{1}{\sqrt{2\pi}\, s_s}\exp\left(-\frac{1}{2}\left(\frac{\sigma_s - \mu_s}{s_s}\right)^2\right)$$

where μ_s and s_s are the mean and standard deviation of the strength. Thereby, the probability of failure P_f, Eq. (7.32) becomes

$$P_f = 1 - \int_{\bar{\sigma}_\ell}^{\infty} F_\ell(\sigma_s)\, p_s(\sigma_s)\, d\sigma_s$$

as the load $\sigma_\ell > \bar{\sigma}_\ell$. After some algebra the results can be written, Mansour (1972)

$$P_f = \Phi\left(-\frac{\beta}{\eta}\sqrt{1+\eta^2}\right) + \frac{1}{\sqrt{1+\eta^2}}\Phi\left(\frac{\beta}{\eta}\right)\exp\left(-\frac{1}{2}\beta^2\right) \tag{7.56}$$

where

$$\eta = \frac{s_s}{s_\ell}$$

and

$$\beta = \frac{\mu_s - \bar{\sigma}_\ell}{\sqrt{s_s^2 + s_\ell^2}}$$

Normally, the safety index β is larger than 3 and thus the result can be approximated by

$$P_f \simeq \frac{1}{\sqrt{1+\eta^2}}\exp\left(-\frac{1}{2}\beta^2\right)$$

using Eq. (7.48).

The corresponding first order reliability method (FORM) can be carried out as follows. The basic variables σ_ℓ and σ_s are replaced by the standard normal distributed variables Z_1 and Z_2, defined through

$$F_\ell(\sigma_\ell) = \Phi(Z_1)$$

and

$$F_s(\sigma_s) = \Phi(Z_2)$$

implying

$$\sigma_\ell = s_\ell \sqrt{-2\ln\left(1 - \Phi(Z_1)\right)} + \bar{\sigma}_\ell$$

and

$$\sigma_s = s_s Z_2 + \mu_s$$

Thereby, the limit state function

$$G(Z_1, Z_2) = \sigma_s - \sigma_\ell = 0$$

can be written

$$G(Z_1, Z_2) = \eta Z_2 + \beta\sqrt{1 + \eta^2} - \sqrt{-2\ln\left(1 - \Phi(Z_1)\right)} = 0$$

The Hashofer-Lind safety index β_{HL} is obtained from Eq. (7.47)

$$\beta_{HL} = \min_{G(\underline{Z})=0} \sqrt{\underline{Z}^T \underline{Z}}$$

or

$$\beta_{HL} = \min_{Z_1 \in R} \left\{ \left[\frac{\beta}{\eta}\sqrt{1 + \eta^2} - \frac{1}{\eta}\sqrt{-2ln\left(1 - \Phi(Z_1)\right)} \right]^2 + Z_1^2 \right\}^{\frac{1}{2}}$$

As an example consider the container ship treated in Example 4.7.1 .

The most probable largest sagging wave bending moment during its operational life time was found to be 4.5 10^9 Nm.

Assume a stationary sea state where this moment was equal to the most probable largest value during 3 hours operations (i.e. 1000 peaks). Then the corresponding standard deviation s_M becomes

$$s_M = 4.5\ 10^9\ \text{Nm}/\sqrt{2\ln 1000} \simeq 1.2\ 10^9\ \text{Nm}$$

assuming Rayleigh distributed peaks (i.e. ignoring the influence of the non-linearities on the tail of the distribution). The still water sagging bending moment is taken as the rule value, see Example 4.7.1.

$$\bar{\sigma}_M = 2.1\ 10^9\ \text{Nm}$$

This is very conservative as the still water moment normaly is in hogging for all loading conditions for a container vessel. To get an estimate for the strength of the hull girder, it is assumed that hull girder collapse will take place if the stress in the deck reaches the nominal yield stress equal to 235 MN/m^2.

In order to compare the strength and the stress, it is assumed that the section modulus W of the deck is given by

$$W = \frac{M^{rule}_{sag,\,rule} + M^{rule}_{sag,stillwater}}{180\mathrm{MN/m^2}} = \frac{4.3\ 10^9 + 2.1\ 10^9}{180\ 10^6} \simeq 35\ \mathrm{m}^3$$

i.e. the most probable largest bending moment will yield a stress of 180 MN/m^2. Hence,

$$s_\ell = \frac{s_M}{W} \simeq 35\ \mathrm{MN/m^2}$$

$$\bar{\sigma}_\ell = \frac{\bar{\sigma}_M}{W} \simeq 60\ \mathrm{MN/m^2}$$

$$\mu_s = 235\ \mathrm{MN/m^2}$$

and finally, the standard deviation of the yield stress s_s is taken to be

$$s_s = 15\ \mathrm{MN/m^2}$$

Thereby,

$$\eta = \frac{s_s}{s_\ell} = 0.4286$$

$$\beta = \frac{235 - 60}{\sqrt{15^2 + 35^2}} = 4.596$$

and, numerically,

$$\beta_{HL} = 4.069$$

for $Z_1 = 3.69$ and $Z_2 = -1.70$, corresponding to a design point of $\sigma_{sD} = \sigma_{\ell D} = 210\ \mathrm{MN/m^2}$. The exact probability of failure becomes

$$P_f = \Phi(-11.67) + 0.919\,\Phi(9.80)\,\exp\left(-\frac{1}{2}\,4.596^2\right) = 2.4\ 10^{-5}$$

whereas

$$\Phi(-\beta) = 0.2\ 10^{-5}$$

and

$$\Phi(-\beta_{HL}) = 2.4\ 10^{-5}$$

The approximation

$$P_f \simeq \Phi(-\beta_{HL})$$

is thus very accurate. The reason is that the failure surface is nearly linear in Z_1, Z_2 for a large range around the design point.

The probability of failure calculated above relates to each peak in the load. Thus for the complete 3 hours operation ($N = 1000$ peaks):

$$P_f(3\text{hours}) = 1 - P(\text{no failure}) = 1 - (1 - P_f)^N$$
$$= 1 - (1 - 2.4\ 10^{-5})^{1000} = 0.023$$

assuming as usual that the probability of failure for each peak is statistically independent of each other. This might seem a rather high probability of failure, but the sea state is only expected once in the life time of the vessel.

Rather than calculating the probability of failure for each peak in the load and then summing up as above for all 1000 peaks, one can also consider failure during the most probable largest peak. Its probability distribution is given by

$$F_{\ell N}(\sigma_\ell) = \left[F_\ell(\sigma_\ell)\right]^N$$

and thus the only difference in the FORM analysis is that the load transformation now becomes

$$\left[F_\ell(\sigma_\ell)\right]^N = \Phi(Z_1)$$

or

$$\sigma_\ell = s_\ell \sqrt{-2\ln\left(1 - (\Phi(Z_1))^{1/N}\right)} + \bar{\sigma}_\ell$$

The numerical result for the present example becomes

$$\beta_{HL} = \min_{G(\underline{Z})} \sqrt{\underline{Z}^T \underline{Z}} = 2.088$$

and

$$P_f(3\ hours) \simeq \Phi(-\beta_{HL}) = 0.018$$

This result is less than the value 0.023 obtained previously, because only failure at the largest load peak is considered. The design point becomes $Z_{1D} = 1.53$, $Z_{2D} = -1.42$ corresponding to $\sigma_{\ell D} = \sigma_{sD} = 214\text{MN/m}^2$.

Finally, it should be mentioned that if the strength is deterministic, $s_s = 0$, then Eq. (7.56) reduces to

$$P_f = \exp\left(-\frac{1}{2}\left(\frac{\mu_s - \bar{\sigma}_\ell}{s_\ell}\right)^2\right)$$

which of course is equal to

$$P_f = P(\sigma_\ell - \bar{\sigma}_\ell > \mu_s) = 1 - F_\ell(\mu_s - \bar{\sigma}_\ell)$$

Hence, for the numerical example

$$P_f = \exp\left(-\frac{1}{2}\left(\frac{235 - 60}{35}\right)^2\right) = 3.7\ 10^{-6}$$

implying that in the present example the uncertainty in the strength is also important.

Example 7.3.4

In the following a proof that the stochastic variable X, Eq. (7.54), is normal distributed is given.

$$P(X < x) = F_x(x) = \int\int_{X<x} p_{u_1}(u_1)\, p_{u_2}(u_2)\, du_1\, du_2$$

$$= \int\int_{X<x} du_1\, du_2$$

as U_1 and U_2 are statistically independent and uniformly distributed between 0 and 1. From the definition of X

$$X = \sqrt{-2\ln U_1}\,\cos(2\pi U_2)$$

the bound $X < x$ can be transformed to bounds on u_1 and u_2

$$h(x) \le u_2 \le 1 - h(x)$$
$$0 \le u_1 \le \exp\left(-\frac{1}{2}x^2\right)$$

where

$$h(x, u_1) \equiv \frac{1}{2\pi} Arccos\left(\frac{x}{\sqrt{-2\ln u_1}}\right)$$

Hence

$$F_x(x) = \int_0^{e^{-\frac{1}{2}x^2}} \int_{h(x,u_1)}^{1-h(x,u_1)} du_2 du_1$$

$$= 1 - \frac{1}{\pi}\int_0^{e^{-\frac{1}{2}x^2}} h(x, u_1) du_1$$

Differention yields

$$\frac{dF_x(x)}{dx} \equiv p_x(x) = \frac{1}{\pi}\int_0^{\exp\left(-\frac{1}{2}x^2\right)} \frac{du_1}{\sqrt{-2\ln u_1 - x^2}}$$

as $Arccos\ 1 = 0$.

Taking $v^2 \equiv -2\ln u_1 - x^2$ gives

$$p_x(x) = \frac{1}{\pi}\int_0^{\infty} e^{-\frac{1}{2}(v^2 + x^2)} dv$$

$$= \frac{1}{\pi} e^{-\frac{1}{2}x^2} \int_0^{\infty} e^{-\frac{1}{2}v^2} dv = \frac{1}{\sqrt{2\pi}} e^{-\frac{1}{2}x^2} = \varphi(x)$$

showing that $p_x(x)$ is a standard normal distributed.

7.3.4 Fatigue Failure under Stochastic Loading

To estimate the probability of fatigue failure in a maritime structure as a result of long-term wave loads, the wave-induced stress amplitudes σ_u are assumed to be Weibull distributed with the following probability density function, Section 4.4:

$$p_p(\sigma_a) = \frac{\beta}{\alpha^\beta}\sigma_a^{\beta-1}\exp\left(-(\sigma_a/\alpha)^\beta\right) \tag{7.57}$$

where β and α are parameters in the distribution. If β is equal to 1 it becomes the exponential distribution, and if β is equal to 2 it is a Rayleigh distribution. The parameter α is a measure for the stress level. α and β can be related to the most probable largest stress amplitude $\tilde{\mu}_N$ among the total N stress peaks through Eq. (3.159).

The number of stress peaks in which the stress amplitude lies between σ_a and $\sigma_a + d\sigma_a$ is $Np_p(\sigma_a)d\sigma_a$. Thus, the *cumulative fatigue damage coefficient D* defined by Eq. (7.22) can be written

$$D = \int_0^\infty \frac{N\,p_p(\sigma_a)d\sigma_a}{A(2\sigma_a)^{-m}} = \frac{2^m N}{A}E[\sigma_a^m] \tag{7.58}$$

by use of the S-N curve, Eq. (7.19). The expectation $E[\sigma_a^m]$ in Eq. (7.58) can be integrated yielding

$$D = \frac{(2\alpha)^m N}{A}\Gamma(1 + m/\beta) \tag{7.59}$$

using Eq. (7.57) and the definition, Eq. (3.40), of the Gamma function $\Gamma(x)$.

If failure is expected for $D = D_{cr}$ the lifetime L in years can be determined by

$$L = \frac{D_{cr}}{D}T \tag{7.60}$$

where D is given by Eq. (7.59) and T is the operational life time corresponding to N peaks in the wave load.

Theoretically, failure will occur when $D_{cr} = 1$. Therefore, it is possible, by specifying a maximum acceptable D, to take into account the consequence of a failure as well as the possibility of detecting and repairing a fatigue crack. Typically, $0.1 \le D \le 1$ where $D = 0.1$ applies to significant structural elements, difficult to inspect during operation.

A Level 2 or Level 3 analysis can be performed using the failure surface

$$G(\underline{X}) = D_{cr} - D = 0 \tag{7.61}$$

where the basic variables are

$$\underline{X} = \{D, \alpha, \beta, A, m\}$$

The probability distributions of the basic variables are very difficult to determine and the Level 2 method using mean values and standard deviations of these variables might be the most appropriate method. However, normally, Eqs. (7.59)-(7.60), are applied with the uncertainty in the strength taken into account by using the S-N curve corresponding to the mean minus two standard deviations.

References

Abramowitz, M. and Stegun, I.A. (1970). "Handbook of Mathematical Functions", Dover Publications.

Andersen, P. and He, W. (1985). "On the Calculation of Two-Dimensional Added Mass and Damping Coefficients by Simple Green's Function Technique", Ocean Engineering, Vol. 12. No. 5, pp. 425-451.

Beck, R. F., Reed, A. M. and Rodd, E. P. (1996). "Application of Modern Numerical Methods in Marine Hydrodynamics", Trans. SNAME, Vol. 104, pp. 519-537.

Betts, C.V., Bishop, R.E.D. and Price, W.G. (1977). "A Survey of Internal Hull Damping", RINA, Vol. 119.

Bishop, R. E. D. and Price, W. G. (1979). "Hydroelasticity of Ships", Cambridge University Press, London.

Breitung, K. (1984). "Asymptotic Approximations for Multinormal Integrals", J. Eng. Mechanics Div., ASCE, Vol 110, pp 357-366.

Breslin, J. and Andersen, P. (1994). " Hydrodynamics of Ship Propellers", Cambridge Ocean Technology Series 3, Cambridge University Press.

Buckley, W. H., (1983). "A Study of Extreme Waves and their Effect on Ship Structures", Ship Structure Committee, SSC-320, U.S. Coast Guard, Washington, D.C.

Buckley, W. H., (1988). " Extreme and Climatic Wave Spectra for Use in Structural Design of Ships". Naval Engineers' Journal, Vol. 100, No. 5, pp. 36-57.

Caldwell, J. B. (1965). "Ultimate Longitudinal Strength", Trans. RINA, Vol. 107, pp. 411-430.

Cartwright, D. E. and Longuet- Higgins, M. S. (1956). "The Statistical Distribution of the Maxima of a Random Function", Proc. Royal Soc. London, Serie A, Vol. 237, pp. 212-232.

Castillo, E. (1988). "Extreme Value Theory in Engineering", Academic Press, San Diego, California.

Chakrabarti, S. K. (1987). "Hydrodynamics of Offshore Structures", Springer-Verlag.

Clarke, J. D. (1986). "Wave Loading in Warships", Proc. Advances in Marine Structures, eds. Clarke and Smith, Elsevier Applied Science, London.

Clauss, G., Lehmann, E. and Ostergaard, C. (1992). "Offshore Structures Conceptual Design and Hydromechanics", Springer-Verlag, Vol. I.

Collatz, L. (1963). "Eigenwertaufgaben mit technischen anwendungen", Akademische Verlagsgesellschaft, Leipzig.

Cornell, C. A. (1969). "A Probability-Based Structural Code", J. Am. Concrete Institute, Vol. 66, pp. 974-985.

Cowper, G.R. (1966). "The Shear Coefficient in Timoshenko's Beam Theory", J. Appl. Mech., Vol. 33.

Danish Code of Practice DS 449 (1984). "Pile-Supported Offshore Steel Structures", 1st Edition, DIF, NP-162-T.

Faltinsen, O. M. (1988). Written Discussion to Report I.2, "Enviromental Forces", ISSC' 88, Lyngby, Denmark.

Flokstra, C. (1974). "Comparison of Ship Motion Theories with Experiments for a Containership", International Shipbuilding Progress, Vol. 21.

Fricke, W., Petershagen, H. and Paetzold, H. (1997, 1998). " Fatigue Strength of Ship Structures", Part I, GL-Technology No. 1/97; Part II, GL-Technology No. 1/98, Germanischer Lloyd. Hamburg.

Friis Hansen, P. (1994). "Reliability Analysis of a Midship Section", PhD thesis, Dept. Ocean Eng., Tech. Univ. of Denmark, Lyngby, Denmark.

Friis Hansen, P. and Nielsen, L. P. (1995). "On the New Wave Model for the Kinematics of Large Ocean Waves", Proc. OMAE' 95, Vol. IA, pp. 17-24.

Gerritsma, J. and Beukelman, W. (1964). "The Distribution of the Hydrodynamic Forces on an Heaving and Pitching Ship Model in Still Water", Fifth Symposium on Naval Hydrodynamics, ACR-112, US Office of Naval Research, Washington, D. C., pp. 219-251.

Gower, J. (1994). "Canadian West Coast Giant Waves", Mariners Weather Log, Spring, pp. 4-8.

Guldhammer, H.E. and Meldahl, A. (1971). "Skibsbygning og styrkelære", Bogfondens forlag.

Hackmann, D. (1979). Written discussion to Jensen and Pedersen (1979).

Halden, K.O. (1980). "Early Design-Stage Approach to Reducing Hull Surface Forces Due to Propeller Cavitation", Trans. SNAME, Vol. 88.

Hansen, A. M. (1996). "Strength of Midship Sections", Marine Structures, Vol. 9, pp 471-494.

Hasofer, A. M. and Lind, N. (1974). "Exact and Invariant Second Moment Code Format", J. Eng. Mechanics Div., ASCE, VOL. 100, pp 111-121.

Hogben, N., Dacunha, N. M. C. and Olliver, G. F. (1986). "Global Wave Statistics". British Maritime Technology, Unwin Brothers Ltd., UK.

Hughes, O. F. (1988). "Ship Structural Design" SNAME, USA.

Inglis, R. B. and Price, W. G. (1982). "A Three-Dimensional Ship Motion Theory - Comparison Between Theoretical Predictions and Experimental Data of the

Hydrodynamic Coefficients with Forward Speed", Trans. RINA, Vol. 124, pp. 141-158.

Jensen, J.J. and Madsen, N.F. (1977). "A Review of Ship Hull Vibration", Shock and Vibr. Digest, Vol. 9, Nos. 4-6.

Jensen, J. J. and Pedersen, P. T. (1978). "On the Calculation of the Joint Probability Density of Slightly Non-linear Stochastic Processes", GAMM-Tagung, Lyngby, ZAMM Vol. 58, pp. T 481-T 484.

Jensen, J. J. and Pedersen, P. T. (1979). "Wave-induced Bending Moments in Ships - a Quadratic Theory", Trans. RINA, Vol. 121, pp. 151-165.

Jensen, J.J. and Pedersen, P.T. (1981). "Bending Moments and Shear Forces in Ships Sailing in Irregular Waves", J. Ship Research, Vol. 25, pp.243-251.

Jensen, J. J. (1983). "On the Shear Coefficient in Timoshenko's Beam Theory". J. Sound and Vibration, Vol. 87, pp. 621-635.

Jensen, J. J., Banke, L. and Dogliani, M. (1994). "Long-term Predictions of Wave-induced Loads using a Quadratic Strip Theory", Proc. Int. Conf. on Ship and Marine Research (NAV'94), Rome.

Jensen, J. J., Baatrup, J. and Andersen, P. (1995). "Probabilistic Damage Stability Calculations in Preliminary Ship Design", Proc. PRADS'95. Seoul, Korea, pp. 1.565-1.577.

Jensen, J. J. (1996). "Second-order Wave Kinematics Conditional on a Given Wave Crest", Applied Ocean Research, Vol. 18, pp. 119-128.

Jensen, J. J. and Dogliani, M. (1996). "Wave-induced Ship Hull Vibrations in Stochastic Seaways", Marine Structures, Vol. 9, pp. 353-387.

Kumai, T. (1962). "On the Three-Dimensional Correction Factor for the Virtual Inertia Coefficient in the Vertical Vibration of Ships", JSNA-Japan, Vol. 112, December.

Longuet-Higgins, M. S. (1963). "The Effect of Non-linearities on Statistical Distributions in the Theory of Sea Waves", Journal of Fluid Mechanics, Vol. 17, pp. 459-480.

Longuet-Higgins, M. S (1964). "Modified Gaussian Distribution for Slightly Non-linear Variables", Radio Science, National Bureau of Standards, Vol. 68D(9), pp. 1049-1062.

Madsen, H. O., Krenk, S. and Lind, N. C. (1986). "Methods of Structural Safety", Prentice Hall Inc., Englewood Cliffs, New Jersey.

Madsen, N.F. (1978). "On the Influence of Three-Dimensional Effects and Restricted Water Depth on Ship Hull Vibration", ISP, Vol. 25, No. 286.

MAN B&W (1992). "An Introduction to Vibration Aspects of Two-Stroke Diesel Engines in Ships". MAN B&W Diesel A/S, P.9207-222.

Mansour, A. E. (1972). "Probabilistic Design Concepts in Ship Structural Safety and Reliability", Trans. SNAME, Vol. 80, pp 64-97.

Mansour, A. E. and Jensen, J. J. (1995). "Slightly Non-linear Extreme Loads and Load Combinations", Journal of Ship Research, Vol. 39, No. 2, pp. 139-149.

Melchers, R.E. (1999). "Structural Reliability Analysis and Prediction", John Wiley & Sons, Second Edition.

Naess, A., Storli, H. and Storm, L. E. (1996). "Statistical Predictions of Long Return Period Design Values", Proc. OMAE' 96, Vol. II, pp. 17-22.

Naonosuke et al. (1993). "R&D of a Displacement-Type High-Speed Ship". (Part 1-4). Proc. FAST'93. Yokohama, Japan, pp. 317-359.

Ochi, M. K. (1990). "Applied Probability and Stochastic Processes". John Wiley and Sons, New York.

Ogilvie, T. F. (1964). " Recent Progress Toward the Understanding and Prediction of Ship Motions", 5th Symp. Naval Hydrodynamics, Bergen, Norway, pp. 3-80.

Pedersen, P.T. (1983) "Beam Model for Torsional-Bending Response of Ship Hulls", Trans. RINA, Vol. 125, pp. 171-182.

Phillips, O. M. (1958) " The Equilibrium Range in the Spectrum of Wind-generated Waves", Journal of Fluid Mechanics, Vol. 4, pp. 426-434.

Pickands, J. (1975). "Statistical Inference Using Extreme Order Statistics", The Annals of Statistics, Vol. 3, No. 1, pp. 119-131.

Pierson, W. J., and Moskowitz, L. (1964). "A Proposed Spectral Form for Fully Developed Wind Seas Based on the Similarity of S. A. Kitaigorodskii", Journal of Geophysical Research, Vol. 69 (24), pp. 5181-5203.

Price, W. G. and Bishop, R. E. D. (1974). "Probabilistic Theory of Ship Dynamics", Chapman and Hall, London.

Rice, O. (1945). "Mathematical Analysis of Random Noise". Bell Syst. Tech. J., Vol. 24, pp. 46-156.

Rutherford, S. E. and Caldwell, J. B. (1990). "Ultimate Longitudinal Strength of Ships: A Case Study", Trans. SNAME, Vol. 98, pp 441-471.

Salvesen, N., Tuck, E. O. and Faltinsen, O. (1970). "Ship Motions and Sea Loads", Trans. SNAME, Vol. 78, pp. 250-287.

Schalck, S. and Baatrup, J. (1990). "Hydrostatic Stability Calculations by Pressure Integrations", Ocean Engineering, Vol. 17, No. 1-2, pp. 155-169.

Simonsen, B. C. (1997). "Mechanics of Ship Grounding", PhD thesis, Dept. Naval Arch. Offshore Eng., Tech. Univ. of Denmark, Lyngby, Denmark.

Smith, C. S. (1977). "Influence of Local Compressive Failure of Ultimate Longitudinal Strength of a Ship's Hull", Proc. PRADS, Tokyo, pp 72-79.

Söding, H. (1982). "Lekstabilität im Seegang", Report 429 of the Institut für Schiffbau, Hamburg.

Tasai, F. (1959). "On the Damping Force and Added Mass of Ships Heaving and Pitching Reports of Research Insitute for Applied Mechanics, Japan, Vol. VII, No. 26.

Townsin, R.L. (1969). "Virtual Mass Reduction Factors. J'values for Ship Vibration Calculations Derived from Tests with Beams including Ellipsoids and Ship Models". RINA, Vol. 111.

Tromans, P. S., Anaturk, A. R. and Hagemeijer, P. (1991). "A New Model for the Kinematics of Large Ocean Waves - Application as a Design Wave", Proc. First ISOPE Conf., Vol. III, Edinburgh, UK, pp. 64-71.

Yamamoto, Y., Sugai, K., Inoue, H., Yoshida, K., Fujino, M. and Ohtsubu, H. (1986), "Wave Loads and Response of Ships and Offshore Structures from the Viewpoint of Hydroelasticity", International Conference on Advances in Marine Structures, 20-23 May, Admirality Research Establisment, Dunfermline, Scotland.

Yao, T. and Nikolov, P. I. (1991). "Progressive Collapse Analysis of Ship's Hull under Longitudinal Bending", J. Soc. Naval Architects Japan, Vol. 170, pp 200-214.

Yao, T. and Nikolov, P.I. (1992). "Progressive Collapse Analysis of Ship's Hull under Longitudinal Bending (2nd report)", J.Soc. Naval Architect Japan, Vol. 172, pp 437-446.

Vinje, T. and Haver, S. (1994). "On the non-Gaussian Structure of Ocean Waves". Proc. BOSS'94, Vol. II, Boston, USA, Pergamon Press, pp. 453-479.

Wang, Z. , Jensen, J. J. and Xia, J. (1998). "On the Effect of Green Water on Deck on the Wave Bending Moments", Proc. 7th Int. Symposium on Practical Design of Ships and Mobile Units, PRADS'98, The Hague.

Watanabe, I., Ueno, M. and Sawada, H. (1989). "Effects of Bow Flare Shape to the Wave Loads of a Container Ship", J. Society of Naval Architects of Japan, Vol. 166, pp. 259-266.

Winterstein, S. R. (1985). "Non-Normal Responses and Fatigue Damage", J. Engineering Mechanics, ASCE, Vol. 111, No. 10, pp. 1291-1295.

Winterstein, S. R. (1988). "Nonlinear Vibration Models for Extremes and Fatigue". J. Engineering Mechanics, ASCE, Vol. 114, NO. 10, pp. 1772-1790.

Xia, J., Wang, Z. and Jensen, J. J. (1998). "Non-linear Wave Loads and Ship Responses by Time-Domain Strip Theory", Marine Structures, Vol. 11, No. 3, pp. 101-123.

Ximenes, M-C., Adhia, G. J., Abe, A. (1997). "Design and Construction of a Floating Storage and Offloading Vessel Escravos LPG FSO". Trans. SNAME, Vol. 105, pp. 455-489.

Zhao, R., Faltinsen, O. M. (1993). "Water Entry of Two-dimensional Bodies", J. Fluid Mech., Vol. 246, pp. 593-612.

Index

A

B

C

D

E

F

G

H

I

J

K

L

M

Z